Frontispiece by Ruth Weisberg

Foundations of Measurement

VOLUME III

Representation, Axiomatization, and Invariance

R. Duncan Luce
University of California, Irvine

David H. Krantz
Columbia University

Patrick Suppes
Stanford University

Amos Tversky

DOVER PUBLICATIONS, INC.
Mineola, New York

Bibliographical Note

This Dover edition, first published in 2007, is an unabridged and slightly corrected republication of the work originally published by Academic Press, Inc., San Diego and London, in 1990.

International Standard Book Number: 0-486-45316-2

Manufactured in the United States of America
Dover Publications, Inc., 31 East 2nd Street, Mineola, N.Y. 11501

. . . For whatever ratio is found to exist between intensity and intensity, in relating intensities of the same kind, a similar ratio is found to exist between line and line, and vice versa. For just as one line is commensurable to another line and incommensurable to still another, so similarly in regard to intensities certain ones are mutually commensurable and others incommensurable in any way because of their continuity. Therefore, the measure of intensities can be fittingly imagined as the measure of lines, since an intensity could be imagined as being infinitely decreased or infinitely increased in the same way as a line.

Again, intensity is that according to which something is said to be "more such and such," as "more white" or "more swift." Since intensity, or rather the intensity of a point, is infinitely divisible in the manner of a continuum in only one way, therefore there is no more fitting way for it to be imagined than by that species of a continuum which is initially divisible and only in one way, namely by a line. And since the quantity or ratio of lines is better known and is more readily conceived by us—nay the line is in the first species of continua, therefore such intensity ought to be imagined by lines and most fittingly by those lines which are erected perpendicularly to the subject.

—Nicole Oresme, De configurationibus qualitatum et motuum (14th century A.D.*)*

Table of Contents

19. Nonadditive Representations

20. Scale Types

21. Axiomatization

22. Invariance and Meaningfulness

Preface

Sometime toward the end of 1969 or the beginning of 1970 we came to realize that the material covered by this treatise was of such extent that the originally projected single volume would require two volumes. Because most of the work in Volume I was by then complete, we concentrated on finishing it, and it was published in 1971. There (pages 34–35) we listed the chapter titles of the projected Volume II, which we expected to complete by 1975. Fifteen years later, when we finally brought the work to a close, our plan had changed in several respects. First, the total body of material far exceeded a reasonably sized second volume so, at the urging of the publisher, the manuscript expanded into two volumes, for a total of three. Second, the chapter on statistical methods was not written, largely because the development of statistical models for fundamental measurement turned out to be very difficult. Third, we decided against a summary chapter and scattered its contents throughout the two volumes. Fourth, Volume III came to include two chapters (19 and 20) based on research that was carried out in the late 1970s and throughout the 1980s. Volume II discusses or references many results that have been obtained since 1971 by the large number of persons working in the general area of these volumes.

As in Volume I, we attempt to address two audiences with different levels of interest and mathematical facility. We hope that a reader interested in the main ideas and results can understand them without having to read proofs, which are usually placed in separate sections. The proofs themselves are given somewhat more fully than would be appropriate for a purely mathematical audience because we are interested in reaching those in scientific disciplines who have a desire to apply the mathematical results of measurement theories. In fact, several chapters of Volume II contain extensive analyses of relevant experimental data.

Volume II follows the same convention used in Volume I of numbering definitions, theorems, lemmas, and examples consecutively within a chapter. Volume III deviates from this with respect to lemmas, which are numbered consecutively only within the theorem that they serve. The reason for this departure is the large number of lemmas associated with some theorems.

Volumes II and III can be read in either order, although both should probably be preceded by reading at least Chapters 3 (Extensive Measurement), 4 (Difference Measurement), and 6 (Additive Conjoint Measurement) of Volume I. Within Volume II, Chapter 12 (Geometrical Representations) should probably precede Chapters 13 (Axiomatic Geometry), 14 (Proximity Measurement), and 15 (Color and Force Measurement). Within Volume III, much of Chapter 21 (Axiomatization) can be read in isolation of the other chapters. Chapter 22 (Invariance and Meaningfulness) depends only on the definability section of Chapter 21 as well as on Chapters 10 (Dimensional Analysis and Numerical Laws), 19 (Nonadditive Representations), and 20 (Scale Types). Chapter 19 should precede Chapter 20.

Acknowledgments

A number of people have read and commented critically on parts of Volume III. We are deeply indebted to them for their comments and criticisms; however, they are in no way responsible for any errors of fact, reasoning, or judgment found in this work. They are Theodore M. Alper, Ching-Yuan Chiang, Rolando Chuagui, Newton da Costa, Zoltan Domotor, Kenneth L. Manders, Margaret Cozzens, Jean-Claude Falmagne, Peter C. Fishburn, Brent Mundy, Louis Narens, Reinhard Niederée, Raymond A. Ravaglia, Fred S. Roberts, and Robert Titiev.

In addition, each of us has taught aspects of this material in classes and seminars over the years, and students have provided useful feedback ranging from detecting minor and not-so-minor errors to providing clear evidence of our expository failures. Although we do not name them individually, their contribution has been substantial.

Others who contributed materially are the many secretaries who typed and retyped portions of the manuscript. To them we are indebted for accuracy and patience. They and we found that the process became a good deal less taxing in the past eight years with the advent of good word processing.

With chapters primarily written by different people and over a long span, the entire issue of assembling and culling a reference list is daunting. We were fortunate to have the able assistance of Marguerite Shaw in this endeavor, and we thank her.

Financial support has come from many sources over the years, not the least being our home universities. We list in general terms the support for each of the authors.

Luce: Numerous National Science Foundation grants to The University of California, Irvine, and Harvard University (the most recent being IRI-8602765; The University of California, Irvine, and Harvard University); The Institute for Advanced Study, Princeton, New Jersey, 1969–1972; AT & T Bell Laboratories, Murray Hill, New Jersey, 1984–1985; and The Center for Advanced Study in the Behavioral Sciences, Stanford, California, 1987–1988 (funds from NSF grant BNS-8700864 and the Alfred P. Sloan Foundation).

Tversky: The Office of Naval Research under Contracts N00014-79-C-0077 and N00014-84-K-0615 to Stanford University.

Chapter 18 Overview

This chapter describes briefly and somewhat informally the several topics covered in this volume, relates them to Volumes I and II, and discusses the advantages of the axiomatic representation and uniqueness approach taken in this work.

The theme of Volume I was measurement in one dimension. Each measurement structure assumed both a qualitative ordering of some empirical objects and a method of combining those empirical objects, either by concatenation, or by set-union, or by factorial combination. Measurement involved the assignment of numerical representations to such empirical structures. The ordering of the representing numbers always preserved the qualitative ordering of the empirical objects. Moreover, there was always some numerical combination rule, involving addition or multiplication, or both at once, which could be used to calculate the number assigned to any allowable combination of the empirical objects from the numbers assigned to the various constituents of that combination. In Chapters 3, 5, and 6 the principal numerical combination rule was addition; in Chapter 4, subtraction; in Chapter 8, weighted average, or expectation; and in Chapters 7 and 9, polynomial combination rules with certain restricted forms.

Volume I unified measurement theory by reducing the proofs of most representation theorems to Holder's theorem. This reduction showed that a common procedure underlies different methods for assigning numerical values to empirical objects. One constructs sequences of equally spaced objects, called standard sequences. The object to be assigned a number is bounded between successive terms of a standard sequence. One counts the number of terms needed to reach the target object, and the number needed to reach an object that serves as a unit of measurement. The ratio of these two counts approximates the desired numerical value.

Much of Volume II is devoted to geometrical representations that have two or more dimensions, although Chapters 16 and 17 returned to one-dimensional representations of fallible data. To some degree, additivity continued to play a significant role in that volume; for example, considerable use was made of the additivity of segments along single dimensions.

Chapters 19 and 20 of this volume go substantially beyond the framework of Volume I by considering one-dimensional representations that are essentially nonadditive. Chapter 19 deals with two kinds of structure: an ordering with a binary operation that is monotonic but can violate associativity, and a two-factor conjoint ordering that is monotonic but can violate the cancellation (Thomsen) axiom. (The associativity and cancellation axioms each force additivity.)

Chapter 20 goes further to consider in very general terms the possibilities for measurement when the primitives of the structure are specified only to the extent that one is a weak order and the others are all relations of finite order. However, the structure is restricted by an assumption of homogeneity that implies a certain interchangeability of elements. In contrast to axioms previously encountered, which are formulated in terms of the primitives of the system, homogeneity is formulated somewhat indirectly. For such homogeneous structures, our understanding is moderately complete concerning the possible types of uniqueness that can arise, the so-called scale types. This knowledge, coupled with fairly weak structural constraints, is useful because it yields a detailed understanding of the types of numerical representations that exist for homogeneous concatenation operations and conjoint structures. Moreover, a good deal is also known about numerical representations of structures that have neither a binary operation nor a factorial structure.

Chapter 21 is devoted to issues (largely treated by methods of mathematical logic) concerning the formal axiomatization of measurement structures. It also explores axiomatic issues that have arisen in all three volumes, such as the significance of Archimedean axioms, and general logical topics, including the notion of definability, that are of particular interest to measurement theorists.

The work on scale types in Chapter 20 and on definability in Chapter 21 leads naturally to the study in Chapter 22 of meaningful assertions within a measurement context. It includes, among other things, a modification and extension of parts of Chapter 10. Meaningfulness in the homogeneous case is taken to mean at least invariance under the transformations that describe the scale type. Various arguments are given for the plausibility of this criterion, including the equivalence of several possible definitions of meaningfulness. How the concept of invariance is best restricted to capture meaningfulness in the more restricted sense of definability (which is critically important for nonhomogeneous structures) is an apparently difficult and unresolved problem.

18.1 NONADDITIVE REPRESENTATIONS (CHAPTER 19)

A natural question raised by Chapters 3 and 6 is: What can be said about concatenation and conjoint structures with nonadditive representations? When we wrote Volume I, nearly 20 years ago, virtually nothing was known about such structures aside from the polynomial models of Chapter 7 and the weighted average models (which are closely related to additive conjoint structures) arising both from bisymmetry (Section 6.9) and from the expected-utility axioms (Chapter 8). Since then, the nonadditive theory has been developed rather fully for concatenation operations and conjoint structures.

As in Volume I, if a structure has any numerical representation, then it has one for each strictly increasing function on the range of the given representation. That is, if $\oplus$ is a numerical operation that represents a structure on a subset R of the positive real numbers, and h is a strictly increasing function from R onto R, then the operation $\odot$ defined by

$$x \odot y = h^{-1}[h(x) \oplus h(y)]$$

is an equally good representation. Thus the mere fact that there is a nonadditive representation does not imply that an additive one does not exist. But if an additive one does exist, then the nonadditivity is said to be inessential. In this case, where, for example $\oplus \equiv +$, then any operation involved in an alternative representation is necessarily associative, i.e.,

$$x \odot (y \odot z) = (x \odot y) \odot z.$$

We are interested here in essential nonadditivity, i.e., cases for which no additive representation exists, and so the empirical operation is necessarily nonassociative.

18.1.1 Examples

Nonassociative operations are neither unfamiliar nor bizarre. Averaging is an example of a common operation that is not associative. In geometry, a concept of midpoint algebra has been a basis of some axiomatizations (for further discussion, see Section 19.6.5). Nonassociative algebras arose briefly in the discussion of the Desarguesian property (Section 13.4). The familiar vector product is another nonassociative operation. And, as will be discussed more fully, the theory of nonadditive conjoint structures, which appears to be of potential importance in the behavioral sciences, is reduced mathematically to the theory of nonassociative operations.

To get some sense of the theory, consider the operation $\oplus$ defined on the positive real numbers as follows. Let f be a strictly increasing function from Re^+ onto Re^+ that has the property that $f(x)/x$ is a strictly decreasing function. For x, y in Re^+, define $\oplus$ by

$$x \oplus y = yf\left(\frac{x}{y}\right). \tag{1}$$

The assumptions about f imply that the operation is monotonic in the usual sense: if $x < x'$, then $x \oplus y < x' \oplus y$; and if $y < y'$, then $x \oplus y < x \oplus y'$. As will be shown in Chapter 20, Equation (1) captures all numerical, monotonic operations in a broad class of situations.

18.1.2 Representation and Uniqueness of Positive Operations

The first issue taken up in Chapter 19 is the axiomatization of qualitative concatenation operations that lead to a numerical representation. For positive operations, the basic finding (Theorem 19.3) is that one can simply drop the associativity assumption from the definition of an extensive structure and slightly modify the Archimedean property, and still prove that numerical representations exist, albeit with operations other than $+$. Moreover, this can be done in a constructive fashion not unlike, but more subtle than, the standard-sequence procedure used for extensive structures. The representation can be selected to be continuous in a certain familiar sense (Theorem 19.5). Finally, a natural definition of continuity of the operation in topological terms is given an equivalent algebraic formulation (Theorem 19.6).

The uniqueness result in this case differs in an important way from the extensive case. Since we do not have any special operation, such as $+$, singled out from among the infinity of possible numerical operations that can represent the structure, it is not immediately obvious which representation to select, thereby making it difficult to give a simple characterization of all admissible transformations. However, for exactly one of the possible representations, the group of admissible transformations has a particularly simple formulation, namely, as multiplication by certain positive constants. Put more formally, the admissible transformations form a subgroup of the similarity group (Theorem 19.4). Recall that in the extensive (associative) case, the additive representation has this property, with the admissible transformations being the full similarity group. In the general case, proper subgroups can arise. As in extensive measurement, it remains true that when two representations of a positive concatenation structure agree at a point, they are necessarily the same representation (Corollary, Theorem 19.2); however, it is no longer true that an admissible transformation always exists that maps the number assigned to an object into any other number. The special case of concatenation structures for which the admissible transformations are the similarity group (ratio scales), as in extensive measurement, is of considerable interest. The form of those operations is characterized in Chapter 20, and it is simply Equation (1).

An important distinction must be made between representations that map *onto* Re^+ and those that map *into* it. Typically, theories involving onto representations are appreciably simpler to formulate than are those whose variables are into the positive real numbers. One example of this arises in Chapter 20, where the classification of scale types is quite simple in the onto case. Of course, little real distinction exists when a structure that maps into Re^+ admits a relatively unique extension to one that maps onto Re^+. An example of this is the classical (nineteenth century) extension by the mathematician Dedekind of $\langle \mathrm{Ra}, \geqslant, +, \cdot \rangle$ to $\langle \mathrm{Re}, \geqslant, +, \cdot \rangle$, in which the latter is uniquely determined by the former. The general problem of doing this now goes under the name of Dedekind completion. Concatenation structures that are extendable to Dedekind complete structures are explored in some detail in Section 19.4. The major result for positive concatenation structures is necessary and sufficient conditions for such an extension to exist (Theorem 19.9).

18.1.3 Intensive Structures

The tack taken for constructing representations of positive concatenation structures works for those structures but does not generalize much beyond that case. In particular, it is of no help for intensive structures, which, necessarily, have idempotent operations ($x \bigcirc x \sim x$). The weighted mean is a prototypical example. Among other things, a standard sequence based on concatenating an element with itself simply goes nowhere. A partial theory, based on the strong property of bisymmetry, was developed in Chapter 6. Developed here is the general theory for intensive structures, which treats the operation as inducing a conjoint structure,

$$(a, b) \succsim' (c, d) \quad \text{iff} \quad a \bigcirc b \succsim c \bigcirc d.$$

18.1.4 Conjoint Structures and Distributive Operations

The general theory of binary conjoint structures follows exactly the same pattern as in the additive case. Recall that the tradeoff structure described by the conjoint ordering can be represented as a binary operation on one of the components (Holman, 1971; Section 6.2.4). Under the assumptions of additive conjoint measurement, including the Thomsen condition, the induced operation was shown to form an extensive structure, and that fact was used to develop the additive representation. Dropping the Thomsen condition, Theorem 19.11 establishes that the induced operation looks like two positive concatenation structures on either side of a zero element. Once again, from the representations of the positive concatenation structures one constructs a representation for the conjoint structure (Theorem 19.19). The issue of how the several induced operations relate is more subtle than in the additive case (see Theorem 19.15).

Chapter 10 treated, as a model for interlocked physical variables, a situation involving two concatenation structures, one on a component of a conjoint structure and the other either on the other component or on the entire structure. We assumed the concatenation structures to be extensive, the conjoint structure to be additive, and the three structures to be related by a law of exchange in the former case and a law of similitude in the latter. This work has subsequently been generalized considerably. To get the product of powers representation needed for the vector space of dimensions found in physics, the following will do: instead of two concatenation structures, assume just one on a factor of the conjoint structure; instead of extensive structures, assume a concatenation structure with a ratio scale representation; instead of an additive conjoint structure, postulate a general conjoint structure; and instead of a law of exchange or similitude relating two operations and the conjoint structure, assume a simple qualitative interlock between the concatenation and the conjoint structures, which is called *distributivity*. Some results about these interlocked structures are reported in Theorem 19.18, but the full treatment is postponed to Chapter 20 where a further generalization to ratio-scale structures that are not necessarily based on a concatenation operation becomes possible (Theorem 20.7; for further discussion, see Section 18.2.3).

18.2 SCALE TYPES (CHAPTER 20)

The fact that nice uniqueness results (though different from the additive case) were obtained with generalized operations and conjoint structures has motivated the study of the general question: What admissible transformations can arise in the theory of measurement? More can be said about that than one might first suspect, and Chapter 20 explores what is currently known.

The first issue is what general class of structures to consider. At present, the theory is worked out for general ordered relational structures in which there is a set A of objects, a total ordering $\succsim$ of A, and a number of additional relations R_j of finite order, i.e., subsets of the product $A^{n(j)}$, where $n(j)$ is a finite integer that is called the order of the relation. Such structures are too general to say much of interest about them, and so a second issue is how to impose additional constraints without becoming too specific. The restrictions imposed in earlier chapters have always taken the form of explicit assumptions about the relations, such as the associativity of a binary operation. This chapter takes a different, more abstract approach. Despite the abstractness, quite specific and concrete results are achieved about both scale types and specific structures, such as concatenation structures.

The restriction is formulated not in terms of specific properties of the primitives but in terms of symmetries exhibited by the structure. The term *symmetry* means, intuitively, a transformation of the structure so that it looks the same both before and after the transformation. Consider a square. Rotate it 90 degrees in the plane, or flip it 180 degrees about a diagonal, or flip it 180 degrees about an imaginary line parallel to two sides and half way between; in each case there is no difference between the original and transformed shape unless a distinctive mark is placed somewhere off center on the square, which destroys the symmetry. These and other transformations reflect some of the appreciable symmetry of an unmarked square. Likewise, when dealing with a more abstract structure, we can speak of its symmetries in terms of structure-preserving maps, isomorphisms, of the structure with itself. These representations of the structure onto itself rather than onto a numerical system are called *automorphisms*. The entire set of automorphisms forms a mathematical group under function composition.

18.2.1 A Classification of Automorphism Groups

The main thrust of Chapter 20 is to restrict structures indirectly by means of explicit restrictions on their automorphism groups. Three types of results follow. First, it is possible to arrive at very general results about the types of automorphism groups that can arise under these restrictions; in particular, it becomes clear why the classification of scale types into ordinal, interval, and ratio is natural and almost complete for all structures than can be mapped onto the real numbers. Second, it is possible to show that the structures exhibiting any of these automorphism groups have numerical representations with easily described uniqueness properties. And third, if we limit ourselves to specific classes of structures, such as concatenation ones, the structural impact of these automorphism restrictions is easily described.

The classification that has been proposed and found successful examines two logically distinct aspects of the automorphism group. Consider the fixed points exhibited by an automorphism α, i.e., values a for which $\alpha(a) = a$. The first aspect is a number N such that any automorphism with that many distinct fixed points is forced to be the identity automorphism, for which every point is fixed. In that case, the structure is said to be N-point unique. Thus a ratio scale, such as arises in extensive measurement, is 1-point unique whereas an interval scale, such as arises in additive conjoint measurement, is 2-point unique but not 1-point unique. As was noted, positive concatenation structures are also 1-point unique.

The second aspect of the classification concerns the richness of the automorphism group. Consider two M-tuples of points of the structure that are ordered in the same way. Does there exist an automorphism of the structure that maps one ordered set of points onto the other? If there does for each pair of similarly ordered M-tuples, the structure is said to be M-point homogeneous. The term *homogeneous* arises because, if the structure is at least 1-point homogeneous, no element can be distinguished from the others by its properties. Thus a structure with a maximal element, a minimal element, or a zero element cannot be M-point homogeneous for any $M \geqslant 1$ because each of these points is characteristically different from the remaining points of the structure. In this language, a ratio-scale structure is 1-point homogeneous whereas an interval scale is 2-point homogeneous. The general positive concatenation structure, though 1-point unique, can be either 0- or 1-point homogeneous.

A natural question to raise is what combinations can arise of N-point uniqueness, where N is the least value of uniqueness, and of M-point homogeneity, where M is the largest value of homogeneity. For structures with an infinity of points, it is easy to verify that $M \leqslant N$. The chapter focuses primarily on those structures for which $M \geqslant 1$ and N is finite. A major result (Theorem 20.5) is that for any relational structure meeting these two conditions and having a representation onto all of the real numbers, then necessarily $N \leqslant 2$. Moreover, the representation can be chosen so that interval-scale uniqueness obtains when $M = N = 2$ and ratio-scale uniqueness obtains when $M = N = 1$. In the case $M = 1$ and $N = 2$, the representation can be selected so that the group of admissible transformations is a proper subset of the affine transformations of an interval scale and a superset of the similarity transformations for a ratio scale. Put another way, most of the a priori possibilities simply cannot arise. There is no one-dimensional structure that is finitely unique, homogeneous, and can be mapped onto the real numbers (see discussion in Section 18.1.2) for which the scale type is other than ratio, interval, or something lying in between these two.

In addition to these cases, there are others about which we know a good deal less. There are those that are not N-point unique for any finite N, which are called ∞-point unique. An ordinal or an ordered metric scale is an example that is also M-point homogeneous for every M, and so is said to be ∞-point homogeneous. At the other end are cases that are 0-point homogeneous. These include anything from the most heterogeneous structures to those that are highly regular, for example, structures that are homogeneous on either side of a zero element. Little is known about how best to classify them, though their automorphism groups have been described.

18.2.2 Unit Representations

Both of the results just cited and those for positive concatenation structures have suggested additional properties about the automorphism groups. It is possible to partition the automorphisms into two types according to whether they have at least one fixed point or not. Automorphisms with no fixed points, together with the identity automorphism, are the ones that are represented by the similarity group in the theorem just mentioned. It is shown (Theorem 20.7) that certain properties of the automorphisms with no fixed points correspond exactly in the homogeneous case to the existence of a numerical representation with the same kind of uniqueness described earlier. These are called unit representations because of the role of the similarity group. Moreover, it becomes clear from studying this result that the proper generalization of the concept of Archimedeaness of an operation lies not in some easily described property of the structure itself but in the behavior of the subset of its automorphisms that have no fixed points. The result gives a recipe for determining if an arbitrary homogeneous structure has a numerical representation and provides a systematic way of finding it in terms of the automorphisms of the structure.

An additional result of importance concerns the product of powers representation of conjoint structures so prevalent in the dimensional structure of physics. A general definition is given for the distribution in a conjoint structure of an ordered relational structure defined on one of its factors. It is shown that if the relational structure has a unit representation and the conjoint structure is suitably solvable, then distribution forces the conjoint structure to be additive, and there is a product of powers representation (Theorem 20.8). This result, which is more general than the trinary laws that were studied in Section 10.7, demonstrates that the extensive structures assumed in dimensional analysis can be easily generalized to more general structures, including nonadditive operations, so long as they have ratio-scale representations. Thus new opportunities are opened for adjoining nonadditive scales to physical measurement.

18.2.3 Characterization of Homogeneous Concatenation and Conjoint Structures

The characterization of the possible automorphism groups in the homogeneous case can be brought to bear on more specific relational structures. The two cases that have been carefully examined are concatenation and conjoint structures. The concatenation case is the easier to describe. In essence one shows that for homogeneous, finitely unique concatenation structures, Equation (1) describes the class of representations in which the automorphisms with no fixed points are represented by similarity transformations (Theorem 20.10). Moreover, the scale type is completely determined (Theorem 20.11) by considering the set of values of $\rho > 0$ for which the following equation holds:

$$f(x^{\rho}) = f(x)^{\rho}, \qquad x > 0. \tag{2}$$

The ratio-scale case corresponds to Equation (2) having $\rho = 1$ as its only solution. The interval-scale case corresponds to Equation (2) holding for any $\rho > 0$. And in the intermediate case, Equation (2) holds just for ρ of the form $\rho = k^n$, where k is a fixed positive constant and n ranges over the integers.

The interval-scale case is so stringently limited that f is completely determined except for two constants. This is used to arrive at a theory of utility of risky alternatives that generalizes the classical subjective expected-utility model (Chapter 8) in a possibly useful way (Theorem 20.19).

Since the general conjoint theory is closely related to that for concatenation operations, it comes as no surprise that the above results are used in formulating results for the conjoint case (Theorems 20.23–26). Since the results are complex, we do not attempt to summarize them here. Suffice it to say that unless a zero element (defined as a point that is fixed in every automorphism) is involved, there is a strong tendency for homogeneity plus smoothness of the representation to force additivity. A simple criterion is provided (Theorem 20.27) for ascertaining whether a numerical conjoint structure has an additive representation; it is easier to use than that of Scheffé (1959) (Theorem 6.4).

18.2.4 Reprise

The material of Chapters 19 and 20 is novel in that one does not begin with an explicit operation, such as +, on the real numbers and then establish a representation theorem into that numerical structure. Rather, Chapter 19 shows for concatenation and conjoint structures how to construct both the representation and a suitable operation. Chapter 20 provides a proof schema in terms of properties of a subgroup of the automorphisms that enables a numerical representation to be constructed indirectly, using Hölder's theorem on that subgroup, even when no empirical operation is available. Chapter 22 shows, among other things, that this is just the class of structures needed to generalize the concept of a space of physical dimensions and that invariance under automorphisms corresponds to the usual dimensional invariance.

18.3 AXIOMATIZATION (CHAPTER 21)

18.3.1 Types of Axioms

We begin by noting some of the advantages of describing bodies of empirical knowledge in an axiomatic fashion and of representing such structures in familiar formal systems, usually numerical or geometric. For example, just attending to this volume, the issues that will be treated are (i) how to generalize from structures having additive representations to those with nonadditive ones, (ii) how to understand abstractly the concept of scale type and to describe the corresponding groups of transformations for their real representations, (iii) how to characterize the kinds of axiomatizations possible for different types of structures, and (iv) how to understand the class of propositions that can be meaningfully or invariantly formulated in terms of the primitives of a structure. Little progress had been made toward solving any of these problems or even formulating them clearly until they were treated in an explicitly axiomatic-representational framework. No one had any idea about what classes of numerical operations might arise as representations nor any sense about what scale types there might be other than ratio, interval, and ordinal. It is only within a carefully formulated axiomatic-representational context that these questions have been successfully addressed.

The chapter then turns to questions of a very general nature about types of axioms and what can and cannot be done within certain logical frameworks. Axioms in a measurement system can be grouped, as was discussed at length in Chapter 1, into three broad types, termed *necessary*, *structural*, and *Archimedean*. Typical necessary axioms are transitivity of a binary relation that is to be represented by the usual ordering $\geqslant$ on Re, positivity of a concatenation operation that is to be represented by addition of positive numbers, the Thomsen condition in additive conjoint measurement, and the axiom that any two points lie on some straight line in the classical geometries. Note that the first three examples can be formulated as universal statements in first-order logic. For example, transitivity is formally written as:

$$(\forall a)(\forall b)(\forall c)[(a \succsim b \;\&\; b \succsim c) \rightarrow a \succsim c].$$

the fourth example is also first-order but not universal since it involves an existential assertion:

$$(\forall a)(\forall b)(\exists L)[a \in L \;\&\; b \in L].$$

All of these are necessary in the sense that their truth is implied by the representation that we seek in the respective cases.

The most familiar structural axioms are the solvability ones of conjoint measurement. These, too, are existential, asserting the existence of solutions to certain equivalences; however, they are not necessary for the representation as ordinarily formulated. A structural axiom can, usually, be converted into a necessary axiom by imposing the structural requirements as part of the representation, which is exactly what happens in extensive measurement with a closed operation in which the representation is assumed to be addition, and so is closed. In that case, the imposition of added structure to the representation of an operation seems comparatively, though not totally, innocent. The reason, as we know from Chapter 3, is that at the expense of a more complex proof, it is possible to arrive at the same additive representation for a class of partial operations as for the closed ones. In contrast, the imposition of solvability in conjoint measurement seems considerably more deceptive because an infinity of universal axioms is needed to obtain numerical, additive representations for nonsolvable conjoint structures (see Chapter 9). In such cases, structural axioms are often assumed, despite the fact that they impose constraints on the structure beyond those actually needed for the desired representation, because then a relatively few axioms are sufficient to imply the existence of the representation and, therefore, all of the remaining necessary axioms. Moreover, in many applications the structural axioms are judged harmless in the sense that they appear to hold to a reasonable degree of approximation in the intended domain.

Although the Archimedean axioms are necessary for representations in Re or Re^n, they are classed separately because they are not formulated in a first-order logic and cannot be rewritten in such a logic (see Theorems 21.19 and 21.20). They can usually be formulated as an infinite disjunction of first-order statements. For example, in extensive measurement, the Archimedean axiom can be formulated as

$$(\forall a)(\forall b)[1a \succ b \text{ or } 2a \succ b \text{ or } 3a \succ b \text{ or } \cdots],$$

where $na \succ b$ is an abbreviation for the first-order statement that n concatenations of a with itself exceeds b. Such axioms keep the qualitative structures from having elements that differ only "infinitesimally," in which case representation in a one-dimensional numerical system is impossible.

18.3.2 Theorems on Axiomatizability

All three types of axioms have been used in measurement theory, but are they all needed? The first mathematically rigorous response to this question was obtained by Scott and Suppes (1958) who showed that the algebraic difference representation for finite structures cannot be axiomatized by any finite set of first-order universal statements. More precisely, they showed for any such axioms, one can construct a (finite) set A with a quaternary relation $\succsim$ that satisfies the axioms yet does not have a numerical representation as algebraic differences. These earlier results of Scott and Suppes are extended in Chapter 21 to obtain a negative answer for any finite, first-order axiomatization, universal or not.

In Chapter 21 we develop the background to state and prove in a precise form this negative result on finite axiomatization. As part of such considerations, we develop in explicit form the formalism of first-order logic, which we also refer to as the elementary formalization of theories. As part of these developments, we also introduce the main concepts concerning models of elementary languages, and we state without proof a number of general theorems about elementary logic, which are used in the sequel.

We define a class of measurement structures as a nonempty class of models of an elementary language that is closed under isomorphism and that is homomorphically embeddable in a numerical model of the language. We then state a number of results about the axiomatizability of measurement structures. A typical one is Theorem 21.6: the class of models of any elementary theory that has infinite models and is such that the theory of simple orders is interpretable in it is not a class of measurement structures.

Another section is devoted to the more restricted case, as mentioned above, of elementary theories that can be finitely axiomatized. The kinds of cases considered divide naturally into two: finite models that are axiomatizable by a universal sentence and models that are not finitely axiomatizable. The elementary theories with finite models that are axiomatizable by a universal sentence lead to theorems that follow along the lines of development begun earlier by Scott and Suppes, as mentioned above. Some results on conjoint measurement structures of this sort are given. The deepest result proved in Chapter 21 is an unpublished one of Per Lindstrom concerning the classes of finite models that are not finitely axiomatizable.

Separate subsections are devoted to definability of concepts in a given theory and to the interpretability of one theory in another. In the case of definability, we also give a brief summary of Padoa's method for proving that a particular primitive concept is independent of the other primitive concepts of a theory, i.e., cannot be defined in terms of them. This material plays a role in Chapter 22.

18.3.3 Testability of Axioms

The last part of Chapter 21 is concerned with the testability of axiomatic theories of measurement and of individual axioms, in relation to both finite and infinite data structures. The discussion here does not really depend on much logical apparatus, but it is important for the use of axiomatic systems in the empirical sciences. One approach tries to test individual axioms in isolation from the rest of the structure. For example, in any of the one-dimensional structures we have talked about, one can certainly study transitivity of the order in isolation. Equally, if an operation exists, one can test whether it is monotonic or not. There are distinct issues depending upon whether the data structure is conceived of as finite or infinite. A second approach attempts to fit the representation to a set of data and evaluates the overall adequacy of the representation. Each approach has its merits and proponents; we discuss these.

18.4 INVARIANCE AND MEANINGFULNESS (CHAPTER 22)

We next use the characterization of scale type in terms of automorphism groups, formulated in Chapter 20, to explore questions about various forms of invariance under these transformations. Such questions were discussed extensively in Chapter 10 (especially Sections 10.3 and 10.10) in terms of the controversial concept of dimensional invariance of physical laws. This is the key to dimensional analysis, and it played a significant role in the treatment of mechanics. Invariance arose again in Chapters 12 and 13 in connection with the definition of a geometrical object as something invariant under the automorphisms of the geometry in question (an aspect of the famous Erlanger program) and of a physical law in geometrical models of space-time, including both the classical and relativistic formulations. Within the general domain of measurement, invariance under scale type has been a major feature of the discussion of the meaningfulness of particular statistical quantities, hypotheses, and tests formulated in terms of numerical measures. That too has generated controversy.

Of the various terms used in connection with things that are invariant—geometrical objects, physical laws, and meaningful statements—the last has come to be the generic term; so we subsume all as a form of meaningfulness within an appropriate measurement structure. The aim of the chapter is to define and explain several invariance concepts, describe their significance, show how they are related, demonstrate that dimensional invariance is invariance under automorphisms, and clarify the relevant statistical issues in order to dispose of the existing controversy.

It will become clear, however, that certain questions remain unresolved, in part because the needed concepts are probably not fully evolved. It is possible that invariance under automorphisms may be the appropriate concept for ratio and interval scales and for any scale lying between them, but it is quite clear that it is inappropriate for nonhomogeneous structures without rich automorphism groups. Some of the difficulties relate to the meaning of *definability* in terms of the primitives of the structure together with purely logical concepts. Such issues are described in Sections 21.2.4 and 22.5.

18.4.1 Types of Invariance

A natural concept of invariance within the context of measurement arises by considering relations formulated within a numerical representation. For each representation, one can find the qualitative relation corresponding to a particular numerical relation. If that qualitative relation remains invariant as the numerical homomorphism is altered, we say the numerical relation exhibits *reference invariance*. This concept is more slippery than it first seems. One aspect, illustrated in the case of Hooke's law, is the role of dimensional constants.

At least two other notions of invariance have arisen. One is the invariance of a qualitative relation under endomorphisms, which are homomorphisms of the structure into itself. This is called structure invariance. Another focuses on the notion of admissible transformations within the numerical representation and imposes the condition of invariance of the numerical relation under them. Theorems 22.3–5 explore the conditions under which these three concepts are identical. A sufficient condition is that the structure be homogeneous and 1-point unique.

18.4.2 Applications of Meaningfulness

The two most important areas to which the measurement-theoretic concept of meaningfulness has been applied are statistics and dimensional analysis.

Controversy about meaningfulness considerations vis-à-vis statistical practice has centered largely on whether such considerations are relevant at all, and if so, under what circumstances. Section 22.6 reviews the various positions that have been taken. According to our analysis of the problem, measurement considerations relate to statistical inference in just one way; namely, that any statistical hypothesis formulated in terms of the measures should be meaningful within the measurement system being used. This appears to be the only constraint imposed by measurement considerations. In particular, one may sometimes find a monotonic rescaling that produces population distributions within a specified family, e.g., the Gaussian, and another monotonic rescaling that provides an attractive combination rule relative to conjoint structure or to a concatenation operation. Both rescalings may be unique up to the same class of transformations, e.g., affine ones, but be related only by a monotone transformation unconstrained by theory.

As we have seen in Chapter 10, there are two major problems in understanding how dimensional analysis arises out of fundamental measurement theory. One is to construct from a set of one-dimensional structures describing distinct attributes the multiplicative vector space representing all physical (and possibly other) measures. The other is to improve our understanding of why the condition of dimensional invariance is imposed on physical laws.

Chapter 10 offered one possible way to construct the vector space. The construction is redone in Theorem 22.8 using the considerably more general results of Chapter 20 on distribution of unit structures in conjoint ones. This formulation, unlike the earlier one, is not restricted to extensive operations on components of an additive conjoint structure nor to the specific types of interlocking laws. Second, it is shown that dimensional invariance within that framework, involving invariance under the transformations called similarities, is equivalent to a family of systems related by automorphisms of the qualitative structure represented by a vector space. Such a family constitutes a meaningful (automorphism-invariant) relation within that structure.

Included in this material is a detailed discussion of the significant differences between a measurement-theoretical analysis of an attribute and the uses of indices to substitute for such an analysis. Typical examples are attempts to use physical ratio scales, such as time intervals, as indices for various poorly understood attributes of human performance. We attempt by example to make clear why the scale type of the index does not necessarily carry over to the attribute being indexed.

The last section focuses in a general way on the problems of constructing some representation of a structure and on the uniqueness of that construction. The two issues are closely intertwined. In many cases it is possible, and sometimes easy, to establish the existence of a representation in a nonconstructive fashion. From an empirical point of view, such a result is not entirely satisfactory because it provides no way to approximate the representation to a particular level of accuracy. For that reason, measurement theorists go to some pains to arrive at constructive proofs that can actually be used to devise approximation procedures. Unfortunately, such proofs are often a good deal longer and more tedious than the nonconstructive ones. Moreover, in some instances no such proofs have yet been devised.

Chapter 19 Nonadditive Representations

19.1 INTRODUCTION

Most work in the theory of measurement has been based on additive representations for various kinds of structures. There are good reasons, however, for also studying structures in which the numerical combination rules are intrinsically nonadditive. In this introduction we begin with a discussion of nonadditivity in its various guises. Then we sketch the approach to nonadditivity pursued in this chapter; namely, representation by nonadditive binary numerical operations. Finally, we give an overview of the five major parts of the chapter.

19.1.1 Inessential and Essential Nonadditivities

Conjoint structures are common in the sciences, arising whenever an ordered outcome depends on two or more independent variables and also in other, less obvious ways. We need to distinguish three types of numerical representations for conjoint structures: additive representations, nonadditive representations that can be transformed to additive, and representations that are essentially nonadditive.

In Volume I (e.g., Sections 3.9, 4.4.2, and 6.5.2), we discussed the possibilities of alternative numerical representations for structures that do have additive representations. The most obvious and useful example is the transformation by exponentiation of an additive representation into a multiplicative one. More generally, any strictly monotonic transformation h transforms addition, $x + y$, into the nonadditive binary numerical operation,

$$u \oplus v = h^{-1}[h(u) + h(v)]. \tag{1}$$

That is, instead of representing the two factors by scale values x, y, which add, one could just as well use the scale values $u = h^{-1}(x)$ and $v = h^{-1}(y)$, replacing addition by the numerical binary operation $\oplus$ defined by Equation (1). Any numerical operation obtained by such a transformation has the same key mathematical properties as addition: associativity and commutativity.

The choice of addition or multiplication, rather than some alternative, is from a purely mathematical viewpoint fundamentally a matter of convention; however, powerful pragmatic considerations, such as ease of computation, often dictate the choice. If for some reason a nonadditive alternative is selected, such nonadditivity is said to be *inessential*.

Closely related to the above are cases in which a seemingly natural numerical scale exhibits interactions among variables, but such interaction can be eliminated by a monotonic transformation of the scale. As an example stemming from psychology, consider a two-factor conjoint structure in which the probability Q of solving a particular problem depends both on D, the difficulty of the problem, and on T, the level of training of the problem solver. These two independent variables may interact, in part because very easy problems have a high probability of being solved, even with little training; so there is not much room for Q to increase with T. Likewise, very hard problems have a low Q even at high levels of T; so there is not much room for Q to decrease at lower levels of T. In this example, the interaction can be reduced by transforming the probability scale monotonically to obtain a scale that has more room at the top and bottom. The logit transformation $Q \to \log[Q/(1 - Q)]$ is often used. One can test the axioms of two-factor additive conjoint measurement to determine whether *any* monotonic transformation of Q exists that will eliminate the interaction, resulting in a formula of the form $h(Q) = x(T) + y(D)$. But if the interaction can be eliminated by a monotonic transformation h, then one can also elect to preserve the original scale Q, replacing $x(T)$ by $u(T) = h^{-1}[x(T)]$ and $y(D)$ by $v(D) = h^{-1}[y(D)]$ and replacing $+$ with the binary operation $\oplus$ of Equation (1). Again, it is a matter of convention whether one chooses the "natural" scale Q with the nonadditive but associative binary operation $\oplus$ or the transformed scale $h(Q)$ with addition as the operation.

Not every interaction can be eliminated by a monotonic transformation, however, and indeed there are many cases in which it is doubtful if additivity affords an adequate representation for conjoint structures. Such nonadditivity is called *essential* and corresponds to a nonassociative operation. As was noted in Section 18.1.1, operations that are essentially nonadditive are by no means unusual in the sciences. A striking example is vector product. If $\mathbf{x} \otimes \mathbf{y}$ denotes the vector (or cross) product of vectors $\mathbf{x}$ and $\mathbf{y}$, then neither associativity nor commutativity hold. In their stead are the Jacobi identity,

$$(\mathbf{x} \otimes \mathbf{y}) \otimes \mathbf{z} + (\mathbf{y} \otimes \mathbf{z}) \otimes \mathbf{x} + (\mathbf{z} \otimes \mathbf{x}) \otimes \mathbf{y} = 0,$$

and skew symmetry,

$$\mathbf{x} \otimes \mathbf{y} = -(\mathbf{y} \otimes \mathbf{x}).$$

Another familiar nonassociative example is weighted averages.

Even when an additive representation can be used as a first approximation, the interactions among variables may be of considerable theoretical interest. To illustrate this, consider again the example of difficulty and training factors in problem solving. Suppose that there exists some domain of problems such that the probability Q of correctly solving each problem can be expressed as a product of two factors: the probability S of selecting the correct algorithm times the conditional probability E of executing that algorithm properly after selecting it. Each of the two latter probabilities, in turn, depends on the two empirical variables, difficulty and training. These two dependencies might in turn be described by additive models using logit scales. Thus the complete model might look like this:

$$Q = SE, \tag{2a}$$

$$\log \frac{S}{1-S} = f(T) + g(D), \tag{2b}$$

$$\log \frac{E}{1-E} = k(T) + l(D). \tag{2c}$$

It can be shown (Exercise 1) that the relation specified by Equation (2) between the probability Q of correct solution, the training level T, and the difficulty level D, is essentially nonadditive for general choices of f, g, k, and l. There is no monotone transformation h such that $h(Q)$ can be expressed as the sum of a function of T and a function of D except for special cases in which the functions k and l are related to the functions f and g in a particular way that eliminates the nonadditivity.

Of course, it might be argued that Q is simply the wrong thing to measure; rather, the probabilities S and E should be observed separately. These depend additively on T and D and are more fundamental. That might be possible in the above example: one could ask a person first to select an algorithm for a problem, and if the correct one is selected, then one could ask the person to execute the algorithm and obtain the solution. But there could be other examples in which intervening processes are less observable. What is illustrated here is a way of generating a broad class of examples of two-factor conjoint structures for which additivity fails. The observed measure may depend on several unobservable factors, each in turn depending in a different way on the two empirically specifiable variables. Even if all component dependencies are essentially additive, as in Equation (2a–c), the overall dependency is most likely essentially nonadditive. Moreover, the process might be repeated for some of the component dependencies; why should they be additive any more than the overall relation? In the example above, only the multiplicative relation $Q = SE$ has a clear justification.

The methods to be discussed in this chapter apply to a broad class of nonadditive combination rules in which the dependency of an ordered outcome on two factors can be expressed as a general function of two variables, monotone in each. In the above example, the general equation would take the form

$$Q = H[f(T), g(D)]. \tag{3}$$

This is the equation of *decomposability*, treated in Chapter 7 and here specialized to the case of two variables. In particular, the model of Equation (2) falls under this rubric whenever the functions k and l are monotonically increasing functions of f and g, respectively. More generally, if an observed measure depends monotonically on several unobservable variables, each of which depends on the same two empirically specifiable variables, with all these dependencies covarying monotonically, then the overall relation will satisfy decomposability.

Volume I treated three kinds of essential nonadditivity: weighted-average representations (Section 6.9), polynomial combination rules (Chapter 7 and Sections 2.2.7 and 9.5) and, very briefly, decomposability (Section 7.2). Averaging is a nonassociative binary operation, nonetheless, it can be treated via additive conjoint measurement as was done in Section 6.9. A generalization of that procedure is presented in this chapter (Theorem 21, Section 19.7.2). The polynomial results concern only addition and multiplication, but even so, the analysis in Chapter 7 was restricted to a special subclass of polynomials. The results in Chapter 7 for general decomposable structures are quite weak. The essential innovation discussed in this chapter is to treat two-variable decomposability from the standpoint of binary operations that are not necessarily associative. Because of the nonassociativity, the nonadditivity is essential: the operation cannot be converted into addition by the inverse of Equation (1). This approach is suitable for empirical relational structures that have a binary concatenation operation and for two-factor conjoint structures. It has not yet been generalized to more complex situations.

19.1.2 General Binary Operations

We shall consider numerical representations using a quite general binary operation $\otimes$ in place of $+$. One of the simplest examples is the general linear operation

$$x \otimes y = rx + sy + t, \tag{4}$$

which is used in the representation of bisymmetric structures (see Sections 6.9 and 19.9). Here, r, s, and t are constants, with r and s positive. With $t = 0$ and $r + s = 1$, this corresponds to a weighted average, which is commutative only if $r = s = 1/2$ and is never associative. Another sort of example is the equation

$$x \otimes y = x + y + x^{\alpha} y^{\beta}, \qquad x, y > 0, \quad \alpha, \beta > 0 \tag{5}$$

The operation defined by Equation (5) is commutative if and only if $\alpha = \beta$. A little less obviously, it is associative if and only if α and β are both 0 or both 1. Still another class of examples can be obtained by regarding Equation (2) as the definition of a binary operation.

One consequence of considering arbitrary binary operations is that the representation problem and the uniqueness problem become conceptually more distinct than they were in Volume I. The representation problem is best thought of as one of *discovery*: what operation $\otimes$ (if any) can be used to give a numerical representation for a particular empirical structure? Naturally, if there is some operation that can be used, then there are many, just as alternatives to $+$ are generated by arbitrary monotonic transformations h of Re. But the conditions under which *some* operations can be found are much weaker and easier to discover than are the conditions for a prespecified operation. The uniqueness problem, on the other hand, is best thought of as a problem of *scale construction*: given the form of the desired representation, including a specification of the operation $\otimes$, how firmly is the numerical representation determined?

The results most easily obtained concern the number N of distinct points needed to pin down a representation uniquely. Such a representation is called *N-point unique*. For example, the ratio scales that arise in extensive measurement are 1-point unique since the entire representation is determined by specification of a unit. And interval scales, being determined by the choice of a zero and a unit, are 2-point unique. (As we shall see in Chapter 20, the possible values of N for structures that map onto the real numbers are very limited.) Results about uniqueness are proved constructively: a method is given whereby once the numerical values for N distinct elements are selected, the other scale values can be constructed to any desired degree of approximation.

19.1.3 Overview

Section 19.2 presents some general definitions and examples that are used throughout the chapter. The remainder of the chapter covers five main topics. Clearly, the first task is to establish some general results about nonassociative binary operations. This can be done in an abstract setting without necessarily assuming that the operation is defined for real numbers; and thus we obtain a generalization of the theory of extensive measurement (Chapter 3), in which a binary operation $\bigcirc$, not necessarily associative, is represented by some binary numerical operation $\otimes$, also not necessarily associative. The theory should specialize to that of Chapter 3 when the operation is associative. The concatenation operations considered in this first part are *order preserving* (monotonicity axiom) and *positive*, i.e., $a \bigcirc b$ is always larger than both a and b. In addition, two more technical properties (restricted solvability and Archimedean properties, defined in the next section) are assumed. The acronym PCS (positive concatenation structure) is employed for this generalization of extensive structures.

We do not know of direct applications of PCSs that arise as "naturally" as the associative ones employed in extensive measurement of mass, length, and time duration. Nonetheless, there are good reasons for studying them. The intellectual problem is interesting and not obvious; the results are needed for understanding the numerical PCSs that will be used as representations for various structures; there are important indirect applications to both conjoint measurement and intensive measurement (which are the third and fourth main topics of this chapter); and perhaps, with the machinery in place, some direct applications ultimately will be found.

The striking result of this initial part is that associativity of the concatenation operation $\bigcirc$ can simply be dropped; hardly any other modification of the extensive measurement axioms is needed to obtain nearly the same representation and uniqueness theorems, with a general numerical operation $\otimes$ in place of $+$ or $\cdot$. The most notable change is in the uniqueness theorem. In the context of the additive, associative theory, uniqueness could be described in either of two equivalent ways. The one is to say that once the value of a single element is specified, the representation is unique. The other is to say that any two representations are related by multiplication by a positive constant. For the more general structures considered in Section 19.3, these two notions are no longer equivalent, and it is the former (1-point uniqueness) that arises most naturally. The proof of 1-point uniqueness specifies a constructive procedure for measurement: given the numerical value assigned to a single element in the concatenation structure, we show how to determine the numerical value assigned to any other element. Later in the chapter we consider 2-point uniqueness (in place of interval scales); and, more generally we consider N-point uniqueness in Chapter 20, where we present results concerning permissible transformations and their relation to N-point uniqueness properties.

Complete clarification of the uniqueness issue takes place through the study of the automorphism groups of measurement structures. The reason for their importance is that each automorphism of a structure generates a permissible transformation of any representation (although the converse is not true in general). An important theorem asserts that the automorphism group of any PCS is isomorphic to a subgroup of the additive reals (Section 19.3.4). The consequences of this and related theorems are discussed, though the full development of these consequences appears in Chapter 20, which is about scale types.

The second topic involves studying those PCSs that, like most physical models, can be represented by an (often infinite) interval of real numbers. Of special concern are the conditions under which a given PCS can be densely embedded, much as the rational numbers are in the reals, in a larger PCS that is isomorphic to a real interval and, in particular, such that the automorphisms extend in a natural way. This problem is only partially understood.

The third main topic of the chapter is the analysis of nonadditive conjoint structures. Just as in the additive version, a two-factor conjoint ordering induces an ordering of intervals on each factor, and thence, a concatenation structure can be defined for each factor. The results for PCSs can be applied to these defined concatenation structures to obtain nonadditive, two-factor conjoint measurement. The major departure from the additive case is that this procedure does not generalize readily to three or more factors. With three factors, a PCS can be defined over one factor in terms of a second factor with the level of the third factor held constant. But without additivity, different choices for this constant level lead to different and nonisomorphic PCSs. Hence, additional work is needed to clarify the possibilities for a general nonadditive theory with three or more factors. Presumably it will involve the study of operations with more than two arguments.

The fourth topic is intensive concatenation structures in which the property of $a \bigcirc b$ lying *between* a and b is substituted for positivity. Such structures, with the representation of Equation (4), arise in the measurement theories for density, temperature, and expectations (e.g., expected-utility theory), for which there is a natural concatenation operation that produces a result intermediate in value rather than greater than the larger of the two concatenated objects. Intensive concatenations can be analyzed quite simply in terms of two-factor conjoint structures. This leads to a general nonadditive representation theorem. In addition, one class of intensive structures is shown to have a close connection to positive, nonassociative structures.

The final topic of the chapter is the relation among a number of special properties, such as bisymmetry, that can be added to the axioms of a concatenation structure. These results are mainly of interest in connection with generalizations of subjective expected utility theory (Section 20.4.7).

The reader should be cautioned that proofs in a nonassociative context can be tricky. We are so used to associativity that we frequently invoke it without mention in proofs, and as a result it is easy to sneak it in inappropriately. Care should be taken in the exercises; in particular, parentheses must always be maintained in expressions such as $(a \bigcirc b) \bigcirc c$ and $(a \bigcirc a) \bigcirc a$.

19.2 TYPES OF CONCATENATION STRUCTURE

In this section we define a general notion of concatenation structure that applies throughout the chapter. We also define and discuss various special conditions that concatenation structures can satisfy, and we present some examples of structures that satisfy or violate various of these conditions.

19.2.1 Concatenation Structures and Their Properties

The most general notion of concatenation structure for which we have found use is modeled after the formulation of extensive measurement structures (Definition 3.3). A *concatenation structure* satisfies the same properties as an extensive structure except for associativity, positivity, and the Archimedean axiom. Dropping associativity is, of course, the main topic of the chapter. Positivity will sometimes be assumed as a special condition (as in PCSs, later), but we also study structures in which it is replaced by other properties. The Archimedean axiom is dropped only temporarily because in the nonassociative case there can be various Archmedean axioms; we shall introduce and discuss three of them. The most important conditions that remain are (1) transitivity and connectedness of the ordering $\succsim$, (2) monotonicity of the concatenation $\bigcirc$ with respect to $\succsim$, and (3) a structural condition specifying that $a \bigcirc b$ is defined for all "sufficiently small" elements a and b. These three conditions are listed as the numbered axioms in Definition 1. Some minor technical points are embedded in the specifications of the primitives in that definition.

As in Definition 3.3, the primitive notions consist of a set A, a binary relation $\succsim$ on A, and a partial binary operation $\bigcirc$ from a subset B of $A \times A$ into A. The set B contains those pairs for which concatenation is defined, and we shall say "$a \bigcirc b$ is defined" rather than the more formal "$(a, b) \in B$". Later (Definition 10), we modify property 2 to deal with structures having positive and negative parts.

DEFINITION 1. *Suppose that A is a nonempty set, $\succsim$ is a binary relation on A, and $\bigcirc$ is a partial binary operation on A with nonempty domain B. Let $\mathscr{A} = \langle A, \succsim, \bigcirc \rangle$. $\mathscr{A}$ is called a* concatenation structure *iff the following three conditions are satisfied*:

1. Weak order: *$\succsim$ is a weak order on A.*

2. Local definability: *if $a \bigcirc b$ is defined, $a \succsim c$, and $b \succsim d$, then $c \bigcirc d$ is defined.*

3. Monotonicity:
   (i) *if $a \bigcirc c$ and $b \bigcirc c$ are defined, then $a \succsim b$ iff $a \bigcirc c \succsim b \bigcirc c$;*
   (ii) *if $c \bigcirc a$ and $c \bigcirc b$ are defined, then $a \succsim b$ iff $c \bigcirc a \succsim c \bigcirc b$.*

The following definition serves as a collection point for many of the properties of concatenation structures that come up in this chapter. The examples and discussion in Section 19.2.2 are intended to illustrate and clarify the various conditions introduced in this definition. The Archimedean properties are deferred to Section 19.2.3, and others still later.

DEFINITION 2. *Let $\mathscr{A} = \langle A, \succsim, \bigcirc \rangle$ be a concatenation structure. $\mathscr{A}$ is said to be*:

Closed *iff $a \bigcirc b$ is defined for all $a, b \in A$.*
Positive *iff whenever $a \bigcirc b$ is defined, $a \bigcirc b \succ a, b$.*
Negative *iff whenever $a \bigcirc b$ is defined, $a \bigcirc b \prec a, b$.*
Idempotent *iff $a \sim a \bigcirc a$ whenever $a \bigcirc a$ is defined.*
Intern *iff whenever $a \succ b$ and $a \bigcirc b$ or $b \bigcirc a$ is defined, $a \succ a \bigcirc b \succ b$, and $a \succ b \bigcirc a \succ b$.*

Intensive *iff it is both intern and idempotent.*

Associative *iff whenever one of* $(a \bigcirc b) \bigcirc c$ *or* $a \bigcirc (b \bigcirc c)$ *is defined, the other expression is also defined and* $(a \bigcirc b) \bigcirc c \sim a \bigcirc (b \bigcirc c)$.

Bisymmetric *iff* $\mathscr{A}$ *is closed and* $(a \bigcirc b) \bigcirc (c \bigcirc d) \sim (a \bigcirc c) \bigcirc (b \bigcirc d)$.

Autodistributive *iff* $\mathscr{A}$ *is closed and both* $(a \bigcirc b) \bigcirc c \sim (a \bigcirc c) \bigcirc (b \bigcirc c)$ *and* $c \bigcirc (a \bigcirc b) \sim (c \bigcirc a) \bigcirc (c \bigcirc b)$.[1]

Halvable *iff* $\mathscr{A}$ *is positive and, for each* $a \in A$, *there exists a* $b \in A$ *such that* $b \bigcirc b$ *is defined and* $a \sim b \bigcirc b$.

Restrictedly solvable *iff whenever* $a \succ b$ *there exists* $c \in A$ *such that either* $b \bigcirc c$ *is defined and* $a \succsim b \bigcirc c \succ b$ *or* $a \bigcirc c$ *is defined and* $a \succ a \bigcirc c \succsim b$.[2]

Solvable *iff given a and b there exist c and d such that* $a \bigcirc c \sim b \sim d \bigcirc a$.

Dedekind complete *iff* $\langle A, \succsim \rangle$ *is Dedekind complete, i.e., every nonempty subset of A that has an upper bound has a least upper bound in A.*

Continuous *iff the operation* $\bigcirc$ *is continuous as a function of two variables, using the order topology on its range and the relative product order topology on its domain.*[3]

Archimedean properties are treated separately in Section 19.2.3.

19.2.2 Some Numerical Examples

One artificial but instructive way to obtain a nonadditive numerical structure is to use addition for some pairs and multiplication for others. Specifically, define $\otimes$ for positive x, y as follows:

$$x \otimes y = \begin{cases} x + y, & \text{if either } x \text{ or } y \text{ is } < 3, \\ xy, & \text{otherwise.} \end{cases}$$

The domains of addition and multiplication are graphed in Figure 1. The heavy line in the figure depicts the boundary, at which the operation $\otimes$ is discontinuous. For example, let $x = 4$; as y approaches 3 from below, $x \otimes y$ approaches 7; but it approaches 12 as y approaches 3 from above.

The operation is nonassociative: for example, $(4 \otimes 2) \otimes 2 = 8$, but $4 \otimes (2 \otimes 2) = 16$. The structure $\langle \mathrm{Re}^+, \geqslant, \otimes \rangle$ is a closed, positive concatenation structure. It satisfies restricted solvability but not solvability; for

[1] These two equations are sometimes considered separately; the corresponding properties are called *right* and *left* autodistributivity.

[2] The latter is, of course, impossible in a positive structure and the former, in a negative one.

[3] A more concrete definition of continuity, specialized to PCSs, is given in Section 19.3.5. For definitions, see any standard topology text; e.g., Kelley (1955) defines *order topology* (pp. 57–58), *product topology* (p. 90), and *relative topology* (pp. 50–51).

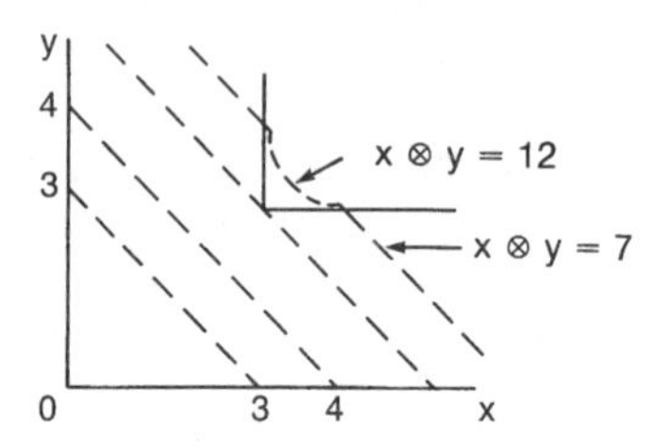

FIGURE 1. Numerical concatenation operation having a region of discontinuity. See text for formula.

example,

$$6 \otimes 3 = 18 > 10 > 8 = 6 \otimes 2,$$

but the equation $6 \otimes x = 10$ has no solution (since 10 falls in the gap created by the discontinuity) though the inequality $10 > y \otimes x$ has a solution x for every y less than 10. Since the ordering is the usual one on the positive reals, the structure is Dedekind complete, but as we have noted, it is not continuous.

The next example is of a one-parameter family of numerical relational structures. They are nonadditive except for three values of the parameter. The domain is the positive real numbers with their usual ordering and $\otimes$ defined by

$$x \otimes y = x + y + 2c(xy)^{1/2},$$

where c is a constant between -1 and 1. For each nonnegative value of c, the structure satisfies the axioms of Definition 1: weak ordering is immediate, monotonicity follows from the fact that the first partial derivatives of $x \otimes y$ are positive (since x and y are positive and c is nonnegative), and local definability is not an issue because the operation in fact is closed. For negative values of c, however, monotonicity fails.

If c is nonnegative, then the operation is positive but neither negative, idempotent, nor intern (and so, not intensive). It is not associative except for $c = -1$, 0, or $+1$. (This can be checked numerically for a particular value of c, e.g., $c = 1/2$; a general proof can be obtained from the fact the partial derivatives of $x \otimes y$ do not satisfy Scheffé's criterion for additivity, Theorem 6.4.) Likewise, it can be shown that the structure is not bisymmetric except for those same exceptional values of c, and it is never autodistributive. It is halvable for nonnegative c since the equation $x \otimes x = y$ has a positive solution $x = y/2(1 + c)$ for every positive y. The structure is obviously continuous and Dedekind complete, and a standard continuity argument shows that it is restrictedly solvable. It is not solvable.

The operation $x \otimes y$ gives the variance of the sum of two random variables **X** and **Y** whose respective variances are x and y and whose correlation is c. One might envisage an application to a situation in which there are two correlated sources of error, whose variances change with changing conditions but whose correlation remains fixed.

It is instructive to consider an example of a finite nonassociative concatenation structure. For this we let $A = \{1, 2, 3, 4, 5\}$, with the usual ordering, and we define a concatenation $\bigcirc$ by the following table:

Values of $a \bigcirc b$

		b 1	2	3	4	5
	1	2	3	5	—	—
	2	3	4	—	—	—
a	3	5	—	—	—	—
	4	—	—	—	—	—
	5	—	—	—	—	—

Note. Dashes mean that $a \bigcirc b$ is undefined.

Observe that, unlike the first example, in which $B = \mathrm{Re}^+ \times \mathrm{Re}^+$, the operation $\bigcirc$ is not defined everywhere. This is necessarily so in a structure that is both positive and finite. The operation is nonassociative since $(1 \bigcirc 1) \bigcirc (1 \bigcirc 1) = 2 \bigcirc 2 = 4$ but $((1 \bigcirc 1) \bigcirc 1) \bigcirc 1 = 3 \bigcirc 1 = 5$. However, it is commutative since the table is symmetric. This structure is in fact the smallest example of a commutative but nonassociative concatenation structure. (Exercise 3 shows that if A has only four elements, commutativity implies associativity.) Note finally that this structure is isomorphic to the structure consisting of $A' = \{1, 3, 5.73, \ldots, 9, 9.12, \ldots\}$ with the concatenation given by $a \bigcirc b = a + b + (ab)^{1/2}$, that is, it is isomorphic to a substructure of the previous example (in which $c = 1/2$). We simply took the first five elements generated by starting with $x = y = 1$ and replaced those numbers by their ranks. The finite structure is, of course, not halvable.

For another example of a nonassociative structure, consider an operation that combines two gambles by forming a probability mixture of them. In the simplest version of this, a gamble is just a probability distribution over a finite set of outcomes. (Later, in Section 20.4.5, we shall consider a quite different method of defining gambles, which leads to concatenation structures with different properties.) For any two finite gambles a, b and any number p such that $0 < p < 1$, the mixture $a \bigcirc_p b$ is defined as the gamble in which gamble a is played with probability p and otherwise b is played. For any suitable p the binary operation $\bigcirc_p$ on finite gambles is closed, idempotent, bisymmetric, and autodistributive. It is closed because the operation, applied to any two finite gambles, yields another gamble. Idempotence, bisymmetry, and autodistributivity all hold with $=$ in place of $\sim$ because the definition of the operation means that the two probability distributions on either side of $=$ are in fact identical. Since the structure is idempotent, it cannot be positive, negative, or halvable.

The ordering over finite gambles is usually based on some sort of human judgment, such as a preference judgment, a riskiness judgment, a selling price, etc. (Borcherding, Brehmer, Vlek, and Wagenaar, 1984; Edwards and Tversky, 1967; Kaplan and Schwartz, 1975; Schoemaker, 1980). For any such ordering, it is highly plausible that the resulting structure is intern.

Solvability, restricted solvability, continuity, and Dedekind completeness depend on the structure of the outcome set over which the finite gambles are defined. If this outcome set is taken to be Re, interpreted as amounts of money or rates of income (including idealized amounts or rates), with the induced ordering the same as the natural ordering of Re, then we expect continuity and Dedekind completeness to be satisfied and the solvability properties to hold also.

The most crucial questions about such mixture structures concern the properties of Definition 1 rather than Definition 2. Both transitivity and monotonicity of preferences have been shown to fail empirically when the set of gambles is identified with finite probability distributions. Failure of monotonicity is demonstrated by the class of examples introduced by Allais (1953). Consider gambles involving the three outcomes 0, 1, 10, where the units are chosen suitably, e.g., thousands of dollars, or, for the wealthier, millions of dollars. Many people prefer gamble $a = 1$ (i.e., 1 with certainty) to the gamble $b = 10 \bigcirc_{.9} 0$, where there is an appreciable risk of getting nothing. If so, then monotonicity implies both that $0 \bigcirc_{.9} a$ is preferred to $0 \bigcirc_{.9} b$ and that $1 \bigcirc_{.9} a$ is preferred to $1 \bigcirc_{.9} b$. However, these latter four gambles have the probability distributions shown in the following table:

		Outcome			
		$0 \bigcirc_{.9} a$	$0 \bigcirc_{.9} b$	$1 \bigcirc_{.9} a$	$1 \bigcirc_{.9} b$
	.90	0	0	1	1
Probability	.09	1	10	1	10
	.01	1	0	1	0

The prediction is that people who prefer a to b will prefer the gamble in column 1 to that in column 2 and the one in column 3 to that in column 4. People who prefer b to a should reverse their preferences for each pair of columns. But what is observed is that many people prefer column 2 to column 1, yet prefer column 3 to column 4. That is, they prefer a sure 1 to a risk of 0, even though they will go for the chance to win 10 when winning is improbable (see Allais and Hagen, 1979; Kahneman and Tversky, 1979). We shall see in Section 20.4.6 that this class of examples cannot necessarily be interpreted as refuting monotonicity of a gamble-mixture concatenation structure. In that section we present a theory in which gambles are not identified with their probability distributions. The concatenation of gambles remains idempotent, but it is no longer bisymmetric or autodistributive. For such structures, monotonicity of concatenation is not violated in the preceding example, and in fact, suitable empirical tests of monotonicity seem not to have been performed.

Even transitivity can fail descriptively, as Tversky (1969) showed. For example, an individual might prefer $a = 2 \bigcirc_{.7} 0$ to $b = 2.5 \bigcirc_{.6} 0$, electing a greater probability of winning at the cost of a smaller win. The same individual may prefer b to $c = 3 \bigcirc_{.5} 0$ for a similar reason. Yet when a and c are compared, the person may now consider the larger difference in amount to be won as sufficiently important that c is chosen despite its lower probability of winning. (Such intransitivities can be avoided if gambles are evaluated singly, but then these evaluations do not necessarily predict choices correctly.) See Section 17.2.5 for a further discussion of this problem.

Another example, mostly negative in its import, has its motivation in studies of sensory thresholds. Let Y be a bounded interval of nonnegative real numbers, representing the possible physical intensities y of a background stimulus, and let X be a similar interval representing the possible physical intensities x of a signal that is presented for possible detection against some background in Y. Define an ordering by $(x, y) \succ (x', y')$ whenever x against background y is more detectable (or more salient—brighter, louder, etc.) than is x' against background y'. Let $\bigcirc$ be ordinary component-wise addition on $X \times Y$, i.e.,

$$(x, y) \bigcirc (x', y') = (x + x', y + y').$$

Let B be the set of pairs for which $\bigcirc$ is defined, i.e., for which $x + x' \prec \sup(X)$ and $y + y' \prec \sup(Y)$.

The operation $\bigcirc$ is, of course, associative because $+$ is. The structure is not closed because addition is limited by the boundedness of X and Y. It is not positive since adding a small increment together with a large background can decrease detectability of a given signal. Neither is it negative since adding a large increment with little background can increase detectability. The simplest version of Weber's law [for an axiomatic treatment, see Narens (1980)] states that the detectability depends only on the ratio of signal to background, x/y, and if this were strictly true, it would imply idempotence:

$$(x, y) \sim (x, y) \bigcirc (x, y) \sim (2x, 2y).$$

Empirically, this fails; generally, $(2x, 2y) \succ (x, y)$, and indeed, a good approximation is often obtained from the linear inhomogeneous Weber law, which asserts that the ordering is monotonic with $x/(y + c)$ for some positive constant c. A more general Weber model assumes that the ordering is monotonic with $f(x + y) - f(y)$ for some increasing concave function f; this reduces to the linear inhomogeneous Weber law for $f(x) = \log(x + c)$. Such models imply that the structure is not intern: if (x, y) and (x', y') are very close together, then their concatenation is more detectable than either alone. More importantly, the monotonicity axiom (Definition 1) is not satisfied, a fact readily predicted by any of the Weber models just introduced. Local definability (Definition 1) fails similarly for this structure.

In this example, weak ordering and associativity are satisfied, but none of the other properties that we have introduced hold. The main point is that in this case the ordering and the "natural" concatenation operation have little to do with one another. The concatenation is available simply because we are dealing with physical intensity measures and hence with real numbers, but it does not mesh well with the ordering based on a sensory detection task. It is far better to drop the concatenation and regard the structure as conjoint. The linear inhomogeneous Weber law is in fact an additive conjoint representation (taking logarithms). The general Weber model, using an arbitrary increasing concave function, does not, in general, have an additive conjoint representation using X and Y as factors; but it does have a nonadditive conjoint representation in X and Y, so we shall return to this example as an illustration of Theorem 19 in Section 19.7.1.

The most prominent "averaging" processes found in both physical and social sciences, such as the earlier gambling example, have usually been assumed to be bisymmetric; but as indicated earlier, we shall see excellent reasons for considering more general idempotent structures. Except for these important empirical cases, we know of very few serious examples of empirical structures that involve both an order and a nonassociative operation. Many physical and geometric operations are not associative, but these structures do not usually possess a monotonic total order. For this reason we have had to resort to numerical examples, with little empirical motivation, or to negative examples, in which the concatenations are unsuitable.

Another major motive exists for studying nonassociative structures: namely, they arise indirectly from nonadditive conjoint structures. Under rather general assumptions, a nonadditive conjoint structure induces on one component a family of nonassociative operations that encode exactly the information included in the conjoint structure. Thus representations for concatenation structures can be used to construct representations for conjoint structures. Thus far we have not found any other general method for treating nonadditive conjoint structures.

Some further comments are in order concerning Dedekind completeness and continuity. These properties were satisfied in earlier examples in which we were dealing with operations defined on intervals of real numbers, or products of such intervals, and the definitions used standard continuous functions. Dedekind completeness will sometimes be assumed as a property of a more abstract structure; its function is to force the structure to be representable by a whole real interval rather than by a ragged set of real numbers. As such, it is invariably an idealization; for instance, in the gambling example, we assumed that amounts of money are indefinitely divisible in order to obtain a mixture operation in which Dedekind completeness is satisfied. This idealization leads to a far simpler mathematical structure and thus facilitates other investigations. Examples are found especially in work on dimensional analysis (Chapter 10) and on automorphisms and scale types (Chapter 20). Perhaps, wherever it is used, analogous but more complex theorems could be formulated without it, but we have not been consistent in carrying out such a program wherever it could be attempted. Continuity is perhaps less of an idealization; it will be made more concrete and discussed in Section 19.3.5 for PCSs.

19.2.3 Archimedean Properties

Archimedean axioms specify the conditions under which the basic process of counting off equal units produces an approximate scale value for any to-be-measured object. In ordinary length measurement, the usual Archimedean axiom specifies that sufficiently many millimeter lengths, laid end-to-end, will reach from earth to the sun. In other words, there is no infinite standard sequence (e.g., a sequence of 1 mm, 2 mm, 3 mm, ...) that is bounded by the length of some fixed object or interval.

The meaning of this axiom is elucidated by cases in which it is violated. For example, it is violated by concatenation of velocity; if $a \bigcirc b$ is the velocity, relative to a fixed reference frame, of an object that is moving with velocity b relative to a frame that, in turn, is moving at velocity a relative to the fixed frame, then, according to special relativity, the sequence $a, a \bigcirc a, (a \bigcirc a) \bigcirc a, \ldots$ never attains its asymptote, the velocity of light. In the non-Archimedean cases that have arisen, most of the usual theory of extensive measurement can nonetheless be obtained. (See Section 3.7 of Volume I.) There is a different sort of example in which some changes are effectively infinitesimal relative to others. This can happen if objects are multidimensional and are ordered lexicographically, so that some dimensions are effective only if the dimensions having precedence are tied. (Some examples of this sort give up solvability as well as the Archimedean property.) In such cases, one has to forego representation in Re; instead, one uses a multidimensional representation or a representation in a nonstandard continuum (Narens, 1974). These examples show the importance of Archimedean axioms. This is discussed in greater detail in Section 21.6.

In Volume I, the Archimedean axiom took various specified forms because in different structures there are different ways of specifying what constitutes "equal units," that is, there are different definitions of standard sequences. For example, in concatenation structures that are positive, it is natural to form an equally spaced sequence by concatenating an element with itself, i.e., $a, a \bigcirc a, (a \bigcirc a) \bigcirc a$, etc. but in idempotent structures, those concatenations are all equal to a. Hence, for these structures, equally spaced sequences are constructed by choosing any pair b, c, of nonequivalent elements and choosing successive elements a_j and a_{j+1} such that

$$a_{j+1} \bigcirc b = a_j \bigcirc c.$$

For nonassociative concatenation operations, this difficulty is exacerbated in two ways. First, there are various ways to concatenate an element with itself. Four copies of an element a can be put together as $((a \bigcirc a) \bigcirc a)) \bigcirc a$, as $(a \bigcirc a) \bigcirc (a \bigcirc a)$, or as $a \bigcirc (a \bigcirc (a \bigcirc a))$. More generally, given any binary rooted tree T with n terminal elements, we can define an element $a(T)$ consisting of n copies of a that have been concatenated according to the structure of T.

Second, different constructions of equally spaced sequences that are equivalent in associative structures are no longer equivalent in more general structures. The method sketched above for idempotent structures could also be used in positive structures. In positive, associative structures, the two constructions lead to Archimedean axioms that are closely related (see Section 3.2), but we do not know in general how they relate in the nonassociative case.

We deal with the first problem rather arbitrarily, by picking out a particular family of binary trees and using them to generate concatenations of equal units. We use the strictly left-branching trees because these can be constructed using the restricted solvability axiom as we have formulated it, with concatenation on the right. (Obviously we could have reversed the order of concatenation in that axiom and used right-branching trees.) We shall use the notation $a(n)$ for the left-branching concatenation of n copies of element a. (The notation na, used in Volume I and in much of the literature, could lead to confusion here, particularly in concatenation structures on Re.) Thus, we define the sequence $a(n)$ inductively by

$$\begin{aligned} a(1) &= a \\ a(n) &= [a(n-1)] \bigcirc a, \quad \text{if the right-hand side is defined} \\ a(n) &\ \text{is undefined otherwise.} \end{aligned}$$

The above definition holds in any concatenation structure for any element $a \in A$. A *standard* (*concatenation*) *sequence* is any $\{a(n) | n \in J\}$, where either $J = \{1, 2, \ldots, n\}$ or $J = \{1, 2, \ldots\}$, with all its elements defined. The sequence is said to be finite or infinite depending on J. (A concatenation structure need not have any infinite standard sequences; of course, $a(1) = a$ is a finite standard concatenation sequence for any element a.)

The use of a particular family of trees could perhaps be avoided by letting $a(n)$ be the minimum of $a(T)$ over all binary rooted trees with n terminal elements. It might be necessary to strengthen the restricted solvability axiom and perhaps to add other assumptions so that an Archimedean axiom formulated in this way would turn out equivalent to one formulated in terms of a particular family of trees.

In this connection, the following example, due to M. Cozzens, makes clear that such a result can only be found under some restrictions. Let $\langle \text{Re}, \geqslant, \odot \rangle$ be a real structure with

$$x \odot y = x + \frac{y}{2}.$$

It is easy to verify that the left-branching definition of a standard sequence yields

$$x_{\mathrm{L}}(n) = \frac{x(n+1)}{2},$$

whereas the right-branching definition yields

$$x_{\mathrm{R}}(n) = \frac{x(2^n - 1)}{2^{n-1}}.$$

The former is unbounded and the latter is bounded by $2x$. Thus it is Archimedean in left-branching but not in right-branching sequences. The structure is not positive since $x + (y/2)$ need not exceed y, but it is weakly positive in the sense that $x \odot x = (3/2)x > x$.

For idempotent structures we capture the notion of equal spacing in a different way, as sketched above: A sequence $\{a_j | j \in J\}$ is called a *difference sequence* if there exist b, c in A with not $(b \sim c)$ such that for $\forall j, j+1 \in J$,

$$a_{j+1} \bigcirc b \sim a_j \bigcirc c,$$

where J is either $\{1, 2, \ldots, n\}$ or $\{1, 2, \ldots\}$, and the sequence is accordingly called finite or infinite.

Note that a standard concatenation sequence can be imagined to be a special case of a difference sequence in which $c = a$ and b is thought of as a (nonexistent) identity element; thus $a_{j+1} \bigcirc b \sim a_{j+1}$. One cannot choose $c = a \bigcirc a$ and $b = a$ since that implicitly assumes associativity.

A major weakness of the concept of difference sequences is that they exist only in structures with many exact solutions to equations. We can be sure they exist in solvable concatenation structures but not otherwise. Thus a more satisfactory concept is the following notion of a *regular sequence* (Narens, 1985): there exist $b, c \in A$ with $c \succ b$ such that for all j, $j + 1 \in J$,

$$a_{j+1} \bigcirc b \succ a_j \bigcirc c \qquad \text{and} \qquad b \bigcirc a_{j+1} \succ c \bigcirc a_j.$$

All three types of sequence will be considered in various places in the remainder of the chapter. Note that the idempotent case is not dealt with until after the treatment of conjoint structures. It is quite possible that still other types of standard sequence may be needed in the development of nonadditive representations for structures different from the ones considered in this chapter.

A sequence of any type is said to be *bounded* if there exist elements $d, d' \in A$ such that for every a in the sequence [i.e., every a_j or every $a(j)$], $d' \succsim a \succsim d$. By an Archimedean axiom we mean an assertion that every *bounded sequence of a particular type is finite*; the type of sequence—standard concatenation, difference, or regular—will be specified in the

Archimedean axioms introduced for various structures. We add to the word *Archimedean* either *in standard sequences*, *in differences*, or *in regular sequences*.[4] For nonidempotent structures, where all three Archimedean properties can be defined, we do not know when one of them implies the other, except in some special cases (Luce, 1987).

We close this section with a theorem of Luce and Narens (1985) that derives the first two Archimedean properties from Dedekind completeness and some level of solvability.

THEOREM 1. *Suppose $\mathscr{A}$ is a Dedekind complete concatenation structure.*

(i) *If $\mathscr{A}$ is positive and left-solvable in the sense that for $b \succ a$, there exists c such that $b = c \bigcirc a$, then it is Archimedean in standard sequences.*

(ii) *If $\mathscr{A}$ is solvable, then it is Archimedean in difference sequences.*

Thus, to obtain representations that are onto Re or Re$^+$ or a closed interval and, hence, represent structures that are Dedekind complete, solvability properties are sufficient to insure Archimedean properties.

With the Archimedean properties, we have in hand all the definitions needed to define PCSs and develop their representation and uniqueness theory. The next three sections cover general and Dedekind-complete PCSs.

19.3 REPRESENTATIONS OF PCSs

19.3.1 General Definitions

We begin with the definitions of a PCS and of a homomorphism from one PCS into another.[5] The latter (Definition 4) is, of course, motivated by

[4] Narens (1985) referred to the concept of being "Archimedean in regular sequences" as "strongly Archimedean." Clearly, it is stronger in the sense of applying when there are no difference sequences, but when difference sequences do exist, as in solvable structures, then it appears to be a weaker concept since as is easily shown (Exercise 4), in a solvable structure, Archimedean in differences implies Archimedean in regular sequences.

[5] Narens and Luce (1976), who first introduced the concept of a PCS, initially called it a "positive concatenation structure." Only later did it become apparent that a general concept of concatenation structure (Definition 1) is needed and that various other concepts, including positivity, should be identified separately (Definition 2). In particular, the phrase *positive concatenation structure* now merely means a concatenation structure that is positive whereas the original concept was specialized by the inclusion of additional properties (Definition 3). Thus, in later publications, they began to use the acronym PCS for the specialized concept in an attempt to avoid confusion. Another term that might have been used is *generalized extensive structure* since the term *extensive structure* is used for an associative PCS.

the idea of a numerical representation for a PCS, that is, a homomorphism of an empirical PCS into a numerical PCS.

DEFINITION 3. *Suppose that* $\mathscr{A} = \langle A, \succsim, \bigcirc \rangle$ *is a concatenation structure.*

1. $\mathscr{A}$ *is said to be a* PCS *iff it is positive, restrictedly solvable, and Archimedean in standard sequences.*
2. *An associative* PCS *is said to be* extensive.
3. *A* PCS *in which* $A \subseteq \mathrm{Re}^+$ *and* $\succsim$ *is the usual ordering* $\geqslant$ *of* Re^+, *restricted to the subset* A, *is said to be* a numerical PCS. (*It is usually denoted* $\mathscr{R} = \langle R, \geqslant, \otimes \rangle$.)

Thus we assume for a PCS essentially all the properties of an extensive structure (Definition 3.3 of Volume I) with the exception of associativity. We say *essentially* because the monotonicity property of Definition 3.3 was stated in a weak form, which is equivalent in the context of the other axioms (including associativity) to the monotonicity property stated in Definition 1.

A homomorphism of $\mathscr{A}$ into $\mathscr{A}'$ will be defined as a function φ from A *into* (but not necessarily *onto*) A' that maps $\succsim$ into $\succsim'$ and maps $\bigcirc$ into $\bigcirc'$. In particular, if $a \bigcirc b$ is defined, then so is $\varphi(a) \bigcirc' \varphi(b)$, but the converse need not hold; for example, in the numerical case, the operation on Re^+ could be defined for all pairs of numbers even though the operation $\bigcirc$ is limited in its domain.

DEFINITION 4. *Let* $\mathscr{A} = \langle A, \succsim, \bigcirc \rangle$ *and* $\mathscr{A}' = \langle A', \succsim', \bigcirc' \rangle$ *be* PCSs, *and let* φ *be a function from* A *into* A'. φ *is a* homomorphism of $\mathscr{A}$ into $\mathscr{A}'$ *iff the following two conditions hold*:

1. $a \succsim b$ *iff* $\varphi(a) \succsim' \varphi(b)$;
2. *if* $a \bigcirc b$ *is defined, then* $\varphi(a) \bigcirc' \varphi(b)$ *is defined and* $\varphi(a \bigcirc b) = \varphi(a) \bigcirc' \varphi(b)$.

A homomorphism in $\mathscr{A}$ *is said to be a* measurement representation *iff the structure* $\mathscr{A}'$ *is a numerical* PCS.

EXAMPLES. The structure in Section 19.2.2 with the concatenation $x \otimes y = x + y + c(xy)^{1/2}$, where x, y are positive, is a numerical PCS for each $c \geqslant 0$. It is extensive when either $c = 0$ or 1. Another example is given by the concatenation $x \otimes' y = (x^2 + y^2 + cxy)^{1/2}$ for the positive reals. These two structures are related by the homomorphism $x \to x^{1/2}$ mapping Re^+ onto itself. If $\otimes$ is interpreted as an addition formula for variances, then $\otimes'$ is just the corresponding formula for standard deviations.

The finite structure of the example on page 29 is not a PCS because it fails to satisfy restricted solvability: we have $5 > 4$, but there is no x such that $5 \geq 4 \bigcirc x$. We noted earlier that this finite structure is isomorphic to a five-element subset of Re^+ with the variance-addition concatenation for $c = 1/2$. Though this isomorphism satisfies the two properties of a homomorphism (Definition 4), it is not one since the domain is not a PCS.

There is, up to PCS-isomorphism, only one five-element PCS that is commutative but has no homomorphism into an extensive (i.e., associative) PCS (Exercise 5).

19.3.2 Uniqueness and Construction of a Representation of a PCS

Our first theorem about PCSs states that the set of homomorphisms of a PCS into a fixed PCS satisfies 1-point uniqueness: if φ and ψ are two homomorphisms of $\mathscr{A}$ into $\mathscr{A}'$ that agree at one point, then they agree everywhere except possibly at one maximal point that is coupled only loosely to other points by concatenation relations. The reason for this uniqueness is that because of the (restricted) solvability and Archimedean properties, the nonmaximal points of a PCS are tightly coupled to one another by concatenations, and a homomorphism must preserve the structure of these concatenations. If a is coupled to b by concatenations, e.g., $b \sim (a \bigcirc a) \bigcirc (a \bigcirc a)$, then $\varphi(a) = x$ and $\varphi(b) = y$ are coupled by corresponding concatenations, i.e., $y = (x \bigcirc' x) \bigcirc' (x \bigcirc' x)$. The theorem shows further that homomorphisms can be ordered: if $\varphi(b) \succsim' \psi(b)$ at one nonmaximal point b, the same direction of inequality holds at every nonmaximal point.

THEOREM 2. *Let φ and ψ be two homomorphisms from a* PCS $\mathscr{A} = \langle A, \succsim, \bigcirc \rangle$ *into a* PCS $\mathscr{A}' = \langle A', \succsim', \bigcirc' \rangle$. *If there is any nonmaximal element b for which $\varphi(b) \succsim' \psi(b)$, then $\varphi(a) \succsim' \psi(a)$ for every nonmaximal $a \in A$.*

COROLLARY. *The representation of a* PCS *is* 1*-point unique in the sense that if for some nonmaximal a, $\varphi(a) \sim' \psi(a)$, then the same is true for all nonmaximal a.*

The proof of this theorem is given in Sections 19.5.2 and 19.5.3.

Assuming that a representation exists, we next outline how to calculate one homomorphism for an arbitrary element c, $\varphi(c)$, given the value of $\varphi(b)$ for some particular b. The details are worked out in Section 19.5.4. Choose some "small" a in A (in the formal proof, we use sequences $\{a_j\}$ that converge to zero in a well-defined sense), and construct the standard concatenation sequence $a(k)$ until b is just attained. This is possible because of the Archimedean property. Let n be the number of copies used, i.e.,

$$a(n+1) \succ b \succsim a(n).$$

Do the same for c, using m copies of a,

$$a(m+1) \succ c \succsim a(m).$$

These inequalities are mapped by φ into inequalities in $\mathscr{A}'$ that use the corresponding standard $\bigcirc'$-concatenation sequences based on $\varphi(a)$:

$$\begin{aligned} \varphi(a)(n+1) &\succ' \varphi(b) \succsim' \varphi(a)(n) \\ \varphi(a)(m+1) &\succ' \varphi(c) \succsim' \varphi(a)(m). \end{aligned}$$

In these inequalities, the quantities $\varphi(b)$, n, and m are known but $\varphi(a)$ and $\varphi(c)$ are not. However, we can search in A' for some element d that is only a little larger than $\varphi(a)$; the criterion is that $d(n) \succ' \varphi(b) \succsim' d(n-1)$. This gives $d(m+1)$ as a good upper bound on $\varphi(c)$. Similarly, find e such that $e(n+2) \succ' \varphi(b) \succsim' e(n+1)$. Such an element must be a little below $\varphi(a)$; hence, $e(m)$ gives a good lower bound for $\varphi(c)$. In Section 19.5.4 we formulate what it means for these estimates to converge to $\varphi(c)$, and we show that for $b \succ c$ they do converge. For $c \succ b$, we replace b by some sufficiently large $b(k)$ so that $b(k) \succ c$, which is satisfactory since knowing $\varphi(b)$ means we also know $\varphi[b(k)] = \varphi(b)(k)$.

19.3.3 Existence of a Representation

We turn now to the question of existence of homomorphisms. If we desired to parallel the uniqueness theory, we might take an arbitrary but fixed numerical PCS $\mathscr{R}$ and ask what conditions on $\mathscr{A}$ are sufficient to allow the construction of a homomorphism from $\mathscr{A}$ into $\mathscr{R}$. However, it does not seem possible to develop a useful theory at this level of generality. The existence theorems of this sort that we do have are particular to certain numerical PCSs with special properties. For example, in the specific case in which $R = Re^+$ and $\otimes$ is addition, multiplication, or any other associative operation, the necessary and sufficient condition for the existence of a PCS homomorphism is simply that $\mathscr{A}$ be associative. This is the standard representation theorem for extensive measurement. Representation theorems for other specific cases also exist, for example, the treatment of bisymmetric representations in Volume I (Section 6.9).

Although we do not know how to produce general theorems about the *existence* of homomorphisms between two arbitrary PCSs $\mathscr{A}$ and $\mathscr{R}$, a weaker representation problem can be posed and solved: when is there *some* numerical PCS that serves as a representation of $\mathscr{A}$? Since we do not specify the numerical operation in advance, we are in a situation in which if one representation exists, so do many others. Recall that in the associative case, any strictly increasing function h led to an alternative numerical representation, with the operation $+$ replaced by the numerical operation $\oplus$ that is related to $+$ by: $x \oplus y = h^{-1}[h(x) + h(y)]$. Here we have the same ordinal uniqueness: any strictly increasing h is a permissible transformation; if $\otimes$ is the representing operation, then the new operation $\otimes'$ is given by

$$x \otimes' y = h^{-1}[h(x) \otimes h(y)].$$

The representation and uniqueness theorem of this subsection is essentially the same as the theory of monotone decomposability in Volume I (Section 7.2), restricting the latter to the case of functions of two variables. We simply recast those results in a different notation and present them more precisely here, with an additional elaboration concerning continuity of the representing operation and of the permissible transformations (Section 19.3.5). Just as monotonicity and the existence of a countable order-dense[6] subset were sufficient in Section 7.2, the PCS conditions, including monotonicity and the Archimedean property, are sufficient here. We show how to construct some representing numerical operation for every PCS.

THEOREM 3. *Suppose* $\mathscr{A} = \langle A, \succsim, \bigcirc \rangle$ *is a* PCS.

(i) *There exists a numerical* PCS $\mathscr{R}$ *such that there is a homomorphism* φ *of* $\mathscr{A}$ *into* $\mathscr{R}$.

(ii) *If* φ *is a homomorphism of* $\mathscr{A}$ *into* $\mathscr{R}$, *then* φ' *is another such representation iff there exists a strictly increasing function h from* $\varphi(A)$ *onto* $\varphi'(A)$ *such that, for all* $a \in A$,

$$\varphi'(a) = h[\psi(a)]$$

and such that the numerical operations $\otimes$ *and* $\otimes'$ *are related as follows*:

[6] Recall B is order dense in A if for each $a, c \in A$ with $a \succ c$, there exists $b \in B$ such that $a \succsim b$ and $b \succsim c$. B is countable if it can be placed in 1:1 correspondence with a subset of the integers.

$x \otimes' y$ is defined iff $h(x) \otimes h(y)$ is defined, and when they are defined,

$$x \otimes' y = h^{-1}\{h(x) \otimes h(y)\}.$$

Several explanatory remarks are in order concerning this theorem.

First, in its essence, the theorem says that all PCSs are such that the usual conditions for the existence of a purely ordinal representation are met (Theorems 2.2 and 2.3), which means there is a countable, order-dense subset. Thus it merely asserts that the objects in set A can be given numerical labels in a way that preserves order. If this can be done at all, it can be done in many ways: any strictly increasing function defined on the set of numerical labels gives a different set of numerical labels that is just as good, and any two such sets of labels must be related by a strictly increasing function since both preserve order.

Second, the numerical operation is just the empirical operation $\bigcirc$ transferred to work on the labels. Since the original structure is a PCS, the resulting numerical structure must also be a PCS. The structures are exactly alike except that the empirical structure converts equivalence into equality of the representing labels. From this standpoint the existence of some numerical PCS to represent an arbitrary empirical PCS is neither surprising nor deep, once one sees how to prove the existence of a countable order-dense subset.

Third, the permissible transformations are easy to specify because they do not operate within a single numerical PCS; rather, they map one set of numerical labels into an entirely different and unrelated set of such labels. This contrasts greatly with the situation that arises when a numerical PCS is specified in advance and we are concerned about permissible transformations within that PCS. In that situation we expect ratio-scale uniqueness, but we do not get it in the nonassociative case without additional structural assumptions (see Theorem 20.11). Instead, we get the 1-point uniqueness result of Theorem 2, which, although useful for purposes of scale construction, does not tell us anything about permissible transformations and alternative scales within the same representing PCS. We do, however, obtain a simple characterization of all permissible transformations (Theorem 4).

Fourth, although not deep, Theorem 3 can be valuable if it leads in some cases to the *discovery* of interesting nonassociative numerical operations. To that end, we devote some care to its proof. One way to prove it, as noted in the first remark above, is simply to show that the axioms of a PCS imply the existence of a countable dense subset, which is what Narens and Luce (1976) did. One can then appeal to the general theory of ordinal measurement, e.g., Theorem 2.2 of Volume I. (Exercises 7, 8, and 9 develop that proof.) Alternatively, one can give an appreciably longer proof (as we shall do in Section 19.5.4) in which a particular countable dense subset is constructed and then used to show in detail how the numerical labels and the numerical operation can be determined approximately from finite sets of order and concatenation relations. A third approach, which in principle applies to structures more general than concatenation ones, has been suggested by Niederée (1987). Its full development remains to be published. It begins with a collection of statements that can be made in terms of the primitives of the structure about two elements, one of which is fixed as a unit while the other varies over the entire domain. For the case at hand, these statements are the inequalities $a(m) \succsim b(n)$ and $a(m) \prec b(n)$. He applies model-theoretic methods, which in the associative case reduces to Dedekind's classical method for constructing the real numbers from the rational ones, to these classes of statements to literally construct the "numbers" needed for the representation. The axioms, of course, are required to establish that the resulting construction is isomorphic to a subset of the real numbers. The method has the advantage that it directly constructs what is needed and does not presuppose the real numbers. Moreover, it introduces an intrinsic topology into measurement that serves to yield a qualitative meaning to the concept of degree of approximation. It is unclear at present how to assign a numerical measure of accuracy or to speak of rate of convergence.

We sketch the construction of Section 19.5.4 because instances may arise when it can provide a practical way of approximating values of the representing homomorphism φ and of discovering the corresponding numerical operation $\otimes$. The basic idea is to construct a sequence of finite approximations to the representation, each based on a finite standard concatenation sequence that is finer-grained than the one used in the previous approximation and also extends to larger elements of A. Thus we first take some a_1 and form a finite standard concatenation sequence $a_1(j)$, stopping the sequence at some arbitrary point. Then we take a smaller a_2 and form a finer standard concatenation sequence, stopping at some higher point. It is shown that we can do this in such a way that for any $a \in A$, no matter how small, there is eventually a standard concatenation sequence $a_k(j)$ such that $a_k(1)$ is smaller than a. Furthermore, we can arrange the stopping points of successive standard sequences in such a way that they pass successive "milestones" that diverge either to infinity or, if there is one, to the maximal element of A. Thus, for any nonmaximal element $b \in A$ and for any small element $a \in A$, we can find some k such that the kth milestone exceeds b and also such that $a_k(1)$ is smaller than a. The finite concatenation sequence $a_k(j)$ brackets b with grain size smaller than a.

The collection of the finite concatenation sequences whose construction we have just sketched yields a countable, order-dense subset. Rather than appeal to the general theory of ordinal representations, however, we construct a specific representation φ using the special properties of this subset. We define φ inductively for successive finite concatenation sequences (constructed as before). At each stage, we assign equally spaced numerical values to any portion of a standard sequence that falls between two successive elements for which the φ values were constructed earlier. (There is no special reason to use equal spacing, but it serves the purpose as well as any other definite method of assigning values.) Starting with the first finite standard sequence $a_1(1), \ldots, a_1(n)$, we choose the equally spaced numerical values $1, \ldots, n$. That is, $\varphi[a_1(j)] = j$. Next, if m elements from $a_2(j)$ fall strictly between two successive elements of the a_1 sequence, the interval between the two values previously defined for those a_1 elements is divided into $m + 1$ equal subintervals by the values assigned to these m elements from the a_2 sequence. The initial portion of the a_2 sequence, which falls below a_1, is assigned a sequence of values equally spaced between 0 and $\varphi[a_1(1)] = 1$; and the final portion of the a_2 sequence, which falls above $a_1(n)$, is assigned a sequence of values equally spaced between n and $n + 1$. The a_3 sequence is then handled similarly, treating the a_1 and a_2 sequences as a single set of elements with previously defined φ values and continuing to subdivide intervals between successive values equally.

It can be shown that for any nonmaximal b in A and for any small positive ϵ, this construction eventually yields an interval from $\varphi[a_k(j)]$ to $\varphi[a_k(j + 1)]$ that has a length less than ϵ and that must include the value of $\varphi(b)$. Formally, φ is defined for any $b \in A$ by taking maxima or minima over φ values defined inductively as above. For practical purposes, however, the inductive construction gives any desired degree of approximation. From close examination of the details in Section 19.5.4, it is even possible, once b has been located relative to a set of milestones in A, to determine what value of k suffices to approximate $\varphi(b)$ by an interval of length at most ϵ.

Given the approximation to φ, one can also obtain an approximation to the required operation $\otimes$. To find the approximate value of $\varphi(a) \otimes \varphi(b)$, one needs merely to construct the concatenation sequence, based on a sufficiently fine element a_k that brackets $a \bigcirc b$.

19.3.4 Automorphism Groups of PCSs

As was noted, we have not yet characterized the class of admissible transformations associated with any particular representation of a PCS. We have shown that such representations are 1-point unique, but we have not described all of them as we did for the structures of Volume I. For example, we know that for the class of extensive structures represented on

$\langle \text{Re}^+, \geqslant, + \rangle$, the group of admissible transformations is the similarity group, i.e., multiplication by all positive constants. Here, it is less clear how to formulate such a result since we do not have a canonical numerical operation. Thus we need a characterization of these transformations that is intrinsic to the structure itself.

Observe that if φ and ψ are both isomorphisms from a totally ordered PCS $\mathscr{A}$ *onto* the same numerical PCS $\mathscr{R}$, and if h is the increasing function that takes φ into ψ, then $\varphi^{-1}h\varphi = \varphi^{-1}\psi$ is an isomorphism of $\mathscr{A}$ onto itself. (See Exercise 10.) Such a self-isomorphism is called an *automorphism* of the structure, and as is easily seen, the set of all automorphisms forms a group under function composition (Exercise 11). We shall fully explore the nature of automorphism groups of quite general measurement structures in Chapter 20. For the moment, however, our only interest in them is to describe as fully as possible the uniqueness of representations of PCSs. The following nice result of Cohen and Narens (1979) formulates the desired information.

THEOREM 4. *The automorphism group of a PCS is an Archimedean ordered group.*

There are three comments to be made.

First, where does the order on the automorphism group come from? Obviously, it must derive from that of the PCS. Indeed, the key observation is that for a PCS (and many other 1-point unique structures) any two automorphisms α and β have the property that if for some $a \in A$, $\alpha(a) \succ \beta(a)$, then the same inequality holds for all $a \in A$. Thus it is natural to define the ordering $\succsim'$ on the automorphism group by

$$\alpha \succsim' \beta \quad \text{iff} \quad \alpha(a) \succsim \beta(a) \quad \text{for all } a \in A.$$

This is the order referred to.

Second, this result coupled with Hölder's theorem means that the automorphism group of a PCS is isomorphic to a subgroup of the similarity group, i.e., $x \to rx$, $x > 0$, $r > 0$. One such subgroup is, of course, the ratio-scale case, which corresponds to the entire group of positive real numbers, but there are other possibilities. Two of the most important are the trivial group consisting of only the identity, in which case the representation is said to be *absolute*, and the discrete group of the form $x \to k^n x$, where $k > 0$ is fixed and n ranges over the (positive and negative) integers. Examples of PCSs with such automorphism groups can be found in Cohen and Narens (1979).

Third, in Section 20.2.6, we use Theorem 4 to formulate a generalized concept of Archimedean structures: ones with automorphism groups that are Archimedean (hence, isomorphic to a subgroup of similarities). In Section 20.4.2 we give a simple, yet complete characterization of all concatenation structures with large (i.e., homogeneous) Archimedean automorphism groups.

19.3.5 Continuous PCSs

In Definition 2, a concatenation structure was defined to be continuous in terms of the continuity of the operation $\bigcirc$ using the order topology on its range and the relative product order topology on its domain.

THEOREM 5. *Suppose $\mathscr{A}$ is a PCS.*

(i) *There exists a homomorphism φ that is a continuous function using the order topology of $\mathscr{A}$ and the relative topology of the homomorphic image $\mathscr{R}$.*

(ii) *If $\mathscr{A}$ is continuous, then under φ of part* (i), *$\mathscr{R}$ is a continuous* PCS.

(iii) *Suppose φ and ψ are homomorphisms related by the strictly increasing function h,* i.e., *$\psi = h\varphi$, and φ is continuous. Then ψ is continuous iff h is bicontinuous.*

This topological result, really a corollary to Theorem 3, is a bit deeper than the theorem itself because it assures us that the representations are continuous using the normal or relative (i.e., open set) topology of subsets of real numbers rather than a special (order) topology for each set of labels. In fact, if only the order topologies are considered, any order-preserving function from one simply ordered set (the equivalence classes of A) onto another (the set of labels) is bicontinuous. The representation merely copies the topology of A onto the set of labels, just as it copies the operation $\bigcirc$ onto the set of labels. By contrast, continuity in the relative topology means that familiar concepts of proximity of real numbers can be used.

Because the definition of continuity given in Definition 2 is rather abstract, stated as it is in terms of the order topology on A and the product topology on $A \times A$, it may be useful to formulate a more concrete alternative.

DEFINITION 5. *Let $\mathscr{A} = \langle A, \succsim, \bigcirc \rangle$ be a* PCS *with no minimal element. It is said to be* lower semicontinuous *iff whenever $a \bigcirc b$ is defined and $\succ c$, then both of the following are true*:

(1) *There exists $a' \in A$ such that $a \succ a'$ and $a' \bigcirc b \succ c$.*

(2) *There exists $b' \in A$ such that $b \succ b'$ and $a \bigcirc b' \succ c$.*

It is upper semicontinuous *iff whenever* $a \bigcirc b$ *is defined and* $\prec c$, *then both of the following are true*:

(1) *If* $a'' \succ a$ *and* $a'' \bigcirc b$ *is defined, then there exists an* a' *between* a'' *and* a *such that* $c \succ a' \bigcirc b$.

(2) *If* $b'' \succ b$ *and* $a \bigcirc b''$ *is defined, then there exists a* b' *between* b'' *and* b *such that* $c \succ a \bigcirc b'$.

Both lower and upper semicontinuity are defined in two parts because they actually assert that right-concatenation $a \to a \bigcirc b$ and left-concatenation $b \to a \bigcirc b$ are semicontinuous. The two parts of each definition are symmetric, with the roles of the first and second variables interchanged.

The definition of upper semicontinuity is more complicated than that of lower because when $c \succ a \bigcirc b$, we need some additional condition to assert that there exists some larger a' or b' such that $a' \bigcirc b$ or $a \bigcirc b'$ is defined. In the case of lower semicontinuity, the corresponding problem does not arise because of local definability and the fact that structure $\mathscr{A}$ was assumed not to have a minimal element. (A PCS that has a minimal element turns out to be discrete, as shown by Lemma 5 in Section 19.5.2, in which case continuity is vacuous).

We next state the connection between the properties specified here and the notion of continuity expressed in Definition 2.

THEOREM 6. *Let* $\mathscr{A}$ *be a PCS with no minimal element. It is continuous in the order topology* (*Definition* 2) *iff it is both lower and upper semicontinuous* (*Definition* 5).

Note that the combination of upper and lower semicontinuity is equivalent to the continuity of concatenation in each variable separately. But in Definition 2 the continuity of $\bigcirc$ was specified as continuity in both variables simultaneously. Ordinarily, continuity in two variables is a much stronger condition than continuity in each separately; but for strictly increasing functions, they are equivalent. This fact gives us the equivalence announced in Theorem 6.

The right- or left-concatenations are actually asserted to be continuous using the *relative order topologies on their domains*. To understand this better, consider a function f whose domain D is a subset of a topological space X and whose range R is a subset of an ordered set Y. The function f is *upper semicontinuous* if, for every y in Y, the set $\{x | x \in D \text{ and } f(x) < y\}$ is open in D. To be open in D means that there is some open set of X whose intersection with D is the set in question. (This topology, induced from that of X on the set D, is called the relative topology on D.) In the present instance, for right-concatenation, the domain D consists of all a for which $a \bigcirc b$ is defined. Thus, asserting that right-concatenation by b is upper semicontinuous means that for every c in A there is an open subset $G(c)$ of A such that if $a \bigcirc b$ is defined, then $c \succ a \bigcirc b$ if and only if a is in $G(c)$. Part of the proof of Theorem 6 shows that this is so if and only if the defining condition of Definition 5 is satisfied. Lower semicontinuity is handled similarly but more simply since we do not need to worry about the possible lower boundedness of the domain over which $a \bigcirc b$ is defined.

19.4 COMPLETIONS OF TOTAL ORDERS AND PCSs

The literature on representational measurement has pursued two somewhat different goals. One approach emphasizes algebraic and counting aspects, and thereby includes numerical representations for finite or countable empirical structures.[7] Another goal is to achieve measurement onto real intervals, to permit use of standard mathematical machinery, for example, functional or differential equations.

This section presents an approach, due to Narens and Luce (1976), that mediates between these two goals. If a structure has "holes," so that its natural representation cannot be onto a real interval, it may nonetheless be possible to fill the holes with ideal elements, while preserving the key properties. Thus, standard mathematics can be used despite the holes. The main theorem we present is due to Cohen (1988). However, more work is needed to understand fully what can be accomplished by this approach.

Algebraic theorems generally include axioms from at least three of the following four types: (i) first-order, universal (e.g., weak ordering or monotonicity); (ii) first-order, existential (e.g., solvability or closure); (iii) second-order (e.g., Archimedean or existence of countable, order-dense subsets); (iv) higher order, e.g., constraints on the automorphism group (see Chapter 20). Measurement onto real intervals often uses the same first-order universal axioms as in comparable algebraic formulations, but replaces types (ii) and (iii) by topological assumptions, such as continuity, Dedekind completeness, topological connectedness, or topological (i.e., uniform) completeness.

[7] One argument, which we do not pursue here, for studying purely algebraic structures is the possibility of understanding how finite algebraic structures approximate infinite ones (Narens, 1974). Finite algebraic models can usually be formulated in terms of either universal algebraic axioms, the number increasing with the number of elements (see Chapter 9), or in terms of a fixed number of universal axioms together with structural ones. Of course, no Archimedean axiom is needed. An important issue is the sense in which these structures converge to an infinite one as the number of elements grows.

As we have seen in several examples in Volume I, the topological postulates usually are strong enough not only to eliminate the finite and denumerable structures but also some of the nondenumerable ones. Indeed, it is usually possible to show that both the structural and the Archimedean properties follow from the topological and universal axioms (Theorem 1 of this chapter provides an example), but not conversely. Put another way, two rather different ideas—a certain density of elements (in particular, very small ones), which is captured by the solvability axioms, and the commensurability of elements or intervals, which is guaranteed by the Archimedean idea—are blurred together in the topological formulations.

What seems to be needed to enhance our understanding of the relations between algebraic and topological structures are additional algebraic conditions that lead in some sense to the topological conditions. The sense that Narens and Luce (1976) proposed was to find algebraic conditions on a structure that make it densely embeddable (much as the rationals are embeddable in the reals) in a Dedekind-complete structure. This latter notion, which was introduced in Definition 2, captures in a general ordinal setting the completeness property that distinguishes the reals from the rationals. Ramsey (1975, 1976) advocated assumption of topological completeness—existence of limits of Cauchy sequences. This depends on uniform or metric structure, rather than on ordinal structure.

The problem, as we deal with it in this section, is rather more than a rewriting of the classical result about the embeddability of the rationals in the real numbers because we study structures with an operation that is not only partial (not closed) but also is not associative. The arguments are more delicate in the absence of associativity.

19.4.1 Order Isomorphisms onto Real Intervals

The first task is to characterize those simple orders that are order-isomorphic to intervals in Re. This is a classical, well-studied mathematical concept (for general summaries, see Kelley, 1955; Rosenstein, 1982). Three necessary conditions are readily identified.

1. The simple order must have a countable order-dense subset since that is the necessary and sufficient condition for an order-isomorphism with some infinite subset of the reals (Theorem 2.2).

2. There must not be gaps. If $a \succ c$ and there is no b such that $a \succ b \succ c$, then there would be a corresponding gap between the values $\varphi(a)$ and $\varphi(c)$ for an isomorphism φ, and so the mapping could not be onto an interval.

3. There must not be "holes," i.e., the simple order must be Dedekind complete.

These conditions are independent of one another (for example, the integers have gaps but no holes, the rationals have holes but no gaps, and a lexicographic ordering of the plane has neither but has no countable order-dense subset; see Section 2.1 of Volume I). They are also sufficient because we know that (i) the first condition guarantees the existence of a continuous isomorphism into Re; (ii) the last two are equivalent to the statement that the simple order is connected in the order topology; and (iii) the image of a connected space by a continuous function is connected, and therefore, in Re, it is an interval. (For the topological assertions, see e.g., Kelley, 1955, pp. 57–58, Problem I.d and p. 100, Problem A.)

We summarize these classical results in the following theorem, which we do not prove.

THEOREM 7. *A total order is isomorphic to a real interval iff it is topologically connected and has a countable order-dense subset. It is topologically connected iff it has no gaps (that is, for each $a \succ c$, there exists b such that $a \succ b \succ c$) and it is Dedekind complete.*

In the case of a PCS with no minimal element, there can be no gaps. If in addition there is no maximal element, then the real interval cannot include its endpoints, and so the interval can be taken to be all of Re^+. This brings us to the following:

COROLLARY TO THEOREM 7. *Suppose $\mathscr{A}$ is a Dedekind complete* PCS *with no maximal and no minimal element. Then there exists a* PCS $\mathscr{R} = \langle \mathrm{Re}^+, \geqslant, \otimes \rangle$ *and a continuous homomorphism φ from $\mathscr{A}$ into $\mathscr{R}$ such that* $\varphi(A) = \mathrm{Re}^+$.

The effect of this corollary is to reduce any questions about the impact of added properties (e.g., associativity) on the representation to the study of corresponding numerical functional equations. This places at our disposal the considerable literature on functional equations (Aczél, 1966, 1984; Dhombres, 1979).

19.4.2 Completions of Total Orders

Having shown that Dedekind-complete structures are the ones that map onto Re^+, we now turn to the question of which PCSs can be densely embedded in Dedekind-complete structures, much as the rationals are dense in the reals. Such a problem is not unlike comparable generalizations that occur in geometry, for example, metric spaces. The reason that the problem is tricky is that even though a PCS with no minimal element has no order gaps, if it has a hole, then the attempt to fill that hole can produce a gap when the concatenation is discontinuous.

To segregate the difficulties that arise in extending concatenation to a PCS completion from the rather straightforward ordinal theory, we present the latter first in this subsection. Since the intended application is to PCSs with no minimal element, we avoid some complications by limiting the formal development to simple orders without gaps. Definition 6 and Theorem 8 (below) generalize readily, however, to orders with gaps, and we indicate the generalizations in brief comments following each. Because the ordinal theory is well known (Rosenstein, 1982), the proof of Theorem 8 is cast as Exercises 12 through 16.

DEFINITION 6. *Let* $\mathscr{A} = \langle A, \succsim \rangle$ *be a total order without gaps. A* completion *of* $\mathscr{A}$ *is a pair* $\langle \boldsymbol{\mathscr{A}}, \varphi \rangle$ *such that:*

1. $\boldsymbol{\mathscr{A}} = \langle \mathbf{A}, \succsim \rangle$ *is a topologically connected simple order,*
2. φ *is an isomorphism from* $\mathscr{A}$ *into* $\boldsymbol{\mathscr{A}}$*,*
3. $\varphi(A)$ *is order-dense in* $\mathbf{A}$*,*
4. $\mathbf{a}$ *is an extremum of* $\boldsymbol{\mathscr{A}}$ *iff* $\mathbf{a} = \varphi(a)$*, where a is an extremum of* $\mathscr{A}$*.*

To generalize this, one can drop the phrase *without gaps* and then add the requirement that $\mathbf{a}, \mathbf{b}$ are the endpoints of a gap in $\mathbf{A}$ if and only if $\mathbf{a} = \varphi(a)$ and $\mathbf{b} = \varphi(b)$, where a, b are the endpoints of a gap in A. This provision guarantees that the completion has exactly those topologically connected components (intervals without gaps) that reflect the gap structure of A.

THEOREM 8.

Existence of a completion. *If* $\mathscr{A}$ *is a simple order without gaps, then there exists a completion of* $\mathscr{A}$.

Extension of homomorphisms. *Suppose that for* $i = 1, 2$, $\mathscr{A}_i$ *is a simple order without gaps, that* $\langle \boldsymbol{\mathscr{A}}_i, \varphi_i \rangle$ *is a completion of* $\mathscr{A}_1$*, and that* β *is a homomorphism from* $\mathscr{A}_1$ *into* $\mathscr{A}_2$*. Then there exists a homomorphism* β' *from* $\boldsymbol{\mathscr{A}}_1$ *into* $\boldsymbol{\mathscr{A}}_2$ *such that* $\beta'\varphi_1 = \varphi_2\beta$*, i.e., the diagram of Figure 2a commutes. Moreover,* β' *is one-to-one iff* β *is one-to-one and* β' *is unique and onto whenever* β *is onto.*

Uniqueness of the completion. *Completions are unique up to isomorphism (Figure 2b), and every automorphism of the simple order has a unique extension to its completion (Figure 2c).*

This theorem generalizes without change to the case in which A has gaps, using the strengthened concept of completion sketched after Definition 6. The proof, however, would be complicated by bookkeeping for the gaps.

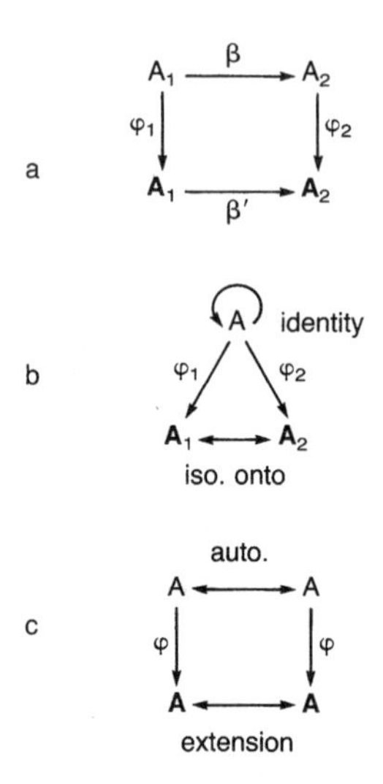

FIGURE 2. Homomorphism extension and uniqueness for completions of simple orders.

Though we leave the proof to exercises, we sketch the basic ideas here. The construction of a completion is exactly the same as Dedekind's construction of the real numbers from the rationals in which each real number is taken to be the set of all smaller rationals. For example, $2^{1/2}$ is identified as the set of all rationals r such that $r^2 < 2$. Those subsets of A that will be identified as nonmaximal elements of the completion $\mathbf{A}$ are called cuts. A *cut* is any nonempty subset with the following properties: (i) it has a nonempty complement, (ii) every element of the complement is larger than every element of the subset, and (iii) the complement has no minimum. Thus an element $\mathbf{a}$ of $\mathbf{A}$ corresponds to the subset $\{b \mid a \succsim b\}$, which is a cut for any nonmaximal a. (Note that the fact that the complement has no minimum corresponds to the absence of gaps.) If there is a hole in A, it is filled in the completion by the set of all elements of A that fall "below the hole" (just as $2^{1/2}$ consists of all rationals with squares less than 2). The homomorphism φ maps each element a into the corresponding cut just defined. The ordering on the set of cuts is just set inclusion since for $a \succ b$ the cut $\varphi(a)$ includes the cut $\varphi(b)$.

The uniqueness of the completion stems from the fact that the definition by cuts captures the underlying structural relation between an order and its completion. Even if a completion $\langle \mathscr{A}, \varphi \rangle$ were defined some other way (as can be done for the reals in terms of the rationals), each element of $\mathbf{A}$ is fully determined by the set of all $\varphi(a)$ less than that element. In the proof, this idea is used to prove what is an even stronger property: that any homomorphism of simple orders can be extended to their completions. The extension formula is the obvious one: for **a** in $\mathbf{A}_1$, take the subset of A_1 mapped by φ_1 into elements bounded above by **a**, map that subset into A_2 by β and thence into the completion $\mathbf{A}_2$ by φ_2, and then take the least upper bound as the value of $\beta'(\mathbf{a})$. If β maps onto, then this is the only possible extension, and it maps onto $\mathbf{A}_2$.

From this extension we derive uniqueness by supposing that there are two different completions and taking β to be the identity map of A into itself. The corresponding β' is an isomorphism between the two completions. Similarly, any automorphism β of A extends to a unique automorphism β' of **A**.

19.4.3 Completions of Closed PCSs

The problem of embedding a PCS $\mathscr{A}$ into a Dedekind complete PCS is, in principle, simply a problem of extending Definition 6 so that for the completion $\langle \mathscr{A}, \varphi \rangle$, the first statement of the definition becomes:

1′. $\mathscr{A} = \langle \mathbf{A}, \succsim, \bigcirc \rangle$ is a Dedekind complete PCS.

The key question is, what conditions on a PCS make this possible?

First, it is important to recognize that such an embedding is not possible for an arbitrary PCS. The following example, and a general graphical procedure for generating such examples, was given in Cohen (1988). [A somewhat similar example given by Narens (1985, p. 150) fails to satisfy monotonicity.] Let Ra^+ denote the positive rational numbers, and let

$$X = \mathrm{Ra}^+ \cup [2^{1/2}\mathrm{Ra}^+ \cap (2,\infty)].$$

The operation is defined by:

$$x \bigcirc y = \begin{cases} x + y, & \text{if } x < 2 \\ x + [(4 - x)/2^{1/2}]y, & \text{if } 2 \leqslant x \leqslant 3,\ y < 2^{1/2} \\ 4 + (x - 1)(y - 2^{1/2}), & \text{if } 2 \leqslant x \leqslant 3,\ y > 2^{1/2} \\ x + 2y, & \text{if } x > 3. \end{cases}$$

The problem in completing the operation occurs at $y = 2^{1/2}$ with $2 \leqslant x \leqslant 3$. Observe that approaching $2^{1/2}$ from either above or below, the value approached by $x \bigcirc y$ is 4. Thus, if $\otimes$ denotes the intended completion, then necessarily $x \otimes 2^{1/2} = 4$, which means $\otimes$ is not monotonic for the x variable in the range $2 \leqslant x \leqslant 3$.

Second, it clearly would be desirable to have structural conditions that are both necessary and sufficient for the embedding, in which case we would fully understand the problem. The next result, though providing necessary and sufficient conditions for a closed operation,[8] cannot be described as fully satisfactory as "structural conditions." What it does do, however, is provide a proof method for establishing sufficient conditions. This is demonstrated in the two corollaries to Theorem 10.

By Theorem 3, there is no loss of generality in assuming the PCS is defined on an order-dense subset of Re^+. In addition, we shall restrict the formulation to closed operations.

DEFINITION 7. *Suppose* $\mathscr{R} = \langle R, \succsim, \bigcirc \rangle$ *is a* PCS *with* $R \subseteq \mathrm{Re}^+$. *For* $x, y \in \mathrm{Re}^+$, *let*

$$P_{xy} = \{u \bigcirc v | u, v \in R, \quad u \geqslant x, \quad v \geqslant y\} \qquad \textit{and}$$

$$Q_{xy} = \{u \bigcirc v | u, v \in R, \quad u \leqslant x, \quad v \leqslant y\}.$$

Then the operations $\bigcirc^+$ *and* $\bigcirc^-$ *are defined by*

$$x \bigcirc^+ y = \inf P_{xy} \qquad \textit{and} \qquad x \bigcirc^- y = \sup Q_{xy}.$$

Observe that the operations $\bigcirc^+$ and $\bigcirc^-$ are defined for all $x, y > 0$; however, they need not be monotone. As we shall see in the proof, they bound any monotone extension that exists, and if they exhibit the properties stated in Theorem 9, then the extension exists.

THEOREM 9. *Suppose* $\mathscr{R} = \langle R, \succsim, \bigcirc \rangle$ *is a closed* PCS *defined on a dense subset* R *of* Re^+. *Then* $\mathscr{R}$ *has a Dedekind completion* $\mathscr{R} = \langle \mathrm{Re}^+, \geqslant, \otimes \rangle$ *iff for each* $x, y, z \in \mathrm{Re}^+$ *with* $y > z$,

(i) *if* $x \bigcirc^+ y = x \bigcirc^+ z$ *and* $x \bigcirc^- y = x \bigcirc^- z$, *then* $x \bigcirc^+ y > x \bigcirc^- y$;
and

(ii) *if* $y \bigcirc^+ x = z \bigcirc^+ x$ *and* $y \bigcirc^- x = z \bigcirc^- x$, *then* $y \bigcirc^+ x > y \bigcirc^- x$.

[8] Cohen (1988) comments that the result can be generalized to partial operations, but we have not attempted to do so.

The earlier example fails the strict inequality $x \bigcirc^+ 2^{1/2} > x \bigcirc^- 2^{1/2}$ for $2 \leqslant x \leqslant 3$ because both terms are 4. This result, due to Cohen (1988), is mainly useful in going from more structural conditions to the conclusion that a Dedekind completion exists. We illustrate this in Corollary 1 following the next result, which responds to a natural question: to what extent is a Dedekind completion, if one exists, unique up to isomorphism?

To this end, consider the following concept of a gap in a concatenation structure that was suggested to us by R. Niederée (personal communication). Suppose $\mathscr{A} = \langle A, \succsim, \bigcirc \rangle$ is PCS and U_1 and U_2 are cuts of A (see Section 19.4.2), then $\mathscr{A}$ is said to have a *gap* at (U_1, U_2) iff there exist $c, d \in A$ such that for all $a_i \in U_i$, $b_i \in -U_i$,

$$b_1 \bigcirc b_2 \succsim d \succ c \succsim a_1 \bigcirc a_2 .$$

THEOREM 10. *If a PCS is strictly ordered, closed, and without gaps, then it has at most one Dedekind completion.*

The easy proof is left as Exercise 17. Hint: show that $\bigcirc^+ \equiv \bigcirc^-$.

With this result available, we may now illustrate the use of Theorem 9.

COROLLARY 1. *Suppose $\mathscr{R} = \langle R, \geqslant, \bigcirc \rangle$ is a* PCS *on a dense subset of* Re$^+$ *that satisfies the following two properties*:

(i) *$\mathscr{R}$ has no gaps*;

(ii) *For every $x, y, z \in \mathrm{Re}^+$ with $y > z$, there exist $u, v, w, w' \in R$ such that $y > w > w' > z$, $u > x > v$, $u \bigcirc w' < v \bigcirc w$, and $w' \bigcirc u < w \bigcirc v$.*

Then $\mathscr{R}$ has a unique Dedekind completion which itself has no gaps.

The proof is left as Exercise 18.

Other sufficient conditions have been given by Narens and Luce (1976) and Narens (1985). In the latter book, the following was shown directly but without the restriction to a closed operation.

COROLLARY 2. *Suppose a closed PCS satisfies lower semicontinuity*[9] *and is Archimedean in regular sequences, then it has a Dedekind completion that is lower semicontinuous and Archimedean in regular sequences.*

The proof is left as Exercise 19.

A further issue is whether the automorphisms of the completion are natural extensions of those of the PCS. There are really two questions: does

[9] Narens (1985) called it "embeddability."

each automorphism of the PCS extend to one of the completion, and does each automorphism of the completion extend one of the PCS? We do not have a general answer about the extensions of automorphisms; however, under somewhat more restricted conditions, a result is known (Theorem 20.15 in Section 20.4.3).

19.5 PROOFS ABOUT CONCATENATION STRUCTURES

19.5.1 Theorem 1 (p. 37)

(i) *If a Dedekind-complete, totally ordered concatenation structure is positive and left-solvable (for $b \succ a$ there exists c with $b = c \bigcirc a$), then it is Archimedean in standard sequences.*

(ii) *If it is solvable, then it is Archimedean in difference sequences.*

PROOF.

(i) Suppose not. Let $\{a(i)\}$ be a bounded, infinite standard sequence based on a. Then $a(1) = a$, $a(2) = a \bigcirc a$, $a(3) = a(2) \bigcirc a$, and so on. By positivity, $a(1) \prec a(2) \prec a(3) \prec \cdots$. By Dedekind completeness, let $\bar{a}$ be the l.u.b. of $\{a(i)\}$. Then $\bar{a} \succ a(1) = a$. By hypothesis, let c be such that $\bar{a} = c \bigcirc a$. By positivity, $\bar{a} \succ c$. Therefore, by the choice of $\bar{a}$, let $a(j)$ be such that $\bar{a} \succ a(j) \succ c$. By positivity,

$$a(j + 1) = a(j) \bigcirc a \succ c \bigcirc a = \bar{a},$$

contradicting the choice of $\bar{a}$.

(ii) Suppose $\mathscr{A}$ is solvable and not Archimedean in differences. We can suppose that there exists a bounded infinite sequence $\{a(i)\}$ and elements $p, q \in A$ such that $a(i + 1) \bigcirc p = a(i) \bigcirc q$. Without loss of generality, we can suppose $a(1) \prec a(2) \prec \cdots$, which by monotonicity implies $q \succ p$. By Dedekind completeness, let $\bar{a}$ be the l.u.b. of $\{a(i)\}$. By solvability, let u be such that $\bar{a} \bigcirc p = u \bigcirc q$. Since $q \succ p$, it follows from monotonicity that $\bar{a} \succ u$. Since $\bar{a}$ is the l.u.b. of the sequence, let j be such that $\bar{a} \succ a(j) \succ u$. Then by monotonicity,

$$a(j + 1) \bigcirc p = a(j) \bigcirc q \succ u \bigcirc q = \bar{a} \bigcirc p,$$

and so by monotonicity, $a(j + 1) \succ \bar{a}$, contradicting the choice of $\bar{a}$. ◇

19.5.2 Lemmas 1–6, Theorem 2

The development in this and the next subsection is similar to the proofs in Sections 2.2 and 16.7; the delicate point is to show that associativity, though used in the former proofs, can be bypassed.

Throughout this section we assume that $\mathscr{A} = \langle A, \succsim, \bigcirc \rangle$ and $\mathscr{A}' = \langle A', \succsim', \bigcirc' \rangle$ are totally ordered PCSs and that φ and ψ are homomorphisms of $\mathscr{A}$ into $\mathscr{A}'$. Definitions and results that apply to an arbitrary PCS will be stated for $\mathscr{A}$ and carried over without comment to $\mathscr{A}'$, substituting $\succsim'$ for $\succsim$ and $\bigcirc'$ for $\bigcirc$. Similarly, definitions and results pertaining to a general PCS homomorphism are stated for φ and carried over without comment to ψ.

Recall that $a(1)$ is defined to be just a and that $a(n+1)$ is defined as $a(n) \bigcirc a$ whenever the concatenation in question is defined.

For any $a, b \in A$, we define a nonnegative integer $N(a, b)$ as

$$N(a, b) = \begin{cases} 0, & \text{if } a \succ b; \\ \sup\{n \mid a(n) \text{ is defined, and } b \succsim a(n)\}, & \text{if } b \succsim a = a(1). \end{cases}$$

By the Archimedean property, $N(a, b)$ is always finite. The function $N(a, b)$ provides a kind of linkage between the elements a and b in terms of concatenation: it specifies the maximum number of copies of a that can be concatenated (in left-branching fashion) without exceeding b. The inequality linkage is not as tight as an equation of form $b \sim a(n)$, but it is far more general and thus more useful.

The following lemma shows that N is monotonically decreasing in its first variable and increasing in its second variable. Since it is a step function, the monotonicity is not strict (except for discrete PCSs, as will be seen).

LEMMA 1.

(i) *If* $a \succsim b$, *then* $N(b, c) \geqslant N(a, c)$.

(ii) *If* $b \succsim c$, *then* $N(a, b) \geqslant N(a, c)$.

PROOF. Induction, using local definability and monotonicity. ◇

The next lemma shows the way in which homomorphisms preserve concatenation-based linkages of the form $N(a, b)$.

LEMMA 2. *Suppose* φ *is a homomorphism from* $\mathscr{A}$ *to* $\mathscr{A}'$.

(i) *If* $a(n)$ *is defined, so is* $\varphi(a)(n)$, *and*

$$\varphi(a)(n) = \varphi[a(n)].$$

(ii) $N[\varphi(a), \varphi(b)] \geqslant N(a, b)$.

(iii) *If* $b \bigcirc a$ *is defined, then* $N(a, b) = N[\varphi(a), \varphi(b)]$.

PROOF.

(i) Induction: the statement is clearly true for $n = 1$. If it is true for some $n \geqslant 1$, then suppose $a(n+1)$ is defined. Then $a(n) \bigcirc a$ is defined, and by the definition of homomorphism, $\varphi[a(n)] \bigcirc' \varphi(a)$ is defined and equal to $\varphi[a(n) \bigcirc a]$. Thus $\varphi(a)(n+1)$ is defined and equal to $\varphi[a(n+1)]$.

(ii) $a \succ b$ iff $\varphi(a) \succ' \varphi(b)$; and hence $N(a, b) = 0$ iff $N[\varphi(a), \varphi(b)] = 0$. If $b \succsim a(n)$, then [using (i)],

$$\varphi(b) \succsim' \varphi[a(n)] = \varphi(a)(n).$$

Hence, $N[\varphi(a), \varphi(b)] \geqslant N(a, b)$.

(iii) If $b \bigcirc a$ is defined, then so is $a[N(a, b) + 1]$; and hence, by the choice of N, $a[N(a, b) + 1] \succ b$. By part (i), $\varphi(a)[N(a, b) + 1] \succ' \varphi(b)$. Therefore, $[N(a, b) + 1] > N[\varphi(a), \varphi(b)]$, and so $N(a, b) \geqslant N[\varphi(a), \varphi(b)]$. By part (ii), the converse inequality holds and so it is an equality. ◇

For any fixed a the function $f(b) = N(a, b)$ is a sort of crude measurement scale, making no discriminations for b's below a and imperfect discriminations for b's above a (since N is not strictly increasing). To obtain better discrimination, we need to consider $N(a, b)$ for a sequence of a's that become arbitrarily small.

We say that the sequence $\{a_k\}$ *converges to zero* in $\mathscr{A}$ if for any $c \in A$, $c \succ a_k$ for all but finitely many values of k.

For any PCS with no minimal element, Lemma 4 below shows how to construct a sequence that converges to zero and, in the process, establishes a useful equivalent to convergence to zero; namely, that the integer sequence $N(a_k, b)$ diverges to $+\infty$ for every $b \in A$. Before turning to this result, we exhibit a very simple but useful improvement of the previous lemma, based on a sequence that converges to zero.

LEMMA 3. *Suppose that* $\{a_k\}$ *is an infinite sequence of elements of* A *that converges to zero in* $\mathscr{A}$. *For any nonmaximal* $b \in A$, $b \bigcirc a_k$ *is defined for all sufficiently large* k, *whence for such* k,

$$N[\varphi(a_k), \varphi(b)] = N(a_k, b).$$

PROOF. Since b is nonmaximal, choose a such that $a \succ b$. By restricted solvability, choose c such that $a \succsim b \bigcirc c$. For all sufficiently large

k, $c \succ a_k$, hence $b \bigcirc a_k$ is defined. The invariance of $N(a_k, b)$ under homomorphisms then follows from Lemma 2(iii). ◇

LEMMA 4.

(i) *If A has no minimal element, then there exists a sequence $\{a_k\}$ such that for $k = 1, 2, \ldots$,*

$$a_k \succsim a_{k+1} \bigcirc a_{k+1}.$$

(ii) *If a sequence satisfies the inequalities of part* (i), *then it converges to zero; and conversely, any sequence that converges to zero has a subsequence that satisfies the inequalities of part* (i).

(iii) $\{a_k\}$ *converges to zero iff $N(a_k, b)$ diverges to $+\infty$ for every $b \in A$.*

PROOF. Suppose A has no minimal element. Choose a_1 arbitrarily; inductively, if a_k has been chosen, take b_k such that $a_k \succ b_k$, and then take c_k such that $a_k \succsim b_k \bigcirc c_k$. Choose $a_{k+1} \precsim \inf\{b_k, c_k\}$. This sequence satisfies the inequalities of part (i).

The assertion in part (ii) that a sequence convergent to zero has a subsequence that satisfies these same inequalities, requires only a minor modification of the proof just given: at the kth stage of construction, choose a_k to be an element of the given sequence rather than an arbitrary element of A that satisfies the constraint.

Next, we establish the easy direction for the divergence criterion of part (iii). Suppose that $N(a_k, b)$ diverges to $+\infty$. Then for all but finitely many k, $N(a_k, b) > 1$, i.e., $b \succ a_k$. If this holds for every b, then by definition $\{a_k\}$ converges to zero.

We now turn to the only delicate part of the proof. We shall show that any sequence satisfying the inequalities of part (i) satisfies the divergence criterion of part (iii). Since we have just shown that the divergence criterion is sufficient for convergence to zero, this proves part (ii) completely. The necessity of the divergence criterion then follows as well: since a sequence convergent to zero has a subsequence that satisfies the inequalities, that subsequence satisfies the divergence criterion, and it is easily shown the main sequence must also do so.

First, we establish

$$a_k(n) \succsim a_{k+1}(n+1). \tag{6}$$

For $n = 1$ it is true by part (i). Suppose it is true for $n \geqslant 1$. By positivity,

$a_k \succsim a_{k+1} \bigcirc a_{k+1} \succ a_{k+1}$, and so using the induction hypothesis and monotonicity, we obtain

$$\begin{aligned} a_k(n+1) &= a_k(n) \bigcirc a_k \\ &\succsim a_{k+1}(n+1) \bigcirc a_k \\ &= a_{k+1}(n+2). \end{aligned}$$

Next, we show that

$$N(a_{k+1}, b) \geqslant N(a_k, b) + 1. \tag{7}$$

Let $n = N(a_k, b)$, then by definition of N and by inequality (6),

$$b \succsim a_k(n) \succsim a_{k+1}(n+1).$$

Thus,

$$N(a_{k+1}, b) \geqslant n + 1 = N(a_k, b) + 1.$$

A simple induction using inequality (7) yields, for each m,

$$N(a_{k+m}, b) \geqslant N(a_k, b) + 1.$$

Thus $N(a_k, b)$ diverges to ∞. ◇

The preceding lemma shows in particular that any PCS with no minimal element has a sequence that converges to zero. What about a PCS that has a minimal element? Such a structure is very simple: it consists only of elements equivalent to some $a(n)$ where a is the minimal element. This is shown in the next lemma. Such a PCS is called *discrete*.

LEMMA 5. *Either $\mathscr{A}$ has a sequence that converges to zero or A has a minimal element a and for all b, $b \sim a[N(a, b)]$.*

PROOF. From Lemma 4, if there is no minimal element, then there is a sequence that converges to zero. If a is minimal and $b \succ a[N(a, b)]$, choose c such that $b \succsim a[N(a, b)] \bigcirc c$. Since $c \succsim a$, we have $b \succsim a[N(a, b) + 1]$, a contradiction. ◇

Lemmas 2 and 5 show the 1-point uniqueness of homomorphisms very simply for the case in which the PCS $\mathscr{A}$ is discrete. If a is minimal, and $\varphi(a) = \psi(a)$, then

$$\varphi[a(n)] = \varphi(a)(n) = \psi(a)(n) = \psi[a(n)].$$

If, however, $\varphi(a) \succ' \psi(a)$, then the above reasoning shows that $\varphi[(n)] \succ' \psi[a(n)]$.

To obtain the corresponding results for nondiscrete PCSs, we need to show that homomorphisms preserve convergence to zero. This is the next lemma.

LEMMA 6. *$\{a_k\}$ converges to zero iff $\{\varphi(a_k)\}$ converges to zero.*

PROOF. If $\{a_k\}$ converges to zero, then $N(a_k, b)$ diverges to $+\infty$ for all $b \in A$. By Lemma 2(ii), $N[\varphi(a_k), \varphi(b)]$ diverges to $+\infty$ for all $b \in A$. In particular, $N[\varphi(a_{k+l}), \varphi(a_k)]$ diverges to $+\infty$ for every k; and it follows, as in the proof of Lemma 4, that $N[\varphi(a_k), c]$ diverges for every $c \in A'$. Hence $\{\varphi(a_k)\}$ converges to zero.

The converse is simpler: if $\{\varphi(a_k)\}$ converges to zero, then for any $c \in A$ only finitely many $\varphi(a_k)$ are $\succ' \varphi(c)$, and thus only finitely many a_k are $\succ c$. ◇

We can now prove Theorem 2 for the nondiscrete case.

19.5.3 Theorem 2 (p. 39)

Suppose φ and ψ are two homomorphisms from one PCS to another. If there is some nonmaximal element b for which $\varphi(b) \succ' \psi(b)$, then $\varphi(a) \succ' \psi(a)$ for every nonmaximal $a \in A$.

PROOF. Let b satisfy the hypothesis. Since A' has no minimal element, we can use restricted solvability in $\mathscr{A}'$ twice to choose $c \in A'$ such that

$$\varphi(b) \succsim' (\psi(b) \bigcirc' c) \bigcirc' c.$$

We also choose some sequence $\{a_k\}$ that converges to zero in $\mathscr{A}$.

By Lemmas 3 and 6, for all sufficiently large k, $b \bigcirc a_k$ is defined and $c \succ' \psi(a_k)$. By Lemma 2(iii), for all such k,

$$N(a_k, b) = N[\varphi(a_k), \varphi(b)] = N[\psi(a_k), \psi(b)].$$

We simplify notation by denoting this common value just by n. We have

$$\begin{aligned} \varphi(a_k)(n+1) &\succ' \varphi(b) \\ &\succsim' (\psi(b) \bigcirc' c) \bigcirc' c \\ &\succsim' [\psi(a_k)(n) \bigcirc' \psi(a_k)] \bigcirc' \psi(a_k) \\ &\sim' \psi(a_k)(n+2). \end{aligned}$$

This inequality shows, to begin with, that $\varphi(a_k) \succ' \psi(a_k)$ for all sufficiently large k. In addition, it serves as a basis for induction: for $m \geqslant 1$, if $\varphi(a_k)(m+n)$ is defined, then

$$\varphi(a_k)(m+n) \succ' \psi(a_k)(m+n+1).$$

Now, if a is any nonmaximal element $\succ b$, take k sufficiently large that the above holds and, in addition, $a \bigcirc a_k$ is defined. Let $N(a_k, a) = m + n$. Hence,

$$\varphi(a) \succsim' \varphi(a_k)(m+n) \succ' \psi(a_k)(m+n+1) \succ' \psi(a).$$

This establishes the desired result for nonmaximal a with $a \succ b$. For $b \succ a$ we simply choose a_k such that $a \succ a_k$ and repeat the above argument with a_k in place of b, using the fact that $\varphi(a_k) \succ \psi(a_k)$. ◇

19.5.4 Construction of PCS Homomorphisms

The preceding proof establishes that all the values of a PCS homomorphism φ are determined, given one value $\varphi(b)$, but it does not contain an explicit formula for constructing $\varphi(c)$ for an arbitrary c. The construction was sketched in Section 19.3 following the statement of Theorem 2. Here we give a formal statement and a proof that the construction converges to $\varphi(c)$.

The notation and assumptions are the same as in Section 19.5.2, and we use Lemmas 1–6 freely without specific reference.

For any $a, b \in A$ we define a set of "upper" estimates of $\varphi(c)$ as follows:

$$U(a,b,c) = \{d[N(a,c)+1] | d \in A' \text{ and } \varphi(b) \succsim' d[N(a,b)-1]\}.$$

To understand this better, set $d = \varphi(a)$ in this formula. The constraint is satisfied: certainly $\varphi(b) \succsim' \phi(a)[N(a,b)-1]$. The corresponding element of $U(a,b,c)$ is just $\varphi(a)[N(a,c)+1]$, and this is certainly $\succsim' \varphi(c)$. Thus, $U(a,b,c)$ contains some elements that are upper estimates of $\varphi(c)$; in fact, if one selects d such that

$$d[N(a,b)] \succ' \varphi(b) \succsim' d[N(a,b)-1],$$

then $d[N(a,c)+1]$ is definitely such an upper estimate. [This last point obviates the need to refer to the unknown $\varphi(a)$ in selecting elements d that generate upper estimates.]

Intuitively, however, d cannot be very much more than $\varphi(a)$ because of the constraint that $N(a, b) - 1$ copies of d do not exceed $\varphi(b)$. Thus we expect the upper estimates not to exceed $\varphi(c)$ by very much. More rigorously, we have the following formula.

LEMMA 7. *If $b \succ c$, then $\varphi(c) = \lim_{a \to 0} \sup U(a, b, c)$.*

Note that the formula requires a bit of interpretation since $a \to 0$ needs definition and, if A' is not Dedekind complete, $\sup U$ may not exist in A'. What we actually prove is the following:

For any $c' \succ' \varphi(c)$, there exists $a_0 \in A$ such that for $a_0 \succ a$, c' is an upper bound for $U(a, b, c)$.

Thus, although $\varphi(c)$ is never an upper bound for any $U(a, b, c)$, every larger element of A' is an upper bound for all $U(a, b, c)$ provided a is sufficiently small. Of course, for representations in Re, $\sup U$ does exist, and the formula has a more straightforward interpretation.

Note further that the formula gives a construction only for $\varphi(c)$ with $b \succ c$. This is easily remedied: for most or all c, we can choose m large enough so that $b(m) \succ c$, and since we know that $\varphi[b(m)] = \varphi(b)(m)$, we use the estimates in $U[a, b(m), c]$ to the same effect. Finally, if we have chosen b to be inconveniently large, so that $b(m)$ is undefined and yet there exists $c \succ b(m-1)$ such that calculation of $\varphi(c)$ is desired, then it is necessary to use the calculation of Lemma 7 to determine, with sufficient accuracy, the value $\varphi(b')$ for some more convenient b' such that $b'(m') \succ c$ for some m'. We do not give the details of this last case.

PROOF OF LEMMA 7. Suppose that $c' \succ' \varphi(c)$. By restricted solvability, take $c'' \in A'$ such that $c' \succ' ((\varphi(c) \bigcirc' c'') \bigcirc' c''$. Choose a_0 such that $c'' \succsim' \varphi(a_0)$ and also $b \succsim (c \bigcirc a_0) \bigcirc a_0$. We show that for $a_0 \succ a$, we have c' as an upper bound for $U(a, b, c)$. Assume otherwise, i.e., there exists $d \in A'$ such that $d[N(a, c) + 1] \succsim' c'$. We have

$$\begin{aligned} d[N(a,c)+1] &\succsim' c' \\ &\succsim' ((\varphi(c) \bigcirc' c'') \bigcirc' c'') \bigcirc' c'' \\ &\succ' \varphi(a)[N(a,c)+3]. \end{aligned}$$

This shows first that $d \succ' \varphi(a)$ and further, by induction, that $d(m) \succ' \varphi(a)(m + 2)$ for every $m \geqslant N(a, c) + 1$. Since, in particular, $N(a, b) - 1 \geqslant N(a, c) + 1$, we have

$$d[N(a,b)-1] \succ' \varphi(a)[N(a,b)+1] \succ' \varphi(b),$$

so d violates the constraint that defines $U(a, b, c)$. Thus, $d[N(a, b) + 1] \notin U(a, b, c)$, and so c' is an upper bound. ◇

A similar formula can be developed for lower bounds on $\varphi(c)$. Define

$$L(a, b, c) = \{d[N(a, c)] | d \in A' \text{ and } d[N(a, b) + 2] \succsim' \varphi(b)\}.$$

We can add the following formula to Lemma 7:

$$\varphi(c) = \lim_{a \to 0} \inf L(a, b, c).$$

The proof (for $b \succ c$) is very similar to that of the upper-bound formula.

In practice, as discussed in Section 19.3, one estimates $\varphi(c)$ by finding d with $d[N(a, b)] \succ' \varphi(b) \succsim' d[N(a, b) - 1]$ and e with $e[N(a, b) + 2] \succsim' \varphi(b) \succ' e[N(a, b) + 1]$. This guarantees that $d[N(a, c) + 1]$ is an element of $U(a, b, c)$ that is actually above $\varphi(c)$ and $e[N(a, c)]$ is an element of $L(a, b, c)$ that is actually below $\varphi(c)$. If these bounds are close enough together, one has a satisfactory interval estimate for $\varphi(c)$. ◇

19.5.5 Theorem 3 (p. 41)

(i) *For every* PCS $\mathscr{A}$ *there exists a numerical* PCS $\mathscr{A}'$ *and a homomorphism* φ *of* $\mathscr{A}$ *into* $\mathscr{A}'$.

(ii) *If* φ *is a representation of a* PCS $\langle A, \succsim, \bigcirc \rangle$, *then* ψ *is another such representation iff there exists a strictly increasing function* h *from* $\psi(A)$ *onto* $\varphi(A)$ *such that, for all* $a \in A$,

$$\varphi(a) = h[\psi(a)],$$

and such that the numerical operations $\otimes$ *and* $\otimes'$ *are related by*

$$x \otimes' y = h^{-1}[h(x) \otimes h(y)], \tag{8}$$

that is, $x \otimes' y$ *is defined iff* $h(x) \otimes h(y)$ *is defined, and when they are defined, Equation* (8) *holds.*

PROOF. We first prove the theorem under the assumption that there is a minimal element, denoted a, in A.

Lemma 5 asserts that $b = a[N(a, b)]$ for every b in A. So the function $\varphi(b) = N(a, b)$ is well defined and order preserving from A onto a finite or infinite sequence of consecutive positive integers. Denote the range of φ by I. Define a partial operation $\otimes$ on I by

$$\varphi(b) \otimes \varphi(c) = \varphi(b \bigcirc c). \tag{9}$$

It is straightforward to verify that $\langle I, \geqslant, \otimes \rangle$ is a numerical PCS and that, by construction, φ is a homomorphism. One can equally well use $h^{-1}[N(a, b)]$, where h^{-1} is any strictly increasing function defined on the positive integers. [We use h^{-1} rather than h in order to obtain the usual formula for an alternative numerical representation ψ, i.e., for this case, $N(a, b) = h(\psi(b))$.] The point is that it suffices for the representing function to preserve the order of $\mathscr{A}$; the operation can then be defined on the image set so that it mirrors completely all the structure of $\mathscr{A}$. Hence the image has all the properties of a PCS because $\mathscr{A}$ has them, and the function is a homomorphism by construction. The numbers are just order-preserving numerical labels for the elements of A.

For this discrete case, the continuity assertions in part (iii) are vacuously true. (Technically, in a discrete topology, all functions are continuous, and so the assertions are true but of no interest.)

Next, assume that there is no minimal element in A. In the discrete case, the entire structure consisted of one finite or infinite standard concatenation sequence based on the minimal element. Here we use a whole family of such sequences based on smaller and smaller elements, yielding sequences with finer and finer grain. For a constructive proof, each sequence must be finite because we cannot complete the numerical construction for an infinite sequence before going to the next finer-grained sequence to obtain a better approximation. But if a sequence is finite, it does not in general cover the whole ordering; thus, as the sequences become finer-grained, they must also go higher and higher. Eventually, any particular element is reached, with as fine-grained a sequence as desired, in only finitely many steps.

Once the family of finite standard concatenation sequences (SCS) is constructed, we show how order-preserving numbers can be assigned to their union in a recursive fashion. Then the numerical assignment for any element of the PCS is obtained simply as the supremum of the values assigned to smaller elements in the union of concatenation sequences or, equivalently, as the infinum of the values assigned to larger elements. Obviously, for any element a an approximation to the value of $\varphi(a)$ is obtained from any of the sequences in the family that continues at least up to a. This construction has the further advantage that the function φ obtained in this way is continuous, using the relative topology on the image $\varphi(A)$.

The above outline is developed in detail in the following titled segments.

Construction of a countable order-dense union of finite standard sequences. In order that our family of SCSs will cover more and more of the ordering, we form a sequence $\{a_k'\}$ of target elements that are strictly increasing and eventually $\succsim$ any given element of A. (It is not hard to prove that such a sequence exists: if there is an infinite standard concatenation sequence, use it; if not, then form an infinite "nonstandard" concatenation sequence in which right-concatenation of identical elements continues until there is no more room; then reduce the size of the standard so that concatenation can continue.)

Next we construct a sequence $\{a_k\}$ that converges to zero and satisfies additional smallness constraints as specified in Equation (10). For each a_k we form the finite standard concatenation sequence $A_k = \{a_k(1), \ldots, a_k(N(a_k, a'_k))\}$, i.e., the sequence that covers the interval up to a'_k. And we let U_k be the union of the first k sequences, A_1 through A_k. The construction goes as follows.

Let $a_1 = a'_1$. Thus $A_1 = \{a_1\} = U_1$ since $N(a_1, a'_1) = 1$. Suppose that a_1 through a_k have been constructed. There exists a solution c to the following set of inequalities:

$$a_k \succsim c \bigcirc c; \tag{10a}$$
$$b \succsim (b' \bigcirc c) \bigcirc c, \tag{10b}$$

for every $b, b' \in U_k$ such that $b \succ b'$; and

$$a'_{k+1} \succsim (a'_k \bigcirc c) \bigcirc c. \tag{10c}$$

We take for a_{k+1} any element that satisfies these conditions.

Equation (10a) simply assures that the a_k satisfy the inequalities of Lemma 4(i) and so converge to zero. The inequalities of (10b), which can be very numerous, assure that each pair of successive elements in U_k is separated by at least one element of A_{k+1} that falls strictly between them. (The inequalities are written for all b, b' with $b \succ b'$, but it suffices that they be satisfied for adjacent pairs.) Equation (10c) assures that the next interval beyond U_k, from a'_k to a'_{k+1}, contains at least one element of A_{k+1} strictly between.

We let U_∞ be the union of all the U_k. Clearly U_∞ is countable and order dense in A. To prove the latter, for any $b \succ b'$ choose c such that $b \succsim (b' \bigcirc c) \bigcirc c$; then for any k with $c \succsim a_k$, we have $a_k[N(a_k, b') + 1]$ strictly between b and b'.

We remark that there are many ways other than the one just given to construct a countable order-dense subset of A, but not all of them have the same useful properties.

Construction of an order-isomorphism into the positive reals. Since U_∞ is countable and order dense, we could use Theorem 2.2 to assert the existence of an order-isomorphism φ from $\langle A, \succsim \rangle$ into $\langle \mathrm{Re}, \geqslant \rangle$. (That theorem is stated for simple orders, but of course it applies to the simple order that is induced by the weak order $\succsim$ on the equivalence classes in A.) The function φ can be taken to map into Re^+. However, it is preferable to specify the construction of φ directly.

We begin the definition by defining φ for $U_1 = \{a_1\}$. For simplicity, just let $\varphi(a_1) = 1$.

Suppose now that φ has been defined on U_k for some $k \geqslant 1$. To extend the definition to A_{k+1}, consider any interval between successive elements of U_k, say $b \succ b'$. For the one or more elements of A_{k+1} that are strictly between b and b', assign φ values that preserve their order and divide the real interval from $\varphi(b)$ to $\varphi(b')$ into equal parts. Similarly, for the smallest elements of A_{k+1} that fall below a_k, assign φ values that divide the real interval between $\varphi(a_k)$ and 0 into equal parts. And for the largest elements of A_{k+1}, those between a'_k and a'_{k+1}, assign φ values that divide the real interval from $\varphi[\sup(U_k)]$ to $1 + \varphi[\sup(U_k)]$ into equal parts.

Clearly the extension of φ to A_{k+1} is still order preserving. Thus φ is defined inductively on all of U_∞. Moreover, because there is at least one element of A_{k+1} between any two successive elements of U_k, and because we have used equal spacing in the interpolation between successive values of φ on U_k, we have the following useful results:

$$\varphi(a_k) \leqslant 2^{-(k-1)}. \tag{11a}$$

Let j be $\geqslant 2$ and k be $\geqslant 0$. If b, b' are successive elements of U_{j+k} such that

$$\sup(U_j) \succsim b \succ b',$$

then

$$\varphi(b) - \varphi(b') \leqslant 2^{-(k+1)}. \tag{11b}$$

Property (11a) is obvious from the construction, but (11b) deserves an inductive proof. For the basis of the induction, let $k = 0$. We have b, b' successive in U_j, and the maximum spacing occurs for $j = 2$ if $b \sim a_1$ and $b' \sim a_2$ and $b \sim b' \bigcirc b'$. Then the difference between $\varphi(b)$ and $\varphi(b')$ is just $1/2$. If $j > 2$, then the maximum spacing is $1/3$, attained if b, b' are the only elements of U_j that are greater than $\sup(U_{j-1})$.

Now suppose that the result holds for some $k \geqslant 0$ and all $j \geqslant 2$. Let b, b' be successive elements of U_{j+k+1} with $\sup(U_j) \succsim b \succ b'$. Choose c minimal in U_{j+k} such that $\sup(U_j) \succsim c \succsim b$. [Such a c exists, of course, since $\sup(U_j)$ is itself an element of U_{j+k}.] Now if $c = \inf(U_{j+k}) = a_{j+k}$, then the difference between $\varphi(b)$ and $\varphi(b')$ is at most $\frac{1}{2}\varphi(a_{j+k})$, and by Equation (11a) this is at most $2^{-(k+2)}$, as required. Otherwise, there exists some $c' \in U_{j+k}$ with $c \succ c'$. By the induction hypothesis, the difference between $\varphi(c)$ and $\varphi(c')$ is at most $2^{-(k+1)}$. Since c was minimal $\succsim b$, we have $b \succsim c'$, and since b, b' are successive in U_{k+l+1} we have also $b' \succsim c'$. Since there must be some element of U_{k+l+1} strictly between c and c', we have either

$$c \succ b \succ b' \succsim c'$$

or

$$c \succsim b \succ b' \succ c'.$$

In the former case we have

$$\varphi(b) - \varphi(b') \leqslant \varphi(b) - \varphi(c') \leqslant \frac{1}{2}[\varphi(c) - \varphi(c')].$$

In the latter case we have

$$\varphi(b) - \varphi(b') \leqslant \varphi(c) - \varphi(b') \leqslant \frac{1}{2}[\varphi(c) - \varphi(c')].$$

So in either case, $\varphi(b) - \varphi(b')$ is at most $2^{-(k+2)}$. This completes the proof of Equation (11b).

We now define φ for all nonmaximal elements of A by use of limits:

$$\begin{aligned} \varphi(a) &= \sup\{\varphi(b) | b \in U_\infty \text{ and } a \succsim b\} \\ &= \inf\{\varphi(b) | b \in U_\infty \text{ and } b \succsim a\}. \end{aligned} \tag{12}$$

Either equation is a valid definition since for nonmaximal elements the sets in question are nonempty and suitably bounded. Clearly, the value of φ is the same as previously defined for elements of U_∞. The equivalence of the two equations follows readily from the spacing inequality (11b) that we have just established. Obviously, we can get upper and lower estimates of $\varphi(a)$ that are as close as we like by taking the sup and inf over the finite set U_k, for sufficiently large k.

It is easy to show that the function φ just constructed is an order-homomorphism of $\langle A, \succsim \rangle$ into $\langle \mathrm{Re}^+, \geqslant \rangle$. If $a \succ a'$, then there is an element b of U_∞ strictly between them, and thus $\varphi(a) > \varphi(b) > \varphi(a')$. Conversely, if $\varphi(a)$ and $\varphi(a')$ are distinct, then there is some $\varphi(b)$ for b in U_∞ between them, and it follows from the definitions of φ in Equation (12) that b is strictly between a and a'.

Construction of the numerical operation; permissible transformations. We now proceed just as in the discrete case: let $\varphi(A)$ be the range of φ, and define a binary operation $\otimes$ on $\varphi(A)$ by Equation (9). It is again straightforward to show that $\mathscr{A}' = \langle \varphi(A), \geqslant, \otimes \rangle$ is a (numerical) PCS: all the needed properties transfer from $\mathscr{A}$ to $\mathscr{A}'$ since $\varphi(a)$ is in effect just a relabeling of the equivalence class of a in A. It is also clear that φ is a homomorphism of $\mathscr{A}$ *onto* $\mathscr{A}'$. The order-preserving property has just been established, and the concatenation-preserving property is Equation (9), the definition of φ.

Concerning permissible transformations [part (ii) of Theorem 3], we note that the uniqueness theorem for ordinal measurement (Theorem 2.3) shows that ψ is another order-isomorphism of $\langle A, \succsim \rangle$ into $\langle \mathrm{Re}^+, \geqslant \rangle$ if and only if $\varphi = h(\psi)$ for some strictly increasing function h from $\psi(A)$ onto $\varphi(A)$. And for any such h an operation $\otimes'$ on $\psi(A)$ can again be defined by Equation (9), using $\psi = h^{-1}(\varphi)$ in place of φ. The relation $x \otimes' y = h^{-1}[h(x) \otimes h(y)\}$, for all x, y in $\psi(A)$, now follows by definition of $\otimes$ and $\otimes'$. $\diamond$

19.5.6 Theorem 4 (p. 45)

The automorphism group of a PCS is an Archimedean ordered group.

PROOF. Let $\langle A, \succsim, \bigcirc \rangle$ be the PCS. With no loss of generality, assume that there is no minimal element and $\succsim$ is a total order. First, suppose $u \in A$ is maximal. Observe that the restriction of the PCS to $A - \{u\}$ has no maximal element since if a were such an element, then since $u \succ a$, by restricted solvability there exists a $b \in A$ such that $a \bigcirc b$ is defined and $u \succ a \bigcirc b$ and by positivity $a \bigcirc b \succ a$, contradicting the assumption that a is maximal. Now, either there exist $a, b \in A$, such that $u = a \bigcirc b$ or not. In the latter case, simply discard u and proceed with the resulting PCS with no maximal element (see below). In the former case, suppose α is an automorphism. Since u is maximal, necessarily $\alpha(u) = u$, and so $a \bigcirc b = u = \alpha(u) = \alpha(a) \bigcirc \alpha(b)$. If $\alpha(a) \succ a$, then by monotonicity, $b \succ \alpha(b)$, which, by Theorem 2 applied to the PCS restricted to $A - \{u\}$, is impossible. A similar argument holds for $\alpha(a) \prec a$. So $\alpha(a) = a$. But then, by Theorem 2 it is impossible for $\alpha(c) \neq c$ for any c; thus only the trivial automorphism exists.

So, we turn to the case of no maximal element. By Theorem 2, $\succsim$ induces a natural ordering $\succsim'$ on the automorphisms, which is clearly a simple order. We show it is monotonic. Suppose that $\alpha, \beta, \gamma \in \mathscr{G}$, and let $\cdot$

denote function composition,

$$\begin{aligned} \alpha \succsim' \beta \quad &\text{iff} \quad \alpha(a) \succsim \beta(a) \\ &\text{iff} \quad \gamma\alpha(a) \succsim \gamma\beta(a) \\ &\text{iff} \quad \gamma \cdot \alpha \succsim' \gamma \cdot \beta; \\ \alpha \succsim' \beta \quad &\text{iff} \quad \alpha[\gamma(a)] \succsim \beta[\gamma(a)] \\ &\text{iff} \quad \alpha \cdot \gamma \succsim' \beta \cdot \gamma. \end{aligned}$$

Finally, we show that $\langle \mathscr{G}, \succsim', \cdot \rangle$ is Archimedean. Suppose $\alpha \succ' \iota_A$, where ι_A is the identity map on A, and for some $\beta \in \mathscr{G}$ and all positive integers n, $\beta \succ' \alpha^n$. So, for any $a \in A$, $\alpha(a) \succ \iota(a) = a$, whence, by restricted solvability, there exists a b such that $\alpha(a) \succ a \bigcirc b$. Thus,

$$\alpha^2(a) \succ \alpha(a \bigcirc b) = \alpha(a) \bigcirc \alpha(b) \succ (a \bigcirc b) \bigcirc b \succ b \bigcirc b.$$

By induction, $\alpha^n(a) \succ b(n)$. Since $\beta \succ' \alpha^n$, $\beta(a) \succ \alpha^n(a) \succ b(n)$, which is impossible since a PCS is Archimedean in standard concatenation sequences. ◇

19.5.7 Theorem 5 (p. 46)

The homomorphism φ of Theorem 3 can be constructed so as to be continuous; in this case, if $\mathscr{A}$ is continuous, then so is its representation; and if $\psi = h\varphi$, then ψ is continuous iff h is bicontinuous.

PROOF. We show that the particular representation φ constructed in Section 19.5.5 is continuous from the order topology of A to the relative topology on $\varphi(A)$. Basically, this is true because φ was constructed so as not to have any gaps in its range [see Equation (11)]. Specifically, suppose a is any nonmaximal element of A, and let (x', x) be any open interval of Re^+ containing $\varphi(a)$. [This is the place where the relative topology is distinguished from the order topology: x, x' are not necessarily elements of $\phi(A)$; rather, an open interval in $\varphi(A)$ is any intersection of an open interval in Re^+ with $\varphi(A)$.] We can find b, b' in U_∞ such that $b \succ a \succ b'$ and such that $\varphi(b)$ differs from $\varphi(b')$ by less than $\inf\{x - \varphi(a), \varphi(a) - x'\}$. Then φ maps the interval (b', b) in A into the interval (x', x) in Re^+.

We note further that any homomorphism, being order preserving, is continuous in the order topology. Moreover, any representation of a PCS can be regarded as a one-to-one function from the set of equivalence classes onto its image in Re^+. This function has an inverse that is continuous in the order topologies, as just noted, and hence also continuous from the relative topology on the image to the order topology on the PCS. (Using a more fine-grained topology on the domain preserves continuity since there are fewer convergent sequences in a finer topology.) For any homomorphism ψ we denote this continuous inverse mapping as ψ^{-1}.

Clearly, if a strictly increasing transformation h is bicontinuous, then the alternative representation $\psi = h^{-1}(\varphi)$ is continuous since it is a composition of continuous functions. Conversely, suppose that we have two continuous representations φ and ψ. The strictly increasing function h from the range of ψ onto the range of φ is bicontinuous since $h = \varphi(\psi^{-1})$ and $h^{-1} = \psi(\varphi^{-1})$ are compositions of continuous functions.

If the operation $\bigcirc$ is continuous, and if we use a continuous homomorphism φ in Equation (9), then the representing operation $\otimes$ is continuous since it can be written as a composition of continuous functions:

$$x \otimes y = \varphi[\varphi^{-1}(x) \bigcirc \varphi^{-1}(y)]. \qquad \diamond$$

19.5.8 Theorem 6 (p. 47)

A PCS *with no minimal element is continuous iff it is both lower and upper semicontinuous.*

PROOF. Let $\mathscr{A} = \langle A, \succsim, \bigcirc \rangle$ be a PCS with no minimal element.

We first show that the definitions of lower and upper semicontinuity in Definition 5 are in fact equivalent to the lower and upper semicontinuity of one-sided concatenation functions in the usual sense. Then we point out why continuity of one-sided concatenations is equivalent, in the present case, to continuity of concatenation in the relative product topology on its domain, i.e., continuity as a function of two variables.

Let f be a function with domain D contained in a topological space X and with range R contained in a totally ordered set $\langle Y, \succsim \rangle$. Then f is lower semicontinuous iff for every y in Y there is an open subset $G(y)$ of X such that $\forall x \in D$, $f(x) \succ y$ iff $x \in G(y)$. (Briefly, the pre-image of any relatively open, lower-bounded interval in R is relatively open in D.) Upper semicontinuity is defined similarly with $\prec$ replacing $\succ$, i.e., the pre-image of any relatively open upper-bounded interval of R is relatively open in D. If both are satisfied, then since the intersection of open sets is open, the pre-image of any relatively open interval of R (possibly bounded on both sides) is again relatively open in D, and this is the definition of continuity of f, using the relative order topology on R and the relative topology on D.

We now apply this standard definition to right-concatenation by some fixed $b \in A$; that is, let the sets X and Y both be the equivalence classes of A relative to $\sim$, use the order topology, let D be the subset $\{a \in A | a \bigcirc b$

is defined}, let $f(a) = a \bigcirc b$, and let R be the subset $f(D) = \{a \bigcirc b | a \bigcirc b$ is defined}. We therefore use part (i) of the definitions of upper and lower semicontinuity in Definition 5, which correspond to right concatenation. The results for left concatenation are exactly the same, using the symmetric part (ii) of the definitions.

For any $c \in A$, let $L(c)$ be the pre-image of the relatively open interval bounded below by c, i.e., the set of all a such that $a \bigcirc b$ is defined and $\succ c$. If $\mathscr{A}$ satisfies lower semicontinuity (Definition 5), then for any $a \in L(c)$ there exists $a' \in L(c)$ with $a \succ a'$. Hence, let $G(c)$ be the set of elements of A that are $\succ a'$ for some a' in $L(c)$. This is the union of open intervals of A, and hence, it is open; and clearly, its intersection with D is exactly $L(c)$. Thus right-concatenation by any fixed b satisfies the lower semicontinuity in the usual sense.

Conversely, if the usual sense of lower semicontinuity is satisfied, then $L(c)$ is the intersection of some open set $G(c)$ with D. Thus, if $a \in L(c)$, there exists $a'' \prec a$ such that every element between them is in $G(c)$. Choose any a' with $a \succ a' \succ a''$. Then $a' \in G(c)$ and, by local definability, $a' \in D$; hence, $a' \in L(c)$, i.e., $a' \bigcirc b \succ c$, as required for Definition 5.

For upper semicontinuity the proof is similar but a bit more complex because of the possibility that there is a maximal a such that $a \bigcirc b$ is defined. Let $U(c)$ be the set of a such that $a \bigcirc b$ is defined and $\prec c$. If there is no maximal element such that $a \bigcirc b$ is defined, then the extra precondition in upper semicontinuity in Definition 5 is always met, and the proof is exactly as before but with signs reversed. If there is such a maximal element, denoted e, and if $e \bigcirc b \succ c$, then again the extra precondition is always satisfied by setting $a'' = e$, and the proof goes the same way. Finally, if $e \bigcirc b \prec c$, then we let $G(c)$ be all of A. This is open, and its intersection with D is $U(c)$.

For the converse, we again assume that $U(c)$ is the intersection of some open set $G(c)$ with D. For any a in $U(c)$ there exists some a''' greater than a such that everything between a''' and a is in $G(c)$. Thus, if a'' exists greater than a with $a'' \bigcirc b$ defined, take a' between $\inf\{a''', a''\}$ and a; then we have $a' \bigcirc b$ defined and $a' \in G(c)$, and hence, $a' \in U(c)$ or $a' \bigcirc b \prec c$, as required.

So far we have established that $\mathscr{A}$ is upper and lower semicontinuous iff each one-sided concatenation function is continuous using the relative topologies on its domain and range. Continuity of $\bigcirc$ in the relative product topology on its domain implies continuity of the one-sided concatenation functions; so it only remains to establish the converse. We emphasize that this depends critically on monotonicity of the one-sided concatenations.

If there are no maximal elements for which any one-sided concatenations are defined, then given any $a \bigcirc b$ in the range of $\bigcirc$ and any relatively open interval containing $a \bigcirc b$, i.e., c', c'' such that $c' \succ a \bigcirc b \succ c''$, we can first choose a' and a'' such that

$$c' \succ a' \bigcirc b \succ a \bigcirc b \succ a'' \bigcirc b \succ c'',$$

and then choose b' and b'' such that

$$c' \succ a' \bigcirc b \succ a' \bigcirc b \succ a \bigcirc b \succ a'' \bigcirc b \succ a'' \bigcirc b'' \succ c''.$$

The Cartesian product of the two intervals a' to a'' and b' to b'' is an open set in the domain of $\bigcirc$ that is mapped into the c' to c'' interval.

Suppose a maximal element d exists, e.g., $d \bigcirc b = \max\{d' \bigcirc b\}$, where d' ranges over A. If $d \bigcirc b \succ a \bigcirc b$, then we proceed as before. If, however, $d = a$, then we select any $a' \prec a$. The choice of a'', b', and b'' is as before. Now the product of the two intervals is an open set in the domain of $\bigcirc$ that is mapped into the c' to c'' interval, in part because for u in the interval $a' \succ u \succ a$, $u \bigcirc b$ is not defined. The case in which b is the maximal extent of $a \bigcirc b'$ is handled similarly. ◇

19.5.9 Corollary to Theorem 7 (p. 50)

A Dedekind-complete PCS *with no maximal or minimal element has a continuous representation onto the positive reals.*

PROOF. We know from Theorem 5 that there is a continuous representation into Re^+. The proof of Theorem 3 establishes that with no minimal element there can be no gaps. The argument given for Theorem 5 shows that the continuous representation constructed in the proof of Theorem 3 must actually be onto an interval in Re^+. Since there is no minimal or maximal element, the interval must be open. The uniqueness part of Theorem 5 then shows how to obtain the representation onto Re^+ using any bicontinuous mapping h of the open interval onto Re^+. ◇

19.5.10 Theorem 9 (p. 54)

Necessary and sufficient conditions for a PCS *on a dense subset of* Re^+ *to have a Dedekind completion is that for* $x, y, z \in \mathrm{Re}^+$, $y > z$,

(i) *if* $x \bigcirc^i y = x \bigcirc^i z$, $i = +, -$, *then* $x \bigcirc^+ y > x \bigcirc^- z$, *and*

(ii) *if* $y \bigcirc^i x = z \bigcirc^i x$, $i = +, -$, *then* $y \bigcirc^+ x > z \bigcirc^- x$.

We first establish two lemmas.

LEMMA 1. *Under the conditions of the theorem, for* $x, y, z \in \mathrm{Re}^+$,

(i) $x \bigcirc^+ y \geqslant x \bigcirc^- y$.

(ii) *If* $\otimes$ *is a monotone extension of* $\bigcirc$, *then* $x \bigcirc^+ y \geqslant x \otimes y \geqslant x \bigcirc^- y$.

(iii) *If* $x > y$, *then* $x \bigcirc^i z \geqslant y \bigcirc^i z$ *and* $z \bigcirc^i x \geqslant z \bigcirc^i y$, $i = +, -$.

PROOF.

(i) For every $u \bigcirc v \in P_{xy}$ and $u' \bigcirc v' \in Q_{xy}$, by definition $u \geqslant x \geqslant u'$ and $v \geqslant y \geqslant v'$, and thus by the monotonicity of $\bigcirc$, $u \bigcirc v \geqslant u' \bigcirc v'$. Thus $\inf P_{xy} \geqslant \sup Q_{xy}$.

(ii) Suppose $\otimes$ is a monotone extension of $\bigcirc$. Select u, v, u', v' as in (i), then,

$$u \otimes v = u \bigcirc v \geqslant x \otimes y \geqslant u' \bigcirc v' = u' \otimes v'.$$

So by the definitions of $\bigcirc^+$ and $\bigcirc^-$, the conclusion follows.

(iii) Select $r \in P_{xy}$, i.e., there exists $u, w \in R$ such that $r = u \bigcirc w$ and $u \geqslant x$ and $w \geqslant y$. Since $x > y$ and R is dense in Re^+, there exists $v \in R$ such that $x > v > y$. By the monotonicity of $\bigcirc$, $u \bigcirc w > v \bigcirc w$. Thus there exists $s = v \bigcirc w \in P_{yz}$ and $r > s$. Therefore, $\inf P_{xz} \geqslant \inf P_{yz}$, proving that $x \bigcirc^+ z \geqslant y \bigcirc^+ z$. The other cases are analogous. ◇

LEMMA 2. *Suppose* $\mathscr{R} = \langle R, \geqslant, \bigcirc \rangle$ *is a* PCS, $\bigcirc$ *is closed, and* R *is dense in* Re^+. *If* $\otimes$ *is a monotone extension of* $\bigcirc$, *then* $\langle \mathrm{Re}^+, \geqslant, \otimes \rangle$ *is a* PCS.

PROOF. $\otimes$ is closed by definition, and $\geqslant$ on Re^+ is a nontrivial, total order. By hypothesis, $\otimes$ is monotonic. So we need only show Archimedean, positivity, and restricted solvability.

Archimedeaness. For $x \in \mathrm{Re}^+$ select $u \in R$ with $u < x$. By the monotonicity of $\otimes$, $u(n) < x(n)$, and so $x(n)$ is unbounded because $u(n)$ is, since $\mathscr{R}$ is a PCS.

Positivity. Consider $x, y \in \mathrm{Re}^+$ and select $u, v \in R$ such that $u < x$ and $v < y$. By parts (i) and (iii) of Lemma 1,

$$x \otimes y \geqslant x \bigcirc^- y > u \bigcirc v > u, v,$$

and so $x \otimes y \geqslant x, y$. Using this and the monotonicity of $\otimes$,

$$x \otimes y > u \otimes y \geqslant y \qquad \text{and} \qquad x \otimes y > x \otimes v \geqslant x.$$

Restricted Solvability. Suppose $x > y$. Select $u, v \in R$ such that $x > u > v > y$. Since $\mathscr{R}$ is a PCS, there exists $z \in R$ such that $x > v \bigcirc z$, whence, by the monotonicity of $\otimes$, $x > v \bigcirc z > y \otimes z$. ◇

PROOF OF THEOREM 9.

Necessity. Suppose $y > z$, $x \bigcirc^+ y = x \bigcirc^+ z$, and $x \bigcirc^- y = x \bigcirc^- z$. Then,

$$\begin{aligned} x \bigcirc^+ y &\geqslant x \otimes y && \text{[Lemma 1(ii)]} \\ &> x \otimes z && (\otimes \text{ is monotonic}) \\ &\geqslant x \bigcirc^- z && \text{[Lemma 1(ii)]} \\ &= x \bigcirc^- y && \text{(hypothesis).} \end{aligned}$$

The other inequality is similar.

Sufficiency. By Lemma 2, it suffices to construct a monotonic extension of $\bigcirc$. Let λ: $\mathrm{Re}^+ \times \mathrm{Re}^+ \to (0,1)$ be any function that is strictly monotonic increasing in each variable. Define the operation $\otimes_\lambda$ by

$$x \otimes_\lambda y = \lambda(x, y)(x \bigcirc^+ y) + [1 - \lambda(x, y)](x \bigcirc^- y).$$

Now, suppose $x > z$, then

$$\begin{aligned} x \otimes_\lambda y - z \otimes_\lambda y &= \lambda(x, y)(x \bigcirc^+ y) + [1 - \lambda(x, y)](x \bigcirc^- y) \\ &\quad - \lambda(z, y)(z \bigcirc^+ y) - [1 - \lambda(z, y)](z \bigcirc^- y) \\ &= \lambda(x, y)[(x \bigcirc^+ y) - (z \bigcirc^+ y)] \\ &\quad + [1 - \lambda(x, y)][(x \bigcirc^- y) - (z \bigcirc^- y)] \\ &\quad + [\lambda(x, y) - \lambda(z, y)][(z \bigcirc^+ y) - (z \bigcirc^- y)]. \end{aligned}$$

By choice of λ, $\lambda(x, y) > 0$, $1 - \lambda(x, y) > 0$, and $\lambda(x, y) - \lambda(z, y) > 0$. By Lemma 1(iii), both $(x \bigcirc^+ y) - (z \bigcirc^+ y) \geqslant 0$, and $(x \bigcirc^- y) - (z \bigcirc^- y) \geqslant 0$, and by part (i), $(z \bigcirc^+ y) - (\bigcirc^- y) \geqslant 0$. If either of the former two is > 0, then $(x \bigcirc_\lambda y) - (z \otimes_\lambda y) > 0$. If not, then by hypothesis $(z \otimes^+ y) - (z \otimes^- y) > 0$, and the conclusion follows. The other inequality is similar. ◇

19.6 CONNECTIONS BETWEEN CONJOINT AND CONCATENATION STRUCTURES

19.6.1 Conjoint Structures: Introduction and General Definitions

Just as the concept of an extensive structure has been generalized to a PCS, we generalize the concept of an additive conjoint structure to that of a general nonadditive one. Basically, it is just a matter of omitting the Thomsen condition (see Definition 8, part 4, below). Once that is done, we shall be concerned with a number of different but related issues concerning both general conjoint and general concatenation structures.

In this section we explore three major ways in which conjoint and concatenation structures are interconnected. First, in Section 19.6.2, we use Holman's definition (Section 6.2.4 and Definition 9 below) to recode the information in a conjoint structure as operations on one of its components; these operations are closely related to PCSs. That discussion continues in Section 19.6.3 as we explain how an important class of order automorphisms of the conjoint structure manifests itself in terms of the induced operation. The next two sections explore intensive structures. In Section 19.6.4 we show how they can be recoded as conjoint structures, which, in turn, induce a nonintensive operation; and the structure of the automorphisms is studied in detail. A special class of these structures is shown in Section 19.6.5 to be closely related to PCSs through a concept called a doubling function. Finally, in Section 19.6.6, we take up the case, so important for the dimensional representation used in physics (Chapter 10), in which some conjoint structures have concatenation operations on one or more of the components. A qualitative property is introduced that describes how the operation distributes in the conjoint structure, and it is shown that the automorphisms of the concatenation structure can be expressed explicitly in terms of the induced operation of the conjoint structure.

Although these results afford considerable insight into the possible conjoint and concatenation structures, in large part they are developed to serve other purposes. One of these is to establish the existence of numerical representations for conjoint structures and then for general concatenation structures. This is taken up in Section 19.7, in which we use the fact, established in this section, that both intensive and conjoint structures induce an operation that can be decomposed into PCSs, and so the representation of Section 19.3 can be invoked. Another result is taken up in the next chapter where we explore the representations of general ratio and interval-scale structures.

Turning to the issues of this section, it is appropriate first to state explicitly the general concept of a conjoint structure and some of the restrictive properties that will be invoked. As in Volume I, we abbreviate $(a, p) \in A \times P$ as ap, except when ambiguity would arise.

DEFINITION 8. *Suppose A and P are nonempty sets and $\succsim$ is a binary relation on $A \times P$. Then, $\mathscr{C} = \langle A \times P, \succsim \rangle$ is a* conjoint structure *iff for each $a, b \in A$ and $p, q \in P$, the following three conditions* (1–3) *are* *satisfied*:

1. Weak ordering: $\succsim$ *is a nontrivial weak ordering.*
2. Independence: $\forall a, b \in A$ *and* $\forall p, q \in P$,
   (i) $ap \succsim bp$ *iff* $aq \succsim bq$,
   (ii) $ap \succsim aq$ *iff* $bp \succsim bq$.

Observe that independence permits the definition of induced weak orders *on* A *and* P, *which are denoted* $\succsim_A$ *and* $\succsim_P$, *respectively.*

3. $\succsim_A$ *and* $\succsim_P$ *are total orders.*

In addition,

4. $\mathscr{C}$ *is said to satisfy the* Thomsen condition *iff for all* $a, b, c \in A$ *and* $p, q, r \in P$,

$$ar \sim cq \quad \text{and} \quad cp \sim br \quad \text{imply} \quad ap \sim bq.$$

5. *For* $a_0 \in A$ *and* $p_0 \in P$, $\mathscr{C}$ *is said to be* solvable relative to $a_0 p_0$ *iff*
   (i) *for each* $a \in A$, *there exists* $\pi(a) \in P$ *such that* $ap_0 \sim a_0\pi(a)$,
   (ii) *for each* $ap \in A \times P$, *there exists* $\xi(a, p) \in A$ *such that* $\xi(a, p)p_0 \sim ap$.

6. $\mathscr{C}$ *is said to be* unrestrictedly A-solvable *iff for each* $a \in A$ *and* $p, q \in P$, *there exists* $b \in A$ *such that* $ap \sim bq$. *The definition of* unrestrictedly P-solvable *is similar. If* $\mathscr{C}$ *is both unrestrictedly* A- *and* P-*solvable, it is said to be* solvable.

7. *Let* J *be an (infinite or finite) interval of integers. Then a sequence* $\{a_j\}_{j \in J}$, $a_j \in A$, *is said to be a* standard sequence on A *iff there exist* $p, q \in P$ *such that*
   (i) *it is not the case that* $p \sim_P q$, *and*
   (ii) *for all* j, $j + 1 \in J$, $a_{j+1}p \sim a_j q$.

The sequence $\{a_j\}_{j \in J}$ *is said to be* bounded *iff for some* $c, d \in A$, $c \succsim a_j \succsim d$ *for all* $j \in J$. $\mathscr{C}$ *is said to be* Archimedean *iff every bounded standard sequence on* A *is finite.*

Note that assumption 3 could read *weak order* instead of *total order*; we assume a total order only as a matter of convenience. By independence, $\pi^{-1}(p) = \xi(a_0, p)$ is unique.

19.6.2 Total Concatenation Structures Induced by Conjoint Structures

In Chapter 6, we constructed additive representations for conjoint structures satisfying the Thomsen condition by showing that Holman's definition of an induced operation leads to extensive structures on each component, and thence to the numerical representation for the conjoint structure. As was described on p. 260 of Volume I, Krantz earlier had proceeded similarly, inducing an operation not on a single component but on the set of equivalence classes of $A \times P$. These defined operations turn out to be similarly useful in the absence of the Thomsen condition, leading in general to nonassociative concatenation structures, and thence to nonadditive representations. For simplicity, we state only the results for the Holman operation with sufficient solvability to get a closed operation. They can be generalized.

DEFINITION 9. *Suppose that $\mathscr{C} = \langle A \times P, \succsim \rangle$ is a conjoint structure that is solvable relative to $a_0 p_0 \in A \times P$. The* (Holman) induced operation on A relative to $a_0 p_0$, *denoted $\bigcirc_A$, is defined by: for each $a, b \in A$,*

$$a \bigcirc_A b = \xi[a, \pi(b)],$$

where π and ξ are defined in part 5 of Definition 8.

Luce and Cohen (1983) introduced the following concept, which broadens that of a closed concatenation structure of Definition 1.

DEFINITION 10. *Let A be a nonempty set, $\succsim$ a binary relation on A, $\bigcirc$ a binary operation on A, and $a_0 \in A$. Then $\mathscr{A} = \langle A, \succsim, \bigcirc, a_0 \rangle$ is said to be a* total concatenation structure *iff the following five conditions hold:*

1. *$\succsim$ is a total order and $\bigcirc$ is monotonic.*
2. *The restriction of $\mathscr{A}$ to $A^+ = \{a | a \in A \ \&\ a \succ a_0\}$ is a* PCS.
3. *The restriction of $\mathscr{A}$ to $A^- = \{a | a \in A \ \&\ a \prec a_0\}$ but with the converse order $\precsim$, i.e., $\langle A^-, \precsim, \bigcirc \rangle$ is a* PCS.
4. *For all $a \in A$, $a \bigcirc a_0 = a_0 \bigcirc a = a$.*
5. *For $a \in A^+$ and $b \in A^-$, there exist p and $q \in A$ such that $p \bigcirc b$ and $q \bigcirc a$ are defined and $p \bigcirc b \succ a$ and $b \succ q \bigcirc a$.*

It is obvious from Theorem 3, which shows any PCS has a real representation, that a total concatenation structure is isomorphic to a real total concatenation structure. Moreover, by the Corollary to Theorem 2, it is 2-point unique with one of the common points always being a_0.

THEOREM 11.

(i) *Suppose $\mathscr{C}$ is a conjoint structure that is solvable relative to $a_0 p_0$ and that $\bigcirc_A$ is the induced Holman operation. Then $\bigcirc_A$ is closed and monotonic, and it is positive over A^+ and negative over A^-. If, in addition, $\mathscr{C}$ is*

Archimedean, then both $\mathscr{I}_{A^+} = \langle A^+, \succsim_{A^+}, \bigcirc_{A^+} \rangle$ *and* $\mathscr{I}_{A^-} = \langle A^-, \precsim_{A^-}, \bigcirc_{A^-} \rangle$ *are Archimedean in standard sequences. Further, if* $\mathscr{C}$ *is both solvable and Archimedean, then* $\mathscr{I}_A$ *is a closed, solvable, total concatenation structure.*

(ii) *Suppose* $\mathscr{A} = \langle A, \succsim, \bigcirc, a_0 \rangle$ *is a closed total concatenation structure. Then there is a conjoint structure* $\mathscr{C} = \langle A \times A, \succsim' \rangle$ *that is solvable relative to* $a_0 a_0$ *and that induces* $\mathscr{A}$. *If* $\mathscr{A}$ *is Archimedean in differences, then* $\mathscr{C}$ *is Archimedean. If* $\mathscr{A}$ *is solvable, Archimedean in standard sequences, and associative, then* $\mathscr{C}$ *is solvable and Archimedean.*

COROLLARY 1. *If* $\mathscr{C}$ *is a conjoint structure that is solvable at a point, then the following statements are equivalent:*

(i) $\mathscr{C}$ *satisfies the Thomsen condition.*

(ii) $\bigcirc_A$ *is associative and commutative.*

COROLLARY 2. *If, in addition,* $\mathscr{C}$ *is a solvable conjoint structure, then the following statements are equivalent:*

(i) $\mathscr{C}$ *satisfies the Thomsen condition.*

(ii) *The induced operations are invariant in the sense that for every* a, $b, c, d \in A$ *and every pair of induced operations* $\bigcirc_A$ *and* $\bigcirc'_A$,

$$a \bigcirc_A b \succsim c \bigcirc_A d \quad \textit{iff} \quad a \bigcirc'_A b \succsim c \bigcirc'_A d.$$

The proofs of the corollaries are left as Exercises 23 and 24.

The preceding theorem, a modification and correction of one due to Luce and Cohen (1983), demonstrates among other things that Archimedean, solvable conjoint structures give rise to concatenation structures with PCS substructures on their positive and negative parts. Since any PCS has a real representation (Theorem 3), these do also. This is stated formally as Theorem 19 in Section 19.7.1.

19.6.3 Factorizable Automorphisms

One way to study how the induced operations relate to a solvable conjoint structure is to ask about the relations of the two classes of automorphisms: one for induced structures $\mathscr{I}_A = \langle A, \succsim_A, \bigcirc_A, a_0 \rangle$, which preserves $\succsim_A$, $\bigcirc_A$, and a_0, and the other for $\mathscr{C} = \langle A \times P, \succsim \rangle$, preserving $\succsim$ and possibly $a_0 p_0$.

A major distinction to be made about the order automorphisms of $\mathscr{C}$ is whether or not they are factorizable in the following sense.

DEFINITION 11. *Suppose $\mathscr{C} = \langle A \times P, \succsim \rangle$ is a conjoint structure, and α is an order automorphism of $\mathscr{C}$. Then α is* factorizable *iff there exist functions θ and η, where θ is a 1:1 mapping of A onto A and η is a 1:1 mapping P onto P, such that $\alpha = \langle \theta, \eta \rangle$, i.e.,*

$$\alpha(ap) = \theta(a)\eta(p).$$

In a conjoint structure, the identity of the independent factors is an essential feature of the concept, one which we believe should be preserved by automorphisms. This view entails that we not consider all order automorphisms, but only those that are factorizable. This view is implicit in dimensional analysis, where the only automorphisms ever considered in the definition of a dimensionally invariant function, the so-called similarities, are factorizable.

One possible source of confusion on this matter is the fact that we do not usually write conjoint structures in a way that fits the formal definition of a relational structure (see Section 20.2.3). When this is remedied, the only remaining automorphisms are factorizable. Suppose, with no real loss of generality, that A and P are disjoint and let $X = A \cup P$. To recast $\mathscr{C}$ as a relational structure, we let $\mathscr{C}' = \langle X, A, R \rangle$, where A is a unary relation (i.e., a subset of X) and R is a quaternary relation. The intention is to use $X - A$ for P and to have the quadruple (a, p, b, q) in R exactly when a, b are in A and p, q are in $X - A$, and $ap \succsim bq$ in the binary relation $\succsim$ on $A \times P$. We leave it as Exercise 25 to show that the automorphisms of $\mathscr{C}'$ are exactly the factorizable ones of $\mathscr{C}$.

As we shall see in Section 20.2.7, it turns out that the assumption of factorizable automorphisms, interval scale representability, and a representation onto Re^+ forces the Thomsen condition and so the usual multiplicative representation of dimensional theory.

Our first result is important. It shows that if the conjoint structure is sufficiently endowed with factorizable automorphisms, then the induced operations are all basically the same.

THEOREM 12. *Let $\mathscr{C} = \langle A \times P, \succsim \rangle$ be solvable at points $a_0 p_0$ and $a_0' p_0'$ and suppose that $\alpha = \langle \theta, \eta \rangle$ maps $a_0 p_0$ to $a_0' p_0'$. Let π and π' denote the mappings from A to P relative to these points (Definition 8.5). α is an automorphism of $\mathscr{C}$ iff $\eta\pi = \pi'\theta$ and either (and so both) of θ or η are isomorphisms of the induced Holman structures.*

It seems natural to investigate further the following concepts of left and right multiplications of the induced concatenation structures. Except for multiplication by the identity, multiplications are not automorphisms of the concatenation structure; the question is, can they be automorphisms of the conjoint structure?

DEFINITION 12. *Suppose* $\mathscr{A} = \langle A, \succsim, \bigcirc \rangle$ *is a concatenation structure, then for each* $a \in A$, *define the* left multiplication[10] a^L *by* $a^L(b) = a \bigcirc b$, *for all* $b \in A$ *for which the right hand side is defined. Define the* right multiplication *analogously by* $b \bigcirc a$.

In a conjoint structure, any pair of multiplications in the Holman structures induced on each factor generates a factorizable transformation. In general, however, this transformation is not an automorphism. We show next that when there is solvability at a point and such pairs of multiplications do in fact yield automorphisms, then the Thomsen condition is satisfied. (Since concatenation is commutative in this case, left and right multiplication are identical.)

THEOREM 13. *Suppose that* $\mathscr{C}$ *is a conjoint structure that is solvable relative to* $a_0 p_0$ *and that for every* $a \in A$, $p \in P$ *there is a* $b \in A$ *such that* $bp \sim ap_0$.

(i) $\mathscr{C}$ *satisfies the Thomsen condition iff, for every* $c \in A$, $r \in P$, $\langle c \bigcirc_A \iota_A, \iota_P \bigcirc_P r \rangle$ *is an automorphism of* $\mathscr{C}$.

(ii) *Suppose* θ *is* $1:1$ *from* A *onto* A *and* η *is* $1:1$ *from* P *onto* P. *Under the conditions of part* (i), $\langle \theta, \eta \rangle$ *is an automorphism of* $\mathscr{C}$ *iff for some automorphism* θ^* *of* $\mathscr{I}_A$,

$$\theta = \theta(a_0)^L \theta^* \qquad \text{and} \qquad \eta = \eta(p_0)^R \pi \theta^* \pi^{-1}.$$

19.6.4 Total Concatenation Structures Induced by Closed, Idempotent Concatenation Structures

In Section 6.9 we developed a representation for each bisymmetric structure by the device of recoding it as an additive conjoint structure. Although bisymmetry was essential in proving the additivity of the conjoint structure, it did not play any role in the definition of the ordering itself. So we can use the same device in more general contexts.

DEFINITION 13. *Suppose* $\mathscr{A} = \langle A, \succsim, \bigcirc \rangle$ *is a closed concatenation structure. Then the* conjoint structure induced by $\mathscr{A}$ *is* $\mathscr{C} = \langle A \times A, \succsim' \rangle$, *where for all* $a, b, c, d \in A$,

$$ab \succsim' cd \qquad \text{iff} \qquad a \bigcirc b \succsim c \bigcirc d.$$

It is, of course, necessary to verify that $\mathscr{C}$ is a conjoint structure; this is left as Exercise 26.

[10] Luce and Narens (1985) used the terms *left* and *right translations*, but this was a poor choice because of a different, more appropriate use of the term *translation* given in Definition 14.

THEOREM 14. *Suppose $\mathscr{A}$ is a closed concatenation structure and $\mathscr{C}$ is the conjoint structure induced by $\mathscr{A}$. Then the following statements are true:*

(i) *If $\mathscr{A}$ is solvable, then $\mathscr{C}$ is unrestrictedly solvable. If for each $a \in A$ there exist $b, c \in A$ such that $a = b \bigcirc c$, then the converse holds.*

(ii) *$\mathscr{C}$ is Archimedean iff $\mathscr{A}$ is Archimedean in differences.*

(iii) *Suppose $\mathscr{A}$ is idempotent and α is a mapping of A onto A. Then, α is an automorphism of $\mathscr{A}$ iff (α, α) is a factorizable automorphism of $\mathscr{C}$.*

Proof of parts (i)–(ii) are left as Exercise 27, and part (iii) is proved in Section 19.8.4.

Observe that by combining Definition 9, which according to Theorem 11 induces a total concatenation structure from a suitably solvable conjoint one, and Definition 13, which induces a conjoint structure from a concatenation one, we can induce a total concatenation structure from any sufficiently solvable concatenation structure. Moreover, in the idempotent case, the automorphisms of $\mathscr{A}$ become factorizable in $\mathscr{C}$, and so by Theorem 12 they are nicely reflected in the induced total concatenation structures. To make clear the exact connections, the following concepts are useful.

DEFINITION 14. *Suppose $\mathscr{A} = \langle A, \succsim, \bigcirc \rangle$ is a concatenation structure and α is an automorphism of $\mathscr{A}$. Then α is said to be a* dilation at *a iff $\alpha(a) = a$, and it is said to be a* translation *iff it is either the identity or is not a dilation for any $a \in A$.*

The choice of terms is motivated by consideration of linear transformations, which (as we shall see in Chapter 20) is not as restrictive a model for automorphisms as one might at first think. The transformation $x \to rx + s$ has a fixed point iff either $r = 1$ and $s = 0$ (the identity map) or $r \neq 1$. Thus the nondilations are $x \to x + s$, which are quite reasonably called "translations." For $r > 1$ the term *dilation* (expanding) seems plausible, and since for $0 < r < 1$ the inverse of $x \to rx + s$, namely, $x \to (x - s)/r$, is expanding, this term has been applied to all nontranslations plus the identity. Note, by Theorem 4, that all the automorphisms of a PCS are translations.

We turn now to the other major case, the idempotent (intensive) concatenation structures, and we see how the dilations and translations appear in the induced structures). The following result is due to Luce and Narens (1985).

THEOREM 15. *Suppose $\mathscr{A}$ is a closed, idempotent, solvable concatenation structure, $\mathscr{T}_a$ is the total concatenation structure induced at a via Definitions 9 and 13, and α is an automorphism of $\mathscr{A}$.*

(i) *α is a dilation at a iff α is an automorphism of $\mathscr{I}_a$.*

(ii) *α is a translation iff α is an isomorphism of $\mathscr{I}_a$ onto $\mathscr{I}_{\alpha(a)}$, where $\alpha(a) \neq a$.*

COROLLARY. *If, in addition, $\mathscr{A}$ is Archimedean in differences, then the set of dilations at a form a commutative subgroup.*

This corollary is easy to prove. An interesting related question is when the set of translations forms an Archimedean ordered group and so, too, is commutative (Hölder's theorem). This topic is currently under investigation, and some results are given in Section 20.2.6.

Theorem 15 provides a good deal of insight into the decomposition of an idempotent structure into a family of induced total concatenation structures that are all isomorphic under the translations of the idempotent structure. Moreover, its dilations are the automorphisms of the induced total concatenation structures. We suspect that such a decomposition will also prove useful in guiding generalizations to structures that, although lacking automorphisms, have some form of local automorphism that permit the structure to be embedded in another structure that is richly endowed with automorphisms.

19.6.5 Intensive Structures Related to PCSs by Doubling Functions

The approach to intensive structures taken in this subsection, first suggested by Narens and Luce (1976), considers only those that have elements that are "double" other elements in the following, somewhat imaginary, sense.[11] Let $*$ denote the intensive operation. Suppose we were able to adjoin an element 0 to the structure in such a way that one could sensibly define $a * 0$ for all a in A and that it would play a role analogous to 0 in computing numerical averages. Then we would say that b is "double" a if $b * 0 \sim a$, because in an intensive structure $b * 0$ lies between b and 0. Of

[11] In a presentation of axiomatic geometry, Szmielew (1983) examined a concept related to the material treated in this section. In particular, let $\bigcirc$ be an operation on A. $\langle A, \bigcirc \rangle$ is said to be a *midpoint algebra* iff $\bigcirc$ is idempotent, commutative, bisymmetric (she calls it bicommutative), and left-solvable. One way such structures arise is to begin with a commutative group $\langle A, \oplus, e\rangle$ where $\oplus$ is the group operation on A and $e \in A$ is the identity. If the group has half-elements, denoted $\frac{1}{2}a$, then define $a \bigcirc b = \frac{1}{2}(a \oplus b)$. This operation can be shown to be a midpoint algebra. Beginning with $\langle A, \bigcirc, e\rangle$, one can define *multiplication by 2 relative to e* as the solution $2a$ to $2a \bigcirc e = a$. This is similar to what we refer to as a "doubling function." The major difference is that we have no identity element but do have an order. Szmielew shows 2 to be an automorphism of $\langle A, \bigcirc, e\rangle$. If one defines $\odot$ by $a \odot b = 2(a \bigcirc b)$, then using the fact that $a \bigcirc e = \frac{1}{2}(a \oplus e) = \frac{1}{2}a$, it is easy to verify that $\odot = \oplus$. Theorem 16 generalizes this to ordered cases that are not commutative, not bisymmetric, and do not have an identity.

course, were we able to do this, then we could speak of a as the half-element of b and think of introducing an operation $\bigcirc$ such that $a \bigcirc a \sim b$. This, then, hints at the possibility of relating a halvable PCS to any intensive one that has an appropriate doubling function. Of course, the problem is that we really do not know how to adjoin 0 to the structure, and so we must invent some less direct way of characterizing a doubling function. This we do in the next definition. Admittedly, the definition of a doubling function is less than transparent; its utility is to be evaluated, however, by the resulting theorems.

DEFINITION 15. *Let A be a nonempty set, $\succsim$ a binary relation on A, and $*$ a partial, intensive operation on A (Definition 2). Suppose that $B \subseteq A$ and δ is a function from B onto A. Then δ is said to be a* doubling function of $\mathscr{A} = \langle A, \succsim, * \rangle$ *iff for all $a, b \in A$, the following hold.*

1. *δ is strictly monotonic increasing.*
2. *If $a \in B$ and $a \succsim b$, then $b \in B$.*
3. *If $a \succ b$, then there exists a $c \in A$ such that $b * c$ is defined and $\in B$, and $a \succ \delta(b * c)$.*
4. *If $a * b$ is defined and $\in B$, then $\delta(a * b) \succ a, b$.*
5. *Suppose that $a_n \in A$, $n = 1, 2, \ldots,$ are such that if $a_{n-1} \in B$, then $a_n \sim \delta(a_{n-1}) * a_1$. For any b, either there is an integer n such that $a_n \notin B$ or $a_n \succsim b$. Such a sequence is called a standard sequence of δ.*

The obvious first question to ask is just how unique is a doubling function. Narens and Luce (1976) reported some partial results that led them to suspect that a structure might have many doubling functions; however, Michael A. Cohen in a theorem published by Luce and Narens (1985) showed that, except possibly for one point, there is at most one doubling function. Before stating this result precisely, we need another result that is used in its proof.

THEOREM 16. *Suppose A is a nonempty set, $\succsim$ is a binary relation on A, and $\bigcirc$ and $*$ are two closed operations on A such that for each $a, a', b, b' \in A$, $a \bigcirc b \sim a' \bigcirc b'$ iff $a * b \sim a' * b'$. Define the function θ by $\theta(a \bigcirc b) = a * b$. Then, $\langle A, \succ, \bigcirc \rangle$ is a* PCS *with half element function θ [i.e., $a \sim \theta(a) \bigcirc \theta(a)$] iff $\langle A, \succ, * \rangle$ is an intensive concatenation structure with doubling function θ^{-1} defined on all of A.*

We now turn to the uniqueness of the doubling function.

THEOREM 17. *Suppose $\mathscr{A}$ is an intensive concatenation structure with a doubling function. Either that doubling function is unique or there is just one* *other doubling function and*

(i) *their domains differ by just one point* b,

(ii) *the functions agree on their common domain, and*

(iii) *the double of* b *is maximal in* $\mathscr{A}$.

COROLLARY. *Suppose* $\mathscr{A}$ *is an intensive concatenation structure with a doubling function, and* $\mathscr{B}$ *is the* PCS *related to it, as in Theorem* 16. *If* $\mathscr{A}$ *has no maximal element, then the automorphism groups of* $\mathscr{A}$ *and* $\mathscr{B}$ *are identical.*

Proofs of these results are outlined as Exercises 28–30.

It is essential to recognize that some important intensive structures are not equipped with doubling functions, and so they differ significantly from positive concatenation structures. Consider $\mathscr{R} = \langle \text{Re}, \geqslant, \otimes \rangle$, where for some real t, $0 < t < 1$, $x \oplus y = tx + (1-t)y$. Clearly, $\mathscr{R}$ is intensive. For any real r, s, with $r > 0$, α defined by $\alpha = rx + s$ is an automorphism of $\mathscr{R}$. This follows from the fact that α is clearly $1:1$, onto, order preserving, and that

$$\begin{aligned}\alpha(x \oplus y) &= r[tx + (1-t)y] + s \\ &= t(rx + s) + (1-t)(ry + s) \\ &= \alpha(x) \oplus \alpha(y).\end{aligned}$$

However, if $\mathscr{R}$ had a doubling function, then by the corollary to Theorem 17 its automorphism group is that of a PCS, which by Theorem 4 must be a subset of the similarity group. This contradicts the fact that in the present case the positive affine group is a subgroup of the automorphism group, and so $\mathscr{R}$ cannot have a doubling function.

One major difference between intensive structures having doubling functions and the example just given is whether the automorphisms have one or two degrees of freedom. We return to this in Section 20.4.2.

19.6.6 Operations That Distribute over Conjoint Structures

In Chapter 10, we encountered important empirical situations in which an additive conjoint structure has an extensive operation on one of its components. Indeed, we concluded that such triples are the basis of the structure of physical quantities. We were especially interested in conditions under which the multiplicative conjoint representation can be stated in terms of a power function of the standard additive representation of the extensive structure. To insure that relation, we invoked what were called laws of similitude and exchange. Since that chapter was written, far more general results have been uncovered. Basically, the same representation can now be derived for any Dedekind complete conjoint structure (i.e., we do not need to assume the Thomsen condition) and for any structure on the component that has a ratio scale representation (a concept that is studied generally in Section 20.2) and that exhibits a distribution property rather than the two somewhat special types of laws. Here we introduce for the special context of concatenation structures the concept of distribution, and we study something about the structure of automorphisms involved. The concept of distribution will be generalized (see Definition 20.5) and the representation result will be studied and proved in the next chapter (see Theorems 7 and 8, Chapter 20).

The following concept was introduced by Narens (1976) in a special context and studied more generally in Narens and Luce (1976) and Luce and Narens (1985). It is considerably generalized in Section 20.2.7.

DEFINITION 16. *Suppose* $\mathscr{C} = \langle A \times P, \succsim \rangle$ *is a conjoint structure and* $\bigcirc$ *is a partial operation on* A. *Then* $\bigcirc$ *is said to distribute over* $\mathscr{C}$ *iff for all* $a, b, c, d \in A$ *for which* $a \bigcirc b$ *and* $c \bigcirc d$ *are defined and for all* $p, q \in P$,

$$\text{if} \quad ap \sim cq \quad \text{and} \quad bp \sim dq, \quad \text{then} \quad (a \bigcirc b, p) \sim (c \bigcirc d, q).$$

THEOREM 18. *Suppose that* $\mathscr{C} = \langle A \times P, \succsim \rangle$ *is a conjoint structure that is solvable relative to* $a_0 p_0 \in A \times P$, *that* $\bigcirc_A$ *is the operation induced on* A *relative to* $a_0 p_0$, *and that* $\mathscr{T}$ *is the set of right multiplications of* $\bigcirc_A$. *Suppose further that* $\mathscr{A} = \langle A, \succsim_A, \bigcirc \rangle$ *is a closed concatenation structure on the* A *component of* $\mathscr{C}$ *and that* $\mathscr{G}$ *is its group of automorphisms. Then the following are true.*

(i) *The operation* $\bigcirc_A$ *does not distribute over* $\mathscr{C}$.

(ii) *If the operation* $\bigcirc$ *distributes over* $\mathscr{C}$, *then* $\mathscr{T}$ *is a subset of the endomorphisms of* $\mathscr{A}$ (*i.e., homomorphisms of* $\mathscr{A}$ *into* $\mathscr{A}$).

(iii) $\mathscr{T} \subseteq \mathscr{G}$ *iff* $\bigcirc$ *distributes over* $\mathscr{C}$ *and* $\mathscr{C}$ *is unrestrictedly A-solvable.*

(iv) $\mathscr{T} = \mathscr{G}$ *iff* $\mathscr{T} \subseteq \mathscr{G}$ *and* $\mathscr{A}$ *is* 1-*point unique.*

This result is included in Luce and Narens (1985).

Since most empirical operations in physics are distributive in this sense in some conjoint structure, part (i) [due to Narens (1981a)] means that the induced operations of a conjoint structure do not coincide with any empirical concatenation operation. Put another way, the empirical information provided by an empirical conjoint structure is distinct from that provided by any distributive empirical operation on one of its components. Thus, for example, the momentum structure $p = mv$, where m is mass and v velocity, can be recoded as an operation $\odot$ over masses, but that operation will not satisfy the property that if $m_1 v = m_1' v'$ and $m_2 v = m_2' v'$, then $(m_1 \odot m_2) v = (m' \odot m_2') v'$.

Of the remaining statements, the most significant is (iv), namely, that the right multiplications of the induced operation coincide with the automorphisms of $\mathscr{A}$ when $\mathscr{C}$ is solvable, $\odot$ is distributive, and $\mathscr{A}$ is 1-point unique. Such an explicit representation of the automorphisms of $\mathscr{A}$ provides us with considerable leverage in developing the representations of such structures (Section 20.2.7).

19.7 REPRESENTATIONS OF SOLVABLE CONJOINT AND CONCATENATION STRUCTURES

We now take advantage of the various interconnections uncovered in Section 19.6 together with the representations of PCSs from Section 19.3 to arrive at general representation and uniqueness theorems for solvable cases.

19.7.1 Conjoint Structures

The existence of a representation of a conjoint structure that is unrestrictedly solvable and Archimedean follows almost immediately from the facts that its induced structure is a total concatenation structure (Theorem 11), that such a structure is made up of two PCSs, and that each PCS has a representation (Theorem 3). We formulate the result explicitly but leave the formal proof as Exercise 31. Exercise 32 presents the generalized Weber model (see Item 5 in Section 19.2.2) as an example of structure fulfilling the hypotheses of this result.

THEOREM 19. *Suppose $\mathscr{C} = \langle A \times P, \succsim \rangle$ is a conjoint structure that is Archimedean and solvable. Then there exist a numerical operation $\oplus$ and functions φ from A and from ψ from P into* Re *such that*

(i) $\varphi(a_0) = 0$, $\psi(p_0) = 0$,

(ii) $x \oplus 0 = x$, $0 \oplus y = y$.

(iii) $ap \succsim bq$ *iff* $\varphi(a) \oplus \psi(p) \geqslant \varphi(b) \oplus \psi(q)$.

The corresponding uniqueness theorem follows from the Corollary to Theorem 2 and the easily verified fact that $\psi = \varphi\pi^{-1}$.

THEOREM 20. *Suppose that, under the conditions of Theorem 19, $\varphi \oplus \psi$ and $\varphi' \oplus \psi'$ are two representations of $\mathscr{C} = \langle A \times P, \succsim, a_0 p_0 \rangle$ with $\varphi(A)$*

$= \varphi'(A)$ *and* $\psi(P) = \psi'(P)$. *If for some* $a \in A$, $\varphi(a) = \varphi'(a)$, *then* $\varphi \equiv \varphi'$ *and* $\psi \equiv \psi'$.

The proof is left as Exercise 33.

This result is misleading in an important way. It suggests that the representation is 1-point unique whereas it really must be viewed as 2-point unique because the point (a_0, p_0) has been mapped into the common value $(0, 0)$. Once that is treated as a free value, it is seen that there may be two degrees of freedom involved.

19.7.2 Solvable, Closed, Archimedean Concatenation Structures

The goal of this subsection is the representation and uniqueness of a broad class of concatenation structures, including at least the intensive ones, that are distinct from the PCSs. As was true for PCSs, there are two possible ways to formulate uniqueness. One is to show that the structure is N-point unique for some N. The other (and better) result is to characterize the nature of the automorphism group as we did with PCSs when we showed their automorphism groups to be isomorphic to subgroups of the additive reals. At present we do not have the better theorem.

THEOREM 21. *Suppose* $\mathscr{A}$ *is a concatenation structure that is closed, solvable, and Archimedean in difference sequences. Then* $\mathscr{A}$ *is isomorphic to a real closed concatenation structure, and* $\mathscr{A}$ *is either* 1- *or* 2-*point unique.*

Observe that for an intensive structure, it is not the closure of the operation that is unrealistic, as it was for PCSs, but solvability, which implies very large elements in the structure.

The proof of Theorem 21 (Section 19.8.7), due to Luce and Narens (1985), entails recoding it as a conjoint structure and using Theorem 19. The conclusion that it is at most 2-point unique is not terribly surprising, at least for idempotent structures since their dilations at each point form a commutative subgroup (Corollary to Theorem 15) and translations have no fixed point. So the set of all dilations appears to be at most 2-point unique. Later, in Section 20.2.5, we shall show that very general real structures, except for those that are N-point unique for every N, are either 1- or 2-point unique.

At present we lack an analogue to Theorem 4 showing an isomorphism between the automorphism group and a real transformation group. A highly plausible conjecture is that it is isomorphic to a subgroup of the positive affine group: $x \to rx + s$, $r > 0$; but that has not been established.

19.7.3 Intensive Concatenation Structures with Doubling Functions

It is clear that combining Theorem 3 on the representation of PCSs and Theorem 16 relating intensive structures with doubling functions to PCSs should result in a representation theorem for such intensive structures. We formulate the concept of such a representation as follows.

DEFINITION 17. *Suppose* $\mathscr{A} = \langle A, \succsim, * \rangle$ *is an intensive concatenation structure with doubling function* δ. *Let* $\oplus$ *be a partial binary numerical operation on* Re^+ *with half-element function* h, *and let* φ *be a mapping from* A *into* Re^+. *Then* φ *is said to be a* $\oplus$*-representation of* a *iff for all* $a, b \in A$,

1. $a \succsim b$ *iff* $\varphi(a) \geqslant (b)$,
2. *if* $a * b$ *is defined*, $\varphi(a * b) = h[\varphi(a) \oplus \varphi(b)]$,
3. *if* a *is in the domain of* δ, $\varphi(a) = h\varphi\delta(a)$.

THEOREM 22. *Suppose* $\mathscr{A}$ *is an intensive concatenation structure with doubling function* δ. *Then there exist* φ: $A \to$ (*into*)Re^+ *and a binary operation* $\oplus$ *on* Re^+ *such that* φ *is a* $\oplus$*-representation of* $\mathscr{A}$. *Moreover, two such* $\oplus$*-representations that agree at one point are identical.*

The formal proof is left as Exercise 34.

19.8 PROOFS

19.8.1 Theorem 11 (p. 78)

(i) *Suppose* $\mathscr{C}$ *is a conjoint structure that is solvable relative to* a_0p_0 *and that* $\bigcirc_A$ *is the induced Holman operation.*

(a) *Then* $\bigcirc_A$ *is closed and monotonic, and it is positive over* A^+ *and negative over* A^-.

(b) *If, in addition,* $\mathscr{C}$ *is Archimedean, then both* $\mathscr{I}_{A^+} = \langle A^+, \succsim_{A^+}, \bigcirc_{A^+} \rangle$ *and* $\mathscr{I}_{A^-} = \langle A^-, \precsim_{A^-}, \bigcirc_{A^-} \rangle$ *are Archimedean in standard sequences.*

(c) *If, in addition,* $\mathscr{C}$ *is both solvable and Archimedean, then* $\mathscr{I}_A$ *is a closed, solvable, total concatenation structure.*

(ii) *Suppose* $\mathscr{A} = \langle A, \succsim, \bigcirc, a_0 \rangle$ *is a closed total concatenation structure.*

(a) *Then there is a conjoint structure* $\mathscr{C} = \langle A \times A, \succsim' \rangle$ *that is solvable relative to* a_0a_0 *and that induces* $\mathscr{A}$.

(b) *If* $\mathscr{A}$ *is Archimedean in differences, then* $\mathscr{C}$ *is Archimedean.*

(c) *If $\mathscr{A}$ is solvable, Archimedean in standard sequences, and associative, then $\mathscr{C}$ is solvable and Archimedean.*

PROOF.
Proof of (i)(a)

Closure. By the solvability of $\mathscr{C}$ relative to $a_0 p_0$, the Holman operation is defined for every pair and so is closed.

Monotonicity

$$\begin{aligned} a \bigcirc_A c \succsim_A b \bigcirc_A c \quad & \text{iff} \quad \xi[a, \pi(c)] \succsim_A \xi[b, \pi(c)] \\ & \text{iff} \quad \xi[a, \pi(c)] p_0 \succsim \xi[b, \pi(c)] p_0 \\ & \text{iff} \quad a\pi(c) \succsim b\pi(c) \\ & \text{iff} \quad a \succsim_A b. \end{aligned}$$

The other case is similar.

Positivity on A^+. Suppose $a \succ_A a_0$, $b \succ_A a_0$. Observe that

$$(a \bigcirc_A b) p_0 \sim \xi[a\pi(b)] p_0 \sim a\pi(b) \succ a_0\pi(b) \sim bp_0,$$

and so by monotonicity, $a \bigcirc_A b \succ_A b$. Since $b \succ_A a_0$,

$$a_0\pi(b) \sim bp_0 \succ a_0 p_0,$$

whence $\pi(b) \succ_P p_0$. Therefore,

$$(a \bigcirc_A b) p_0 \sim a\pi(b) \succ ap_0,$$

and so $a \bigcirc_A b \succ_A a$.

Negativity on A^- is similar.

Proof of (i)(b). Let $\{a(n)\}$ be a standard sequence of $\mathscr{I}_{A^+}$, i.e., for some $a \succ_A a_0$,

$$a(n) \sim_A a(n-1) \bigcirc_A a = \xi[a(n-1), \pi(a)].$$

Thus,

$$a(n) p_0 \sim \xi[a(n-1), \pi(a)] p_0 \sim a(n-1)\pi(a).$$

So $\{a(n)\}$ is also a standard sequence of $\mathscr{C}$. Thus, if it is bounded it must be finite because, by hypothesis, $\mathscr{C}$ is Archimedean.

Proof of (i)(c). Suppose $\mathscr{C}$ is solvable. We first establish that $\bigcirc_A$ is also solvable in the sense that for $b \succ_A a$, there are c, d such that $b \sim_A a \bigcirc_A c \sim_A d \bigcirc_A a$. By the assumed solvability there exists p such that $ap \sim bp_0$.

Let $c = \pi^{-1}(p)$, and so

$$bp_0 \sim a\pi(c) \sim \xi[a, \pi(c)]p_0 \sim (a \bigcirc_A c)p_0.$$

Thus, by monotonicity, $b \sim_A a \bigcirc_A c$, and $c \succ_A a_0$. The existence of d is similar. Thus, solvability holds for $\mathscr{I}_{A^+}$, and so with part (i)(b) we see that $\mathscr{I}_{A^+}$ is a PCS.

To complete the proof that $\mathscr{I}_A$ is a total concatenation structure, we must show compatibility. Suppose $b \in A^+$, $a \in A^-$, and let c and d be the solutions just established. By the fact they are positive and the operation is positive in the positive domain,

$$a \bigcirc_A (c \bigcirc_A c) \succ_A a \bigcirc_A c \sim_A b.$$

The other case of compatibility is similar.

Proof of (ii)(a). Define $\succsim'$ on $A \times A$ by:

$$aa' \succsim' bb' \quad \text{iff} \quad a \bigcirc a' \succsim b \bigcirc b'.$$

We first verify that $\succsim'$ satisfies the axioms of a conjoint structure. It is a weak order because, by assumption, $\succsim$ is a total order. It is independent because $\bigcirc$ is monotonic. And $\succsim'_A$ is a weak order because it is in fact identical to $\succsim$ since by the monotonicity of $\bigcirc$

$$\begin{aligned} a \succsim'_A b \quad &\text{iff} \quad aa' \succsim' ba' \\ &\text{iff} \quad a \bigcirc a' \succsim b \bigcirc a' \\ &\text{iff} \quad a \succsim b. \end{aligned}$$

To complete the proof that this conjoint structure induces the given total concatenation structure, we show that $\bigcirc'_A$ is $\bigcirc$. Observe that since

$$a \bigcirc a_0 = a = a_0 \bigcirc a,$$

it follows that $\pi(a) = a$. Further, $\xi(a, b) = a \bigcirc b$ because

$$\xi(a, b)a_0 \sim' ab \quad \text{iff} \quad \xi(a, b) = \xi(a, b) \bigcirc a_0 = a \bigcirc b.$$

Using these facts,

$$a \bigcirc'_A b = \xi[a, \pi(b)] = \xi(a, b) = a \bigcirc b,$$

proving the identity of the operations.

Proof of (ii)(b). It is clearly sufficient to show that if $\{a(n)\}$ is a standard sequence of $\mathscr{C}$, then it is a difference sequence of $\mathscr{A}$. By definition,

$$a(n)b \sim' a(n-1)c \quad \text{iff} \quad a(n) \bigcirc b = a(n-1) \bigcirc c,$$

which establishes the connection and so the Archimedeanness of $\mathscr{C}$.

Proof of (ii)(c). The solvability of $\mathscr{C}$ is immediate from that of $\mathscr{A}$.

To show Archimedeanness, suppose $\{a(n)\}$ is an increasing conjoint standard sequence, i.e., for some $b \prec c$,

$$a(n)b \sim' a(n-1)c \quad \text{iff} \quad a(n) \bigcirc b = a(n-1) \bigcirc c.$$

With no loss of generality, assume the sequence numbering is such that $a(1) \succ a_0$ and $a(0) \precsim a_0$. Consider the positive part. By solvability, there is d such that $c = d \bigcirc b$, whence by monotonicity and associativity, $a(n) = a(n-1) \bigcirc d$. Let $a' = \min\{a(1), d\}$, then if $\{a'(n)\}$ is the standard sequence based on a', an easy induction and monotonicity show that $a(n) \succsim a'(n)$. Thus, if the original standard sequence is bounded, so is this one and hence it must be finite. The negative terms are dealt with similarly. ◇

19.8.2 Theorem 12 (p. 80)

Suppose that $\mathscr{C}$ is a conjoint structure that is A-solvable relative both to a_0p_0 and to $a_0'p_0'$, that θ is a function from A onto A such that $\theta(a_0) \sim_A a_0'$, and η is a function from P onto P such that $\eta(p_0) \sim_P p_0'$. Then $\langle\theta, \eta\rangle$ is an automorphism of $\mathscr{C}$ iff $\eta = \pi'\theta\pi^{-1}$ and θ is an isomorphism from $\mathscr{I}_A = \langle A, \succsim_A, \bigcirc_A, a_0\rangle$ onto $\mathscr{I}_A' = \langle A, \succsim_A', \bigcirc_A', a_0'\rangle$.

PROOF. Suppose $\langle\theta, \eta\rangle$ is an automorphism. Since by definition of π, for all p in P, $\pi^{-1}(p)p_0 \sim a_0p$, applying the automorphism yields

$$\begin{aligned} a_0'[\pi'\theta\pi^{-1}(p)] &\sim [\theta\pi^{-1}(p)]p_0' \\ &\sim [\theta\pi^{-1}(p)]\eta(p_0) \sim \theta(a_0)\eta(p) \sim a_0'\eta(p), \end{aligned}$$

whence $\eta = \pi'\theta\pi^{-1}$. Next, apply the automorphism to the definition of $\bigcirc_A$, $(a \bigcirc_A b)p_0 \sim a\pi(b)$, which yields

$$\begin{aligned} (a \bigcirc_A b)p_0' &\sim \theta(a \bigcirc_A b)[\pi'\theta\pi^{-1}(p_0)] \\ &\sim \theta(a)[\pi'\theta\pi^{-1}\pi(b)] \\ &\sim \theta(a)[\pi'\theta(b)] \\ &\sim [\theta(a) \bigcirc_A' \theta(b)]p_0', \end{aligned}$$

whence $\theta(a \bigcirc_A b) \sim_A \theta(A) \bigcirc'_A \theta(b)$. Since θ is 1 : 1 and order preserving by the fact $\langle \theta, \pi \rangle$ is an automorphism, this proves θ is an isomorphism from $\mathscr{I}_A$ onto $\mathscr{I}'_A$.

Conversely, suppose θ is an isomorphism,

$$\begin{aligned} ap \succsim bq \quad &\text{iff} \quad a \bigcirc_A \pi^{-1}(p) \succsim_A b \bigcirc_A \pi^{-1}(q) \\ &\qquad\text{(definition of } \bigcirc_A) \\ &\text{iff} \quad \theta(a \bigcirc_A \pi^{-1}(p)) \succsim_A \theta(b \bigcirc_A \pi^{-1}(q)) \\ &\qquad (\theta \text{ an isomorphism}) \\ &\text{iff} \quad \theta(a) \bigcirc'_A \theta\pi^{-1}(p) \succsim_A \theta(b) \bigcirc'_A \theta\pi^{-1}(p) \\ &\qquad (\theta \text{ an isomorphism}) \\ &\text{iff} \quad \theta(a)[\pi'\theta\pi^{-1}(p)] \succsim \theta(b)[\pi'\theta\pi^{-1}(q)] \\ &\qquad \text{(definition of } \bigcirc'_A), \end{aligned}$$

and so $\langle \theta, \pi'\theta\pi^{-1} \rangle$ is an automorphism of $\mathscr{C}$. ◇

19.8.3 Theorem 13 (p. 81)

Suppose $\mathscr{C}$ is a conjoint structure that is A-solvable relative to $a_0 p_0$ and that for every $a \in A$, $p \in P$ there exists some $b \in A$ such that $bp \sim ap_0$.

(i) *$\mathscr{C}$ satisfies the Thomsen condition iff for every $a \in A, p \in P$, $\langle a \bigcirc_A \iota_A, \iota_P \bigcirc_P p \rangle$ is an automorphism of $\mathscr{C}$.*

(ii) *Under these conditions, suppose θ is 1 : 1 from A onto A and η is 1 : 1 from P onto P. Then $\langle \theta, \eta \rangle$ is an automorphism of $\mathscr{C}$ iff for some automorphism θ^* of $\mathscr{I}_A = \langle A, \succsim_A, \bigcirc_A, a_0 \rangle$,*

$$\theta = \theta(a_0) \bigcirc_A \theta^* \qquad \textit{and} \qquad \eta = \pi\theta^*\pi^{-1} \bigcirc_P \eta(p_0).$$

PROOF.

(i) Suppose $\langle a \bigcirc_A \iota_A, \iota_P \bigcirc_P p \rangle$ is an automorphism for every $a \in A$, $p \in P$. For $b, c \in A$, apply it with $p = \pi(c)$ to $bp_0 \sim a_0\pi(b)$ to yield

$$\begin{aligned} (a \bigcirc_A b)\pi(c) &\sim (a \bigcirc_A b)[p_0 \bigcirc_P \pi(c)] && (p_0 \text{ is identity of } \bigcirc_P) \\ &\sim (a \bigcirc_A a_0)[\pi(b) \bigcirc_P \pi(c)] && \text{(application of automorphsim} \\ &\sim a[\pi(b) \bigcirc_P \pi(c)] && (a_0 \text{ is identity of } \bigcirc_A). \end{aligned}$$

Observe that

$$\begin{aligned} a_0\pi(b \bigcirc_A c) &\sim (b \bigcirc_A c)p_0 && \text{(definition of } \pi) \\ &\sim b\pi(c) && \text{(definition of } \bigcirc_A) \\ &\sim [\pi^{-1}\pi(b)]\pi(c) && \text{(identity)} \\ &\sim a_0[\pi(b) \bigcirc_P \pi(c)] && \text{(definition of } \bigcirc_P). \end{aligned}$$

Using these two equivalences,

$$\begin{aligned} [(a \bigcirc_A b) \bigcirc_A c]p_0 &\sim (a \bigcirc_A b)\pi(c) && \text{(definition of } \bigcirc_A) \\ &\sim a[\pi(b) \bigcirc_P \pi(c)] && \text{(by first equivalence)} \\ &\sim a\pi(b \bigcirc_A c) && \text{(by second equivalence)} \\ &\sim [a \bigcirc_A (b \bigcirc_A c)]p_0 && \text{(definition of } \bigcirc_A). \end{aligned}$$

Thus, by monotonicity, $\bigcirc_A$ is associative, and so, by Corollary 1 to Theorem 11, the Thomsen condition holds.

Conversely, suppose the Thomsen condition holds. We first show the triple cancellation property known as the Reidmeister condition (p. 252 of Volume I). Suppose $ax \sim by$, $fy \sim gx$, $gp \sim fq$. So, by definition of $\bigcirc_A$,

$$\begin{aligned} a \bigcirc_A \pi^{-1}(x) &\sim_A b \bigcirc_A \pi^{-1}(y), \\ f \bigcirc_A \pi^{-1}(y) &\sim_A g \bigcirc_A \pi^{-1}(x), \\ g \bigcirc_A \pi^{-1}(p) &\sim_A f \bigcirc_A \pi^{-1}(q). \end{aligned}$$

Using the associativity and commutativity of $\bigcirc_A$ (Theorem 11) freely, we obtain

$$\begin{aligned} [a \bigcirc_A \pi^{-1}(p)] \bigcirc_A [f \bigcirc_A \pi^{-1}(y)] &\sim_A [a \bigcirc_A \pi^{-1}(p)] \bigcirc_A [g \bigcirc_A \pi^{-1}(x)] \\ &\sim_A [a \bigcirc_A \pi^{-1}(x)] \bigcirc_A [g \bigcirc_A \pi^{-1}(p)] \\ &\sim_A [b \bigcirc_A \pi^{-1}(y)] \bigcirc_A [f \bigcirc_A \pi^{-1}(q)] \\ &\sim_A [b \bigcirc_A \pi^{-1}(q)] \bigcirc_A [f \bigcirc_A \pi^{-1}(y)], \end{aligned}$$

and so by the monotonicity of $\bigcirc_A$, $ap \sim bq$ follows. Consider $\theta(c) = \theta(a_0) \bigcirc_A c$, and let $\bigcirc'_A$ be the operation induced by $a'_0 p'_0$, where $a'_0 = \theta(a_0)$. For any $c \in A$,

$$\begin{aligned} (c \bigcirc'_A \theta(b))p'_0 &\sim c\pi'[\theta(b)] && \text{(definition of } \bigcirc'_A) \\ \theta(a_0)\pi'[\theta(b)] &\sim \theta(b)p'_0 && \text{(definition of } \pi') \\ \theta(b)p_0 &\sim \theta(a_0)\pi(b) && \text{(definition of } \pi) \end{aligned}$$

By the Reidmeister condition,

$$[c \bigcirc'_A \theta(b)] p_0 \sim c\pi(b) \sim (c \bigcirc_A b) p_0.$$

Using this

$$\begin{aligned} \theta(a \bigcirc_A b) &= \theta(a_0) \bigcirc_A (a \bigcirc_A b) \\ &= [\theta(a_0) \bigcirc_A a] \bigcirc_A b \qquad \text{(Theorem 11)} \\ &= \theta(a) \bigcirc_A b \\ &= \theta(a) \bigcirc'_A \theta(b). \end{aligned}$$

θ is clearly order preserving, and it is onto since for since $b \in A$ there exists an a such that

$$bp_0 \sim a\pi[\theta(a_0)] \sim [a \bigcirc_A \theta(a_0)] p_0 \sim [\theta(a_0) \bigcirc_A a] p_0 \sim \theta(a) p_0.$$

(ii) Suppose $\langle \theta, \eta \rangle$ is an automorphism of $\mathscr{C}$. Define θ^* as the solution to

$$[\theta^*(a) \bigcirc_A \theta(a_0)] p_0 \sim \theta^*(a)\pi[\theta(a_0)] \sim \theta(a) p_0.$$

Because θ is onto and 1 : 1 and unrestricted solvability holds, θ^* is onto and 1 : 1. It is order preserving because

$$\begin{aligned} a \succsim_A b \quad &\text{iff} \quad ap_0 \succsim bp_0 \\ &\text{iff} \quad \theta(a)\eta(p_0) \succsim \theta(b)\eta(p_0) \\ &\text{iff} \quad \theta(a) \succsim_A \theta(b) \\ &\text{iff} \quad \theta^*(a) \bigcirc_A \theta(a_0) \succsim_A \theta^*(b) \bigcirc_A \theta(a_0) \\ &\text{iff} \quad \theta^*(a) \succsim_A \theta^*(b). \end{aligned}$$

Next, we show that $\theta^*(a \bigcirc_A b) \sim_A \theta^*(a) \bigcirc_A \theta^*(b)$. Observe that by the technique used to prove the Reidmeister condition, we can prove any system of cancellations that follows from an additive representation. In particular, we apply cancellation to the following eight equations:

$$\begin{aligned} \theta^*(a \bigcirc_A b)\pi[\theta(a_0)] &\sim \theta(a \bigcirc_A b) p_0 && \text{(definition of } \theta^*) \\ \theta(a \bigcirc_A b)\eta(p_0) &\sim \theta(a)\eta[\pi(b)] && (\langle \theta, \eta \rangle \text{ on definition of } a \bigcirc_A b) \\ \theta(a) p_0 &\sim \theta^*(a)\pi[\theta(a_0)] && \text{(definition of } \theta^*) \\ \theta^*(a)\pi[\theta^*(b)] &\sim (\theta^*(a) \bigcirc_A \theta^*(b)) p_0 && \text{(definition of } \bigcirc_A) \\ \theta^*(b) p_0 &\sim a_0\pi[\theta^*(b)] && \text{(definition of } \pi) \\ \theta(b) p_0 &\sim \theta^*(b)\pi[\theta(a_0)] && \text{(definition of } \theta^*) \\ \theta(a_0)\eta[\pi(b)] &\sim \theta(b)\eta(p_0) && (\langle \theta, \eta \rangle \text{ on definition of } \pi(b)) \\ a_0\pi[\theta(a_0)] &\sim \theta(a_0) p_0 && \text{(definition of } \pi) \end{aligned}$$

yielding

$$\theta^*(a \bigcirc_A b) p_0 \sim (\theta^*(a) \bigcirc_A \theta^*(b)) p_0.$$

In like manner, define η^* and show that it is an automorphism of $\mathscr{I}_P$. We show that $\langle \theta^*, \eta^* \rangle$ is an automorphism of $\mathscr{C}$, and so, by Theorem 12, $\eta^* = \pi\theta^*\pi^{-1}$. Suppose that $ap \succsim bq$; then

$$\begin{aligned}
\theta^*(a)\pi[\theta(a_0)] &\sim \theta(a) p_0 && \text{(definition of } \theta^*) \\
\pi^{-1}[\eta(p_0)]\eta^*(p) &\sim a_0\eta(p) && \text{(definition of } \eta^*) \\
(a)\eta(p) &\succsim \theta(b)\eta(q) && (\langle \theta, \eta \rangle \text{ is an automorphism}) \\
\theta(b) p_0 &\sim \theta^*(b)\pi[\theta(a_0)] && \text{(definition of } \theta^*) \\
a_0\eta(q) &\sim \pi^{-1}[\eta(p_0)]\eta^*(p), && \text{(definition of} \eta^*)
\end{aligned}$$

and cancellation yields $\theta^*(a)\eta^*(p) \succsim \theta^*(b)\eta^*(q)$.

Conversely, suppose θ and η are of the form given; we then show $\langle \theta, \eta \rangle$ is an automorphism of $\mathscr{C}$. Define u by $\eta(p_0) = \pi\theta^*\pi^{-1}(u)$, and freely use the fact that $\langle \theta^*, \pi\theta^*\pi^{-1} \rangle$ is an automorphism, the definition of $\bigcirc_A$, and the fact that $\bigcirc_A$ is associative, commutative, and monotonic:

$$\begin{aligned}
ap \succsim bq \quad &\text{iff} \quad \theta^*(a), \pi\theta^*\pi^{-1}(p) \succsim \theta^*(b), \pi\theta\pi^{-1}(q) \\
&\text{iff} \quad \theta^*(a) \bigcirc_A \theta^*[\pi^{-1}(p)] \succsim_A \theta^*(b) \bigcirc_A \theta^*[\pi^{-1}(q)] \\
&\text{iff} \quad \theta^*(a) \bigcirc_A \theta(a_0) \bigcirc_A \theta^*[\pi^{-1}(p)] \bigcirc_A \theta^*[\pi^{-1}(u)] \\
&\qquad \succsim_A \theta^*(b) \bigcirc_A \theta(a_0) \bigcirc_A \theta^*[\pi^{-1}(q)] \bigcirc_A \theta^*[\pi^{-1}(u)] \\
&\text{iff} \quad [\theta^*(a) \bigcirc_A \theta(a_0)] \bigcirc_A \theta^*[\pi^{-1}(p) \bigcirc_A \pi^{-1}(u)], p_0 \\
&\qquad \succsim [\theta^*(b) \bigcirc_A \theta(a_0)] \bigcirc_A \theta^*[\pi^{-1}(q) \bigcirc_A \pi^{-1}(u)], p_0 \\
&\text{iff} \quad [\theta^*(a) \bigcirc_A \theta(a_0)], \pi\theta^*[\pi^{-1}(p) \bigcirc_A \pi^{-1}(v)]. \\
&\qquad \succsim [\theta^*(b) \bigcirc_A \theta(a_0)], \pi\theta^*[\pi^{-1}(q) \bigcirc_A \pi^{-1}(u)].
\end{aligned}$$

But,

$$\begin{aligned}
&a_0, \pi\theta^*[\pi^{-1}(p) \bigcirc_A \pi^{-1}(u)] \\
&\quad \sim \theta^*[\pi^{-1}(p) \bigcirc_A \pi^{-1}(u)], p_0 && \text{(definition of } \pi) \\
&\quad \sim (\theta^*\pi^{-1}(p) \bigcirc_A \theta^*\pi^{-1}(u)), p_0 && (\theta^* \text{ is an automorphism of } A) \\
&\quad \sim \theta^*\pi^{-1}(p), \pi\theta^*\pi^{-1}(u) && \text{(definition of } \bigcirc_A) \\
&\quad \sim a_0, [\pi\theta^*\pi^{-1}(p) \bigcirc_P \pi\theta^*\pi^{-1}(u)] && \text{(definition of } \bigcirc_P) \\
&\quad \sim a_0, [\pi\theta^{*\pi^{-1}}(p) \bigcirc_P \eta(p_0)]. && \text{(definition of } u)
\end{aligned}$$

So, by monotonicity,

$$ap \succsim bq \quad \text{iff} \quad \theta(a)\eta(p) \succsim \theta(b)\eta(q). \qquad \diamondsuit$$

19.8.4 Theorem 14, Part (iii) (p. 81)

Suppose $\mathscr{A}$ is a closed, idempotent, concatenation structure, $\mathscr{C}$ is the conjoint structure induced by Definition 12, and α is a mapping of A onto A. Then α is an automorphism of $\mathscr{A}$ iff (α, α) is a factorizable automorphism of $\mathscr{C}$.

PROOF. Suppose α is an automorphism of $\mathscr{A}$, then

$$\begin{aligned} ab \succsim' cd \quad &\text{iff} \quad a \bigcirc b \succsim c \bigcirc d \\ &\text{iff} \quad \alpha(a \bigcirc b) \succsim \alpha(c \bigcirc d) \\ &\text{iff} \quad \alpha(a) \bigcirc \alpha(b) \succsim \alpha(c) \bigcirc \alpha(d) \\ &\text{iff} \quad \alpha(a)\alpha(b) \succsim' \alpha(c)\alpha(d), \end{aligned}$$

and so (α, α) is a factorizable automorphism.

Conversely, if (α, α) is a factorizable automorphism, then α is order preserving since by monotonicity,

$$\begin{aligned} a \succsim b \quad &\text{iff} \quad a \bigcirc c \succsim b \bigcirc c \\ &\text{iff} \quad ac \succsim' bc \\ &\text{iff} \quad \alpha(a)\alpha(c) \succsim' \alpha(b)\alpha(c) \\ &\text{iff} \quad \alpha(a) \bigcirc \alpha(c) \succsim \alpha(b) \bigcirc \alpha(c) \\ &\text{iff} \quad \alpha(a) \succsim \alpha(b). \end{aligned}$$

And α preserves $\bigcirc$ since by idempotency,

$$\begin{aligned} a \bigcirc b \sim c \quad &\text{iff} \quad a \bigcirc b \sim c \bigcirc c \\ &\text{iff} \quad ab \sim' cc \\ &\text{iff} \quad \alpha(a)\alpha(b) \sim' \alpha(c)\alpha(c) \\ &\text{iff} \quad \alpha(a) \bigcirc \alpha(b) \sim \alpha(c) \bigcirc \alpha(c) \\ &\text{iff} \quad \alpha(a) \bigcirc \alpha(b) \sim \alpha(c) \sim \alpha(a \bigcirc b). \qquad \diamondsuit \end{aligned}$$

19.8.5 Theorem 15 (p. 82)

Suppose $\mathscr{A}$ is a closed, idempotent, solvable concatenation structure, $\mathscr{I}_a$ is the induced total concatenation structure at a and α is an automorphism of $\mathscr{A}$.

(i) *α is a dilation at a iff α is an automorphism of $\mathscr{I}_a$.*

(ii) *α is a translation iff α is an isomorphism of $\mathscr{I}_a$ and $\mathscr{I}_{\alpha(a)}$ and $\alpha(a) \neq a$.*

COROLLARY. *If, in addition, $\mathscr{A}$ is Archimedean in differences, then the set of dilations at each point form a commutative group.*

PROOF. α is an isomorphism of $\mathscr{I}_a$ with $\mathscr{I}_{\alpha(a)}$ and $\alpha\pi = \pi\alpha$ because by Theorem 14(iii), (α, α) is a factorizable automorphism of $\mathscr{C}$; and the statement then follows from Theorem 12. Clearly, α is a dilation at a iff α is an isomorphism of $\mathscr{I}_a$ with $\mathscr{I}_a$, i.e., α is an automorphism of $\mathscr{I}_a$. Otherwise, α is a translation.

To show the corollary, suppose α and β are dilations at a. Consider any b in A. If $b \succ a$, then since by Theorems 14(ii) and 11(i) the restrictions of automorphisms off $\mathscr{I}_a$ to its positive part are automorphisms of a PCS, it follows from Theorem 6 and Hölder's theorem that $\alpha\beta(b) = \beta\alpha(b)$. For $b \prec a$, the proof is similar, and of course $\alpha\beta(a) = \alpha(a) = a = \beta(a) = \beta\alpha(a)$. ◇

19.8.6 Theorem 18 (p. 86)

Suppose $\mathscr{C} = \langle A \times P, \succsim \rangle$ is a conjoint structure that is A-solvable relative to a point $a_0 p_0$, $\bigcirc_A$ is the induced operation relative to $a_0 p_0$, $\mathscr{T}$ is the set of right multiplications of $\bigcirc_A$, $\mathscr{A} = \langle A, \succsim_A, \bigcirc \rangle$ is a closed concatenation structure, $\mathscr{E}$ is its set of endomorphisms, and $\mathscr{G}$ is its group of automorphisms. Then:

(i) *$\bigcirc_A$ does not distribute over $\mathscr{C}$.*

(ii) *If $\bigcirc$ distributes over $\mathscr{C}$, then $\mathscr{T} \subseteq \mathscr{E}$.*

(iii) *$\mathscr{T}$ is included in $\mathscr{G}$ iff $\bigcirc$ distributes over $\mathscr{C}$ and $\mathscr{C}$ is unrestrictedly A-solvable.*

(iv) *$\mathscr{T} = \mathscr{G}$ iff property (iii) holds and $\mathscr{A}$ is 1-point unique.*

PROOF.

(i) Suppose, on the contrary, $\bigcirc_A$ does distribute over $\mathscr{C}$. Choose $a \in A$, $a \neq a_0$. By definition, $ap_0 \sim a_0\pi(a)$. Apply the distribution assumption to two copies of this, so that

$$(a \bigcirc_A a)p_0 \sim (a_0 \bigcirc_A a_0)\pi(a) \sim a_0\pi(a) \sim ap_0.$$

By independence in the conjoint structure, $a \bigcirc_A a = a$. Since $\bigcirc_A$ is a positive on A^+ and negative on A^- and $a \neq a_0$, either $a \bigcirc_A a \succ_A a$ or $a \bigcirc_A a \prec_A a$, which is a contradiction.

(ii) The right multiplication a^R is order preserving because $\bigcirc_A$ is monotonic. Applying distributivity to the two definitions

$$(b \bigcirc_A a)p_0 \sim b\pi(a) \qquad \text{and} \qquad (c \bigcirc_A a)p_0 \sim c\pi(a)$$

yields

$$[(b \bigcirc_A a) \bigcirc (c \bigcirc_A a)] p_0 \sim (b \bigcirc c)\pi(a)$$
$$\sim [(b \bigcirc c) \bigcirc_A a] p_0.$$

Thus, by the independence of the conjoint structure,

$$a^R(b \bigcirc c) = a^R(b) \bigcirc a^R(c),$$

proving a^R is an endomorphism.

(iii) Suppose $\bigcirc$ is distributive in $\mathscr{C}$ and $\mathscr{C}$ is unrestrictedly A-solvable. If $\tau \in \mathscr{T}$, then by unrestricted A-solvability it is *onto*; since it is order preserving, it is one-to-one; and by part (ii), it is an endomorphism.

Conversely, suppose $\mathscr{T} \subseteq \mathscr{G}$. Suppose $ap \sim cq$ and $bp \sim dq$. Observe that

$$\begin{aligned} ap \sim cq \qquad \text{iff} \qquad \pi^{-1}(p)^R(a) &= a \bigcirc_A \pi^{-1}(p) \\ &= c \bigcirc_A \pi^{-1}(q) \\ &= \pi^{-1}(q)^R(c). \end{aligned}$$

Similarly,

$$bp \sim dq \qquad \text{iff} \qquad \pi^{-1}(p)^R(b) = \pi^{-1}(q)^R(d).$$

Since $\pi^{-1}(p)^R$ and $\pi^{-1}(q)^R$ are automorphisms and $\bigcirc$ is monotonic,

$$\begin{aligned} \pi^{-1}(p)^R(a \bigcirc b) &= \pi^{-1}(p)^R(a) \bigcirc \pi^{-1}(p)^R(b) \\ &= \pi^{-1}(q)^R(c) \bigcirc \pi^{-1}(q)^R(d) \\ &= \pi^{-1}(q)^R(c \bigcirc d), \end{aligned}$$

which is equivalent to $(a \bigcirc b)p \sim (c \bigcirc d)q$, proving that $\bigcirc$ is distributive.

To show unrestricted A-solvability, suppose $a \in A$ and $p, q \in P$. Since $\pi^{-1}(p)^R \in \mathscr{T}$, its inverse exists. Let

$$a = \pi^{-1}(p)^R\left[\pi^{-1}(q)^R(b)\right].$$

Then,

$$a \bigcirc_A \pi^{-1}(p) = \pi^{-1}(p)^R(a) = \pi^{-1}(q)^R(b) = b \bigcirc_A \pi^{-1}(q),$$

whence $ap \sim bq$, proving solvability.

(iv) Suppose $\mathscr{T} = \mathscr{G}$. If $a^R(c) = b^R(c)$, then $c \bigcirc_A a \sim_A c \bigcirc_A b$, which by monotonicity of $\bigcirc_A$ is equivalent to $a \sim_A b$. Thus, $a^R = b^R$, which proves that $\mathscr{A}$ is 1-point unique.

Conversely, suppose $\tau \in \mathscr{G}$. Denoting $a = \tau(a_0)$, we obtain

$$\tau_a(a_0) = a_0 \bigcirc_A a = a = \tau(a_0),$$

but since $\tau_a \in \mathscr{G}$ and $\mathscr{A}$ is 1-point unique, $\tau = \tau_a$, and so $\mathscr{T} = \mathscr{G}$. ◇

19.8.7 Theorem 21 (p. 88)

Suppose $\mathscr{A} = \langle A, \succsim, \bigcirc \rangle$ is a concatenation structure that is totally ordered, closed, solvable, and Archimedean in differences (Definition 3*). Then $\mathscr{A}$ is either* 1*- or* 2*-point unique, and it is isomorphic to a real closed structure $\langle R, \succsim, \otimes \rangle$ where $R \subseteq \mathrm{Re}^+$.*

PROOF. Let $\mathscr{C} = \langle A \times A, \succsim' \rangle$ be defined as in Definition 13. Note that the orderings induced on the components A of $\mathscr{C}$ by $\succsim'$ are both equal to $\succsim$. We follow the proof in Volume I (p. 298), omitting the proof of double cancellation, which is the only part invoking bisymmetry. It follows that $\mathscr{C}$ is a conjoint structure that is unrestrictedly solvable in each component and Archimedean. By Theorem 20, there are mappings φ and ψ from A into Re and a real operation $\oplus$ such that $\varphi \oplus \psi$ represents $\mathscr{C}$. If we define $\otimes$ on Re^+ by, for all $r, s > 0$,

$$r \otimes s = r \oplus \psi\varphi^{-1}(s),$$

then we have

$$\begin{aligned} u \bigcirc v \succsim x \bigcirc y \quad &\text{iff} \quad uv \succsim' xy \\ &\text{iff} \quad \varphi(u) \oplus \psi(v) \geqslant \varphi(x) \oplus \psi(y) \\ &\text{iff} \quad \varphi(u) \oplus \psi\varphi^{-1}\varphi(v) \geqslant \varphi(x) \oplus \psi\varphi^{-1}\varphi(y), \\ &\text{iff} \quad \varphi(u) \otimes \varphi(v) \geqslant \varphi(x) \otimes \varphi(y), \end{aligned}$$

proving that φ and $\otimes$ yield a representation.

To show that $\mathscr{A}$ is either 1- or 2-point unique, it is sufficient to show that each automorphism of $\mathscr{A}$ that leaves two distinct elements of $\mathscr{A}$ invariant is the identity. Suppose α is the automorphism and a and b are distinct elements for which $\alpha(a) = a$ and $\alpha(b) = b$. Select the representation $\varphi\psi$ of $\mathscr{C}$ with $\varphi(a) = 0$ and $\psi(b) = 0$, and let $\varphi' = \varphi\alpha$ and $\psi' = \psi\alpha$. Observe that $\varphi'(a) = \psi\alpha(a) = \varphi(a) = 0$ and $\psi'(b) = \psi\alpha(b) = \psi(b) = 0$, and

$$\begin{aligned} uv \succsim' xy \quad &\text{iff} \quad u \bigcirc v \succsim x \bigcirc y \\ &\text{iff} \quad \alpha(u) \bigcirc \alpha(v) \succsim \alpha(x) \bigcirc \alpha(y) \\ &\text{iff} \quad \alpha(u)\alpha(v) \succsim' \alpha(x)\alpha(y) \\ &\text{iff} \quad \varphi\alpha(u) \oplus \psi\alpha(v) \geqslant \varphi\alpha(x) \oplus \psi\alpha(y) \\ &\text{iff} \quad \varphi'(u) \oplus \psi'(v) \geqslant \varphi'(x) \oplus \psi'(y), \end{aligned}$$

and so $\varphi' \oplus \psi'$ forms another representation of the conjoint structure $\mathscr{C}$. Since the ranges of φ and φ' and of ψ and ψ' are identical and $\varphi(a) = \varphi'(a)$, Theorem 20 implies $\varphi \equiv \varphi'$. Thus $\alpha = \varphi^{-1}\varphi' = \varphi^{-1}\varphi$ is the identity. ◇

19.9 BISYMMETRY AND RELATED PROPERTIES

There are two motives for including this section in the chapter. The first is that some authors, most notably Ramsay (1976), have argued that the most important generalizations of extensive measurement found in physics are the bisymmetric ones and that dimensional analysis should be generalized to include this class of structures. Averaging is an example. Among those who have worked on bisymmetric structures are Aczél (1966) and Pfanzagl (1968, 1971). The other motive is that under sufficiently strong hypotheses, a number of properties that appear to be weaker than bisymmetry reduce to it. This will prove important in our discussion of generalizations of subjective expected utility (Sections 20.4.6 and 20.4.7).

As we shall come to understand more clearly, Ramsay's suggestion is really not terribly relevant. There are two classes of bisymmetric structures: those having interval scale representations, like weighted averages, and those having ratio scale representations, like those with doubling functions. However, the class of all structures having ratio scale representations onto the positive real numbers is somewhat understood, and these structures are all suitable for generalizations of dimensional analysis (see Sections 20.2.7 and 21.7). Thus the main motive here is to understand how other concepts relate to bisymmetry.

19.9.1 General Definitions

DEFINITION 18. *Suppose $\mathscr{A} = \langle A, \succsim, \bigcirc \rangle$ is a closed concatenation structure. Then the following conditions are said to hold in $\mathscr{A}$, where $a, b, c, d, p, q \in A$.*

1. Thomsen *iff*

$$a \bigcirc p = q \bigcirc d \quad \text{and} \quad q \bigcirc b = c \bigcirc p \quad \text{imply} \quad a \bigcirc b = c \bigcirc d.$$

2. Bisymmetry *iff*

$$(a \bigcirc b) \bigcirc (c \bigcirc d) = (a \bigcirc c) \bigcirc (b \bigcirc d).$$

3. Autodistributive (Aczél, 1966, p. 293) *iff*

$$(a \bigcirc b) \bigcirc c = (a \bigcirc c) \bigcirc (b \bigcirc c),$$
$$d \bigcirc (a \bigcirc b) = (d \bigcirc a) \bigcirc (d \bigcirc b).$$

Separately, the last two equations are referred to, respectively, as right *and* left autodistributivity. *If the property is satisfied for just for some fixed* $c = d$, *then it is referred to as* autodistributivity relative to c.

4. Self distributive *iff*

$$a \bigcirc p = c \bigcirc q \quad \text{and} \quad b \bigcirc p = d \bigcirc q$$
$$\text{imply} \quad (a \bigcirc b) \bigcirc p = (c \bigcirc d) \bigcirc q.$$

5. Symmetrically distributive *iff*

$$a \bigcirc p = p \bigcirc c \quad \text{and} \quad b \bigcirc p = p \bigcirc d$$
$$\text{imply} \quad (a \bigcirc b) \bigcirc p = p \bigcirc (c \bigcirc d).$$

6. Solvable relative to $p \in A$ *iff for each* $a \in A$ *there exist* $b, c \in A$ *such that* $b \bigcirc p = a = p \bigcirc c$. *Denote the right solution to* $a \bigcirc p$ *by* $\pi_p(a)$ *and the left to* $a \bigcirc b$ *by* $\xi_p(a, b)$. *Then the* induced operation $\bigcirc_p$ *is defined by* $a \bigcirc_p b = \xi_p[a, \pi_p(b)]$.

7. Dual bisymmetric *iff* $\mathscr{A}$ *is solvable relative to p and*

$$(a \bigcirc b) \bigcirc_p (c \bigcirc d) = (a \bigcirc_p c) \bigcirc (b \bigcirc_p d).$$

Before examining how several of these concepts relate, it is worth noting that the weighted average, $x \oplus y = rx + (1 - r)y$, is a numerical structure that satisfies properties 1–5 (Exercise 35). However, since all of these are very restrictive properties, most of the examples of concatenation structures that we have encountered do not satisfy any of them.

THEOREM 23. *Suppose* $\mathscr{A} = \langle A, \succsim, \bigcirc \rangle$ *is a closed concatenation structure and* $\mathscr{C} = \langle A \times A, \succsim' \rangle$ *is the conjoint structure of Definition* 13. *Then the following are true.*

(i) $\bigcirc$ *satisfies the Thomsen condition iff* $\mathscr{C}$ *satisfies the Thomsen condition.*

(ii) $\bigcirc$ *is self-distributive iff* $\bigcirc$ *is distributive in* $\mathscr{C}$.

(iii) *If* $\bigcirc$ *is bisymmetric, then* $\mathscr{C}$ *satisfies the Thomsen condition.*

(iv) *If* $\bigcirc$ *is bisymmetric and idempotent, then* $\bigcirc$ *is autodistributive.*

(v) *If* $\bigcirc$ *is autodistributive, then* $\bigcirc$ *is self-distributive and symmetrically distributive.*

If, in addition, $\mathscr{A}$ *is solvable relative to* p, *then the following also hold.*

(vi) $\bigcirc$ *satisfies the Thomsen condition iff* $\bigcirc_p$ *is associative and commutative.*

(vii) *If* $\bigcirc$ *is symmetrically distributive, then* π_p *is an automorphism of* $\mathscr{A}$.

(viii) $\bigcirc$ *is bisymmetric and idempotent iff* $\bigcirc$ *is autodistributive and* $\mathscr{A}$ *is dual bisymmetric.*

Proofs of all save the parts involving symmetrically distributive structures can be found in Luce and Narens (1985); all proofs are left as Exercise 36.

19.9.2 Equivalences in Closed, Idempotent, Solvable, Dedekind Complete Structures

As we noted earlier, several of the concepts defined above reduce to bisymmetry in sufficiently restricted cases. This result is due to Luce and Narens (1985), but the proof will have to be postponed until Section 20.5.4 because it draws on results about scale types.

THEOREM 24. *Suppose* $\mathscr{A} = \langle A, \succsim, \bigcirc \rangle$ *is a concatenation structure that is closed, idempotent, solvable, and Dedekind complete. Then the following are equivalent:*

(i) $\bigcirc$ *is bisymmetric.*

(ii) $\bigcirc$ *is right autodistributive.*

(iii) $\bigcirc$ *is self-distributive.*

19.9.3 Bisymmetry in the 1-Point Unique Case

THEOREM 25. *Suppose* $\langle A, \succsim, * \rangle$ *is an intensive structure with doubling function* θ^{-1}, *and* $\langle A, \succsim, \bigcirc \rangle$ *is the corresponding* PCS *with* $\theta(a \bigcirc b) =$

$a * b$. Then $\bigcirc$ is bisymmetric iff $*$ is bisymmetric and θ is an automorphism (of both structures).

The statement of this result found in Narens and Luce (1976) is incorrect in that it omits the condition that θ is an automorphism. This result generalizes one of Szmielew (1983) on midpoint algebras. The proof is left as Exercise 37.

EXERCISES

1. Prove that the representation given by Equation (2) is nonadditive. (Hint: Recall the criterion of Scheffé, Vol. I, p. 274. (19.1.1)

2. Suppose $\oplus$ is the operation given by

$$x \oplus y = x + y + x^{\alpha}y^{\beta}, \qquad x > 0, \quad y > 0.$$

Show that $\oplus$ is associative iff either $\alpha = \beta = 0$ or $\alpha = \beta = 1$. (19.2.1)

3. If a concatenation structure has only four elements, show that commutativity implies associativity (see also Exercise 5). (19.2.2)

4. In a solvable concatenation structure, prove that being Archimedean in difference sequences implies being Archimedean in regular sequences. (19.2.3)

5. Prove that there is just one PCS on five elements that is commutative and not embeddable in an associative PCS. (See also Exercise 3.) (19.3.1)

6. Suppose $\mathscr{A}$ is a PCS and C is a nonempty, bounded subset of A with $\bar{c}$ a bound. Suppose that for some $a \in A$, $\bar{c} \bigcirc a$ is defined and $\bar{c} \succ a$. Prove there exists $c \in C$ such that $c \bigcirc a$ is defined and $c \bigcirc a \succ c'$ for all $c' \in C$. (Hint: Use the Archimedean property.) (19.3.1)

The following three exercises yield a brief but nonconstructive proof that a PCS has a numerical representation. (19.3.3)

7. Using local definability, monotonicity, Archimedeanness and restricted solvability, prove that each PCS $\mathscr{A} = \langle A, \succsim, \bigcirc \rangle$ with no minimal element has a sequence $\{a_i\}$ with the properties that (i) for each positive integer, $a_{i+1} \bigcirc a_{i+1}$ is defined and $a_i \succ a_{i+1} \bigcirc a_{i+1}$ and (ii) for each $b \in A$, there exists a positive integer j such that $b \succ a_j$.

8. Using all of the properties of a PCS, but especially Archimedeanness, show that for the sequence $\{a_i\}$ described in Exercise 7 the closure of $\{a_i\}$ is both countable and order dense in $\langle A, \succsim \rangle$.

9. Given Exercise 8, let φ be a real homomorphism of $\langle A, \succsim \rangle$ (Theorem 2.2), and define the real operation $\otimes$ on $\varphi(A)$ by

$$x \otimes y = \varphi[\varphi^{-1}(x) \bigcirc \varphi^{-1}(y)],$$

whenever the right side is defined. Show that $\langle \varphi(A), \succsim, \otimes \rangle$ is a representation of $\mathscr{A}$. Deal separately with the case of a PCS with a minimal element by showing that it is a single standard sequence.

10. Suppose φ and ψ are isomorphisms of a totally ordered PCS $\mathscr{A}$ onto a numerical PCS $\mathscr{R}$, and h is the strictly increasing function that maps φ into ψ. Show that $\varphi^{-1}h\varphi = \varphi^{-1}\psi$ is an automorphism of $\mathscr{A}$. Conversely, if α is an automorphism of $\mathscr{A}$, and φ is an isomorphism of $\mathscr{A}$ onto $\mathscr{R}$, then show that $\varphi\alpha$ is also an isomorphism of $\mathscr{A}$ onto $\mathscr{R}$. (19.3.3)

11. Show that the set of automorphisms of a PCS forms a group under function composition. What properties of a PCS are actually used? (19.3.4)

The following five exercises develop a proof of Theorem 8. (19.4.2)

12. Suppose $\mathscr{A} = \langle A, \succsim \rangle$ is a total order. Define $\mathbf{A}$ to consist of those subsets X of A such that (i) X and $A - X$ are nonempty and (ii) if $x \in X$, $y \in A$, and $x \succsim y$, then $y \in X$. Define $\succsim$ on $\mathbf{A}$ as follows: for $X, Y \in \mathbf{A}$, $X \succsim Y$ iff $X \supseteq Y$. Prove that $\mathscr{A} = \langle \mathbf{A}, \succsim \rangle$ is Dedekind complete.

13. Define $\mathbf{A}^*$ to consist of the set of subsets of $\mathbf{A}$ having the form $\mathbf{a} = \{b | b \in A \text{ and } a \succ b\}$. Define $\varphi(a) = \mathbf{a}$. Show that φ is an isomorphism of $\mathscr{A}$ into $\mathscr{A}$.

14. Prove that $\mathbf{A}^*$ is order dense in $\mathbf{A}$.

15. Prove that a is an extremum of $\mathscr{A}$ iff $\mathbf{a} = \varphi(a)$, where $\mathbf{a}$ is an extremum of $\mathscr{A}$.

16. Complete the proof of Theorem 8.

17. Prove Theorem 10. (19.4.3)

18. Prove Corollary 1 to Theorem 10. (19.4.3)

19. Prove Corollary 2 to Theorem 10. (19.4.3)

20. For $x \geqslant 0$, $x_0 > 0$, $y \geqslant 0$, $y_0 > 0$, $\alpha > 0$, $\beta > 0$, $\delta > 0$, let

$$F(x, y) = \frac{\alpha x}{x + \beta y + \delta}$$

induce a conjoint ordering. Show that the Holman operation induced by that ordering relative to x_0, y_0 is given by $x \bigcirc x' = xx'/x_0$. (19.6.2)

21. Suppose a concatenation structure $\mathscr{A}$ meets all of the conditions of a total concatenation structure save, possibly, that the positive and negative components have not been shown to be Archimedean in standard concatenation sequences. Show that if $\mathscr{A}$ is Archimedean in standard difference sequences, then it is in fact a total concatenation structure. (19.6.2)

22. Suppose $\langle A \times P, \succsim \rangle$ is a conjoint structure solvable relative to some point, that satisfies the Thomsen condition, $\bigcirc_A$ is the corresponding Holman operation (Definition 9), and ξ is given by Definition 8. Prove Prove that, for all $a, b \in A$ and $p \in P$, $\xi(a \bigcirc_A b, p) \sim_A \xi(a, p) \bigcirc_A b$. (19.6.2)

23. Prove Corollary 1 to Theorem 11. [Hint: Assuming the Thomsen condition, first prove $(a \bigcirc_A c) \bigcirc_A b \sim (a \bigcirc_A b) \bigcirc_A c$.] (19.6.2)

24. Prove Corollary 2 to Theorem 11. (Hint: Triple cancellation is helpful.) (19.6.2)

25. Suppose $\mathscr{C} = \langle A \times P, \succsim \rangle$ is a conjoint structure and A and P are disjoint sets. Let $\mathscr{C}' = \langle X, A, R \rangle$ be the equivalent relational structure defined in Section 19.6.3. Show that the automorphisms of $\mathscr{C}'$ are equivalent to the factorizable automorphisms of $\mathscr{C}$. (19.6.3)

26. Show that the structure $\mathscr{C}$ introduced in Definition 13 is a conjoint structure. (19.6.4)

27. Prove parts (i), (ii), and (iii) of Theorem 14. (19.6.4)

28. Prove Theorem 16. (19.6.5)

29. Prove Theorem 17 using the following steps (19.6.5):

(i) Suppose δ and δ' are doubling functions with domains B and B'. For $c \in \delta(B \cap B')$, define $f = \delta'\delta^{-1}$. Show that f is strictly increasing, $f(c * d) \succ c, d$, and if $a \prec b$, there exists c such that $f(b * c) \prec a$.

(ii) Show that f is the identity function over $\delta(B \cap B')$.

(iii) Suppose $b' \in B' - B$. Show $B \subset B'$, and use properties of the doubling function to show $B' - B = \{b'\}$.

(iv) Prove that $a' = \delta'(b')$ is maximal in A.

30. Prove the corollary to Theorem 17. (19.6.5)

31. Prove Theorem 19. (19.7.1)

32. The generalized Weber model has the property that the detectability of stimulus x in background y is a strictly increasing function of $f(x + y)/f(y)$, where f is a strictly increasing function of Re^+ onto Re^+ and $f(x+y)/f(y)$ is strictly decreasing to 1. Show that this generates a conjoint structure meeting the conditions of Theorem 19. (19.7.1)

33. Prove Theorem 20. (19.7.1)

34. Prove Theorem 22. (19.7.3)

35. Verify that the weighted average satisfies properties 1–5 of Definition 18. (19.9.1)

36. Prove Theorem 23. (19.9.1)

37. Prove Theorem 25. (19.9.3)

Chapter 20 Scale Types

20.1 INTRODUCTION

20.1.1 Constructibility and Symmetry

What is the key to measurement? Broadly speaking, two points of view have been espoused that, in a sense, complement each other. One is that the empirical structure is sufficiently rich that it is possible to construct, to any degree of approximation, a numerical representation of the qualitative information. We illustrated this perspective in Chapter 2 of Volume I by showing how to construct an additive representation of structures having both an associative operation and a transitive ordering that are tightly interlocked through monotonicity. This was shown to underlie, in some fashion, all of the additive representations treated there. Moreover, as we have seen in Chapter 19, the method can, with some difficulty, be generalized to nonassociative operations that can only be represented by nonadditive numerical operations.

The other point of view is that some considerable degree of symmetry exists in the structures that underlie measurement and that this fact can be exploited. By symmetry, one means that the structure is isomorphic to itself just as a square is under rotations of 90, 180, 270, and 360 degrees. Such self-isomorphisms are called *automorphisms*, and they play a crucial role in this chapter. We have already seen evidence of interest in such symmetry as it is embodied in the uniqueness theorems accompanying (most) representation theorems. Furthermore, as we shall demonstrate in Chapter 22, the similarity transformations that play an essential role in dimensional analysis are, in fact, automorphisms of the entire structure of physical quantities. A far more explicit recognition of the importance of automorphism groups took place in nineteenth century mathematics, especially geometry, in which geometric objects were defined as those collections of points that are invariant under transformations of the automorphism group (see Section 12.2.5). Furthermore, in the past 25 years it has become a dominant theme in the study of ordered algebraic structures; survey references are Fuchs (1963) and Glass (1981), and these give extensive bibliographies to the relevant mathematical literature.

In physics, again in the nineteenth century and throughout this century, the concept of symmetry has been crucial in many developments. The classical formal treatment is Weyl (1952), and in recent years it has been exposited by Rosen (1975, 1983).

In measurement, the earliest recognition of the importance of automorphisms seems to have been the distinction made between ordinal and ratio scales. Physics had, on the one hand, numerous ratio scales that had resulted in its familiar dimensional structure and, on the other hand, some ordinal scales, which were viewed as clearly unsatisfactory and a challenge for upgrading to ratio-scale uniqueness. For example, temperature was successfully upgraded in the nineteenth century whereas hardness has not been upgraded to this day. Economists made the distinction at least as early as the 1920s in terms of ordinal and cardinal scales of utility (Allen, 1956). But probably the most influential explicit recognition of the role of "admissible transformations" of numerical measures—the transformations that correspond to automorphisms of the qualitative structure—was the work of S. S. Stevens (1946, 1951), in which several classes of scales were described and named. Some of his ideas were developed (see Stevens, 1974, pp. 408–409) during the late 1930s through discussions that took place at a Harvard University faculty seminar; some of the key members were G. D. Birkhoff, R. Carnap, H. Feigl, C. G. Hempel, and G. Bergmann.

Stevens focused attention on the rules whereby numbers are assigned to objects and, most importantly, on the degree of uniqueness inherent in the process of assignment:

> In most cases a formulation of the rules of assignment discloses directly the kind of measurement and hence the kind of scale involved. If there remains any ambiguity, we may seek the final and definitive answer in the mathematical group structure of the scale form: in what ways can we transform its values and still have it serve all the functions previously fulfilled? (Stevens, 1951, p. 29)

Stevens' work arose from two major stimuli: (i) his belief in and extensive experimental research on psychophysical measures of loudness, brightness and other kinds of sensory intensity (Stevens and Davis, 1938) and (ii) the position taken by N. R. Campbell (1920/1957, 1928) to the effect that such uniqueness not only is essential to measurement but also is confined to scales that can be constructed by using associative operations, those having additive representations. In a report "Quantitative Estimates of Sensory Events" from a committee appointed by the British Association for the Advancement of Science, Campbell wrote:

> In physical measurement the further condition ... is imposed that the assignment must be unique to this extent, that, when a numeral has been assigned to one member of the group, the numeral to be assigned to any other member is or can be determined by facts within a limited range of "experimental error," arising from the nature of the facts that determine the assignment.
>
> Only one way of fulfilling this condition has ever been discovered: in it use is made of the primary function of numerals to represent number, a property of all groups. The rule is laid down that the numeral to be assigned to any thing X in respect of any property is that which represents the number of standard things or "units," all equal in respect of the property, that have to be combined together in order to produce a thing equal to X in respect of the property. (Ferguson *et al.*, 1940, p. 340)

Campbell and J. O. Irwin also wrote:

> Why do not psychologists accept the natural and obvious conclusion that subjective measurements of loudness in numerical terms (like those of length or weight or brightness) ... are mutually inconsistent and cannot be the basis of measurement? (Ferguson *et al.*, 1940, p. 338)

At that time, the various generalizations of additive operations with which we are now familiar (e.g., conjoint structures, nonassociative operations) were not developed as an integral part of measurement, and so it is not entirely surprising that the scheme to construct additive scales based on an associative concatenation operation was interpreted as the sole source of ratio-scale uniqueness. It is not clear, however, that even if alternative methods had been known, the physicists on the committee would have been willing to accept such systems as forms of measurement. For example, J. Guild commented:

> To insist on calling these other processes [he was referring to sensory procedures based on jnds and judgments of equal intervals] measurement adds nothing to their actual significance but merely debases the coinage of verbal intercourse. Measurement is not a term with some mysterious inherent meaning, part of which may have been overlooked by physicists and may be in course of discovery by psychologists. It is merely a word conventionally employed to denote certain ideas. To use it to denote other ideas does not

broaden its meaning but destroys it: we cease to know what is to be understood by the term when we encounter it; our pockets have been picked of a useful coin.... (Ferguson, 1940, p. 345)

The Ferguson committee failed to reach consensus on the matter; the physicists and psychologists were in considerable disagreement. There was, however, little disagreement among the physicists that uniqueness of measurement is essential and that however regular the psychophysical data, they failed to fulfill the conditions needed for that uniqueness.

20.1.2 Problem in Understanding Scale Types

Although both Campbell and Stevens sensed the crucial importance of groups of transformations, neither seemed fully to appreciate several subtle questions that needed to be resolved to attain a clear classification of measurement representations in these terms. We cite four such problems.

First, in his empirical work on magnitude scaling, Stevens did not remark on the need for isolating qualitative structures from numerical representations and for showing mathematically the existence and uniqueness of the representations. He was unsympathetic to such work, in part because the empirical procedures he developed led directly to numerical responses on the part of subjects and also because he believed that he understood their degree of uniqueness. Most students of measurement are convinced he was wrong in this belief; and such theorems have been developed, quite independent of his beliefs, as we saw in most of the earlier chapters of this work.

Second, Stevens singled out several classes of transformations, introducing for them the terms *ordinal*, *interval*, *log-interval*, *ratio*, and *absolute scales*. These groups were selected because they had arisen in physics: ordinal for such things as hardness, which could be ordered but had not been encompassed by either an extensive or conjoint (derived) theory; interval for temperature and other bisymmetric structures; ratio for extensive measures; and absolute for such things as counting and probability. He believed that the log-interval scales had not arisen in physics (which, as we pointed out on p. 487 of Volume I, is an illusion) but had arisen in psychophysics. He did not address explicitly why one was limited to only these cases. Later developments, including various finite structures and even infinite ones such as semiorders and PCSs, have made clear that these four (or five) groups are not the only possible transformation groups. What, then, are the possible groups of transformations?

Third, neither Stevens nor his contemporaries raised the question of possible candidate representations that exhibit a particular degree of uniqueness. Since he had little interest in representation theorems, it is not surprising that he did not raise the question. That others who had emphasized such theorems did not address the problem until the early 1980s is more surprising.

Finally, it was certainly the case that invariance under automorphisms had played an important role in geometry (Section 12.2.5), in both Newtonian and relativistic physics (Sections 10.14, 13.8, and 13.9), and in dimensional analysis (Section 10.3). Thus it was plausible to generalize the idea to other classes of scientific propositions. Nevertheless, the fact is that no convincing argument has been offered for invoking any of these criteria. It is true that with ratio-scale structures, laws and propositions singled out by invariance seemed to be scientifically interesting, but no deep justification for this had been provided; as we have discussed for dimensional analysis, this troubled some philosophers of physics. Moreover, in some cases, such as probability or any other partial operation, the only automorphism is the identity, and all propositions are invariant; and so the criterion lacks any restrictive bite. Deeper understanding of meaningful propositions is called for beyond an assertion of invariance.

This chapter describes attempts to deal with the second and third questions: the classes of real transformation groups that can arise in measurement and the types of structure that give rise to representations with specified transformation groups. The first question was, of course, dealt with in earlier chapters. The remaining question—what types of propositions should be viewed as meaningful in a measurement structure and how this relates to invariance properties—is dealt with in Chapter 22. It is the least completely resolved of the four issues.

20.2 HOMOGENEITY, UNIQUENESS, AND SCALE TYPE

20.2.1 Stevens' Classification

As was previously noted, Stevens (1946, 1951, 1959) focused attention on five transformation groups that have arisen. It is important to distinguish for each case exactly which domain is involved, either all of the real numbers Re or all of the positive real numbers Re^+. Failure to be careful about this has led to some confusion. With that separation, Table 1 summarizes the key concepts.

If we restrict ourselves to the positive real numbers Re^+, as physicists usually do, then ratio and log-interval are the scale types of most interest, whereas if the domain is all real numbers, difference and interval are the

TABLE 1
Stevens' Classification of Scale Types (Augmented)

Numerical Domain: Re	Numerical Domain: Re^+	Parameters
Ordinal (homeomorphism group) $x \to f(x)$; $f: Re \to$ (onto)Re	$f: Re^+ \to$ (onto)Re^+; f Strictly increasing	Countable
Interval (positive affine group) $x \to rx + s,\ r > 0$	Log-interval (power group) $x \to tx^r,\ r > 0,\ t > 0$	2
Translation if $r = 1$, Dilation if $r \neq 1$ or if the identity		
Difference (translation group) $x \to x + s$	Ratio (similarity group) $x \to tx,\ t > 0$	1
Ratio (similarity group) $x \to rx,\ r > 0$		1
Absolute (identity group) $x \to x$		0

corresponding scale types. A common tendency is to think of ratio scales in terms of representations on Re^+ (and so of the form $x \to rx$, $r > 0$) and of interval scales in terms of representations on Re (and so of the form $x \to rx + s$, $r > 0$). This leads to the misconception that the former is the special case of the latter with $s = 0$. Observe that to get them on the same domain, we must take logarithms in the positive ratio case leading to the difference transformations $x \to x + \log r$, making clear that ratio is the special case of the transformation $x \to rx + s$ with $r = 1$. These difference transformations are often called translations. Alternatively, by taking exponentials one can convert the interval case onto Re^+, with the log-interval transformations $x \to rx^t$, in which case the ratio scale is the special case with $t = 1$. A major distinction, which is exploited below, concerns the existence of fixed points. Recall that a_0 is a fixed point of the transformation α if and only if $\alpha(a_0) = a_0$. No ratio-scale transformation except for the identity has a fixed point whereas every transformation of the interval scale type, save those corresponding to ratio-scale transformations, does have a fixed point.

20.2.2 Decomposing the Classification

So long as we are dealing with just these few cases, we can describe the groups of transformations in terms of the number N of parameters in either of two ways: (1) a transformation exists with arbitrarily specified values at N distinct points provided the values maintain the correct ordering, and (2) given that two transformations agree at N distinct points, they are identical. However, as we discovered in the last chapter, these two concepts need not always coincide. The automorphism group of a PCS can be isomorphic to a proper subgroup of the translation group $x \to rx$, $r > 0$. The transformations $x \to k^n x$, where $k > 0$ is fixed and n ranges over all integers, is an example of such a subgroup. Observe that if for some particular x, say u, $k^n u = k^m u$, then $m = n$, and so the transformations are identical. So the group is 1-point unique in the second sense. It is not, however, true that given any $u, v > 0$ there is an integer n such that $v = k^n u$. Sometimes such an n can be found but not always. So it fails to exhibit one degree of freedom in the first sense.

The concept of a transformation that takes some set of points into an arbitrarily given set of points with the same weak or total order as the first set has been called *homogeneity* (Narens, 1981a, 1981b).[1] It describes something of the richness of the transformation group; it states the size of two arbitrarily selected sets of ordered points that can always be mapped into each other by a member of the transformation group. The other concept, which we have already referred to as a notion of *uniqueness*, has something to do with redundancy in the transformation group. It states the least number of distinct fixed points (points that go into themselves under a transformation) that force that transformation to be the identity map. Put another way, it states the largest number of points at which any two transformations may agree without their being identical.

In Stevens' classification, the degree of homogeneity always equals the degree of uniqueness. But, as we know, there are important cases for which that need not be true. This chapter investigates these two concepts and shows where further refinements are needed.

[1] The term *homogeneity* has a number of meanings in mathematics, but present usage has a venerable history dating to the early Greek mathematicians. Intuitively, it means that each part of the space in question looks like every other part. In geometry much the same notion (but without an order playing a role) is often called a transitive family of transformations (Busemann, 1955). Exactly the same notion of M-point homogeneity as defined below has been used extensively since 1969 in the study of ordered permutation groups (Glass, 1981); however, the results achieved in the measurement literature, which focus on a formal definition of scale type, appear to be distinct from those reported by Glass.

20.2.3 Formal Definitions

To formalize these notions, we develop a general framework of relational structures that applies to both empirical and numerical structures and does not make a strong commitment to a particular set of primitives. Suppose A is a nonempty set (possibly of empirical entities, possibly numbers), J is a nonempty set (usually integers) which we refer to as an *index set*, and for each j in J, S_j is a relation of finite order on A. The latter means that for some integer $n(j)$, called the order of the relation, S_j is included in $A^{n(j)}$. Then $\mathscr{A} = \langle A, S_j \rangle_{j \in J}$ is said to be a *relational structure*. When J is finite or countable, as will usually be the case (but see Section 20.4.5 for an exception), we can write $\mathscr{A} = \langle A, S_0, S_1, \ldots \rangle$. If one relation, say S_0, is a weak (or total) order, we write $\succsim$ instead of S_0, use the notation $\mathscr{A} = \langle A, \succsim, S_j \rangle_{j \in J}$, and speak of the structure as *weakly* (or *totally*) ordered. In what follows, we shall deal only with totally ordered structures so as to avoid the numerous detailed (and not very interesting) complications of working with indifferences that are not equality. If A is a subset of Re and S_0 is $\geqslant$, then we usually write R for A and R_j for S_j and refer to $\mathscr{R} = \langle R, \geqslant, R_j \rangle_{j \in J}$ as an *ordered numerical structure*.

Suppose $\mathscr{A}$ and $\mathscr{A}'$ are relational structures with the same index set J and for each j in J, the order (S_j) = order (S_j'), then $\mathscr{A}$ and $\mathscr{A}'$ are *isomorphic* iff there is a one-to-one mapping φ from A *onto* A' such that for all $j \in J$ and all $a_1, a_2, \ldots, a_{n(j)} \in A$,

$$(a_1, a_2, \ldots, a_{n(j)}) \in S_j \quad \text{iff} \quad [\varphi(a_1), \varphi(a_2), \ldots, \varphi(a_{n(j)})] \in S_j'.$$

If the mapping φ is a function satisfying this condition but is not one-to-one, then φ is called a *homomorphism*. If $\mathscr{A}$ is a totally ordered structure, $\mathscr{R}$ is an ordered numerical structure, and $\mathscr{A}$ and $\mathscr{R}$ are isomorphic, then $\mathscr{R}$ is said to be a *numerical representation* of $\mathscr{A}$. An isomorphism (homomorphism) of a structure $\mathscr{A}$ with itself is called an *automorphism* (*endomorphism*).

DEFINITION 1. *Suppose $\mathscr{A} = \langle A, \succsim, S_j \rangle_{j \in J}$ is a totally ordered relational structure, $\mathscr{G}$ is its group of automorphisms, and M and N are nonnegative integers. Let $\mathscr{H}$ be a subset of $\mathscr{G}$.*

1. *$\mathscr{H}$ is said to be M-*point homogeneous *iff for each $a_i, b_i \in A$, $i = 1, \ldots, M$ such that $a_1 \succ a_2 \succ \cdots \succ a_M$ and $b_1 \succ b_2 \succ \cdots \succ b_M$, there exists an automorphism $\alpha \in \mathscr{H}$ such that $\alpha(a_i) = b_i$, $i = 1, 2, \ldots, M$.*

2. *$\mathscr{H}$ is said to be* N-point unique *iff for every $\alpha, \beta \in \mathscr{H}$ and $a_i \in A$, $i = 1, 2, \ldots, N$, such that $a_1 \succ a_2 \succ \cdots \succ a_N$, if $\alpha(a_i) = \beta(a_i)$, $i = 1, 2, \ldots, N$, then $\alpha(a) = \beta(a)$ for all $a \in A$.*

3. *If $\mathscr{G}$ is M-point homogeneous (N-point unique), then $\mathscr{A}$ is also said to be M-point homogeneous (N-point unique). $\mathscr{A}$ is said to be* homogeneous *if it is M-point homogeneous for some $M \geqslant 1$ and is said to be* unique *if it is N-point unique for some $N < \infty$.*

4. *$\mathscr{A}$ is said to be of scale type (M, N) iff its largest degree of homogeneity is M and its least degree of uniqueness is N. In case the structure is M-point homogeneous for all M, we set $M = \infty$, and in case $\mathscr{A}$ is not unique for any finite N, we set $N = \infty$.*

These definitions were introduced by Narens (1981a, 1981b) in an attempt to classify structures by their scale type. As we noted earlier, the concept of homogeneity has long been used in other branches of mathematiacs, in particular geometry, and during the 1970s the concept of M-point homogeneity was used to classify ordered permutation groups (Glass, 1981). The results reported below about scale types appear not be found in the literature on ordered permutation groups.

As Roberts and Rosenbaum (1985, 1988) have pointed out, some variants of these definitions bear consideration, especially in the finite case. For example, they consider structures that have a value structure $v \colon A \to \mathrm{Re}$ such that $a \succsim b$ iff $v(a) \geqslant v(b)$. Then they distinguish between automorphisms for which value is preserved, i.e., $v[\alpha(a)] = \alpha v(a)$, which they call tight automorphisms, and those for which only the order is preserved, i.e., $v(a) \geqslant v(b)$ iff $v[\alpha(a)] \geqslant v[\alpha(b)]$, which they call loose automorphisms. They then modify the concepts of homogeneity and uniqueness by placing value restrictions on the sets of elements involved. We do not go into this here.

As we shall see, homogeneity is a strong concept that greatly restricts the possible structures. Thus it is of considerable interest to understand when it holds, and a number of the following results bear on it. However, there is one exceedingly simple criterion that must be met: no single element can be distinguished from the others solely in terms of properties of the structure. The reason is that in a homogeneous structure any element can be mapped under an automorphism into any other element, and the structural relations are preserved under automorphisms. For example, in a probability structure, such as in Chapter 5, the assertion that "A is an event such that for all $B \in \mathscr{E}$, $A \succsim B$" is true for the universal event and no other; therefore, that structure is not homogeneous.

In the next subsection, we explore some general constraints that relate the structure and its scale type. Then in Section 20.2.5 we turn to the general question of an arbitrary ordered structure on the real numbers—in essence, to all candidate representations that are not discrete—and we find that the possibilities are strikingly limited, much in the way Stevens believed. In Section 20.4 we combine what we know about concatenation structures and about scale types to arrive at a rather detailed characterization of real homogeneous concatenation structures, and in Section 20.6 we do the same for conjoint structures.

20.2.4 Relations among Structure, Homogeneity, and Uniqueness

At present, we know of just four results of a general sort about how structures, degree of homogeneity, and degree of uniqueness limit each other. The first result discusses homogeneity and uniqueness properties in the presence of structural properties often assumed or provable in special cases. The second establishes a powerful tie between structure and homogeneity. The third and fourth establish somewhat special ties between structure and uniqueness.

THEOREM 1. *Suppose $\mathscr{A}$ is a totally ordered relational structure.*

(i) *If for some integer $M > 1$, the cardinality of $A > M$ and $\mathscr{A}$ is M-point homogeneous, then $\mathscr{A}$ is $(M - 1)$-point homogeneous.*

(ii) *If $\mathscr{A}$ is N-point unique, then $\mathscr{A}$ is $(N + 1)$-point unique.*

(iii) *If $\mathscr{A}$ is of scale type (M, N) and the cardinality of A is $> M$, then $M \leqslant N$.*

The proof is left as Exercise 1.

To state the next result, due to Roberts and Rosenbaum (1985), recall that the order of a relation is the number of elements that are related.

THEOREM 2. *Suppose $\mathscr{A} = \langle A, \succsim S_j \rangle_{j \in J}$ is a totally ordered relational structure that is M-point homogeneous. If for each $j \in J$, $\text{order}(S_j) \leqslant M \leqslant |A|$, then $\mathscr{A}$ and $\langle A, \succsim \rangle$ have the same automorphism group.*

The way in which we shall use this result is embodied in the following corollary.

COROLLARY. *Suppose, in addition, that $\langle A, \succsim \rangle$ is isomorphic to $\langle \text{Re}, \geqslant \rangle$ then either for some $j \in J$, $M < \text{order}(S_j)$ or $\mathscr{A}$ is of scale type (∞, ∞).*

For example, a concatenation structure $\langle A, \succsim, \bigcirc \rangle$ that has a representation onto the real numbers and that is not of scale type (∞, ∞) must have

$M \leqslant 2$ since order $(\bigcirc) = 3$. As we shall see in Section 20.4 the (∞, ∞) case can usually be ruled out, insuring that $M \leqslant 2$.

When it comcs to uniqueness, we do not have any results of comparable generality to these about homogeneity, although quite a bit is known in special cases. Two such results due to Luce and Narens (1985) are given here, and another will be presented in Section 20.4.

To state the first result we need several concepts. Consider a totally ordered relational structure $\mathscr{A}$. Suppose F is a function (which in this context can be thought of as a generalized operation) on A^n. We say F is *$\mathscr{A}$-invariant* iff for each automorphism α of $\mathscr{A}$, and $a_1, a_2, \ldots, a_n \in A$,

$$\alpha F(a_1, a_2, \ldots, a_n) = F[\alpha(a_1), \alpha(a_2), \ldots, \alpha(a_n)].$$

(Such invariance will be studied at some length in Chapter 22.)

If B is a subset of A and $\mathscr{F}$ is a collection of generalized operations, then C is the *algebraic closure of B under $\mathscr{F}$* iff it is the smallest set that includes B and, for each F in $\mathscr{F}$, if $c_i \in C$, then $F(c_1, \ldots, c_n) \in C$. Such a closure exists.

THEOREM 3. *Suppose $\mathscr{A} = \langle A, \succsim S_j \rangle_{j \in J}$ is a totally ordered relational structure, $\mathscr{F}$ is a set of $\mathscr{A}$-invariant operations such that the identity function is in $\mathscr{F}$. If for some integer N, the algebraic closure under $\mathscr{F}$ of each subset of N distinct elements is order dense in A, then $\mathscr{A}$ is N-point unique.*

The next result, though apparently somewhat special, is of considerable importance. To state it, we need to generalize Definition 19.13 to general measurement structures.

DEFINITION 2. *An automorphism of a totally ordered structure is said to be a* dilation *iff it has a fixed point $[\alpha(a) = a]$, and it is said to be a* translation *iff it is either the identity or is not a dilation. The set of dilations is denoted $\mathscr{D}$ and of translations, $\mathscr{T}$.*

Note that $\mathscr{G} = \mathscr{D} \cup \mathscr{T}$ and $\{\iota\} = \mathscr{D} \cap \mathscr{T}$.

The choice of the word *translation* for automorphisms with no fixed points is justified by two facts. First, as we noted earlier, among affine transformations, the translations are just those that have no fixed points. Second, as we shall see in Theorem 5, in very general contexts translations defined in this way do in fact correspond to our usual usage.

THEOREM 4. *Suppose $\mathscr{A}$ is a totally ordered relational structure, and $\mathscr{T}$ is its set of translations. Then the following statements are equivalent:*

(i) *$\mathscr{T}$ is a group under function composition.*

(ii) *$\mathscr{T}$ is 1-point unique.*

(iii) *Each dilation, aside from the identity, and each translation have a point in common.*

COROLLARY. *If $\mathscr{A}$ is N-point unique, its automorphisms commute (for $\alpha, \beta \in \mathscr{G}$, $\alpha\beta = \beta\alpha$), and no point is singular in the sense that it is a fixed point of every automorphism, then $\mathscr{A}$ is 1-point unique.*

The substance in requiring that the translations form a group is that the function composition of two translations is itself a translation; all of the other group properties are automatically satisfied.

20.2.5 Scale Types of Real Relational Structures

When attention is confined to real relational structures, for which the automorphism group is necessarily a group of transformations (homeomorphisms relative to the natural topology) from Re onto Re, we are able to say a surprising amount not only about the scale type but, in the homogeneous case, about the transformation group itself. To state these results we need the following definitions from group theory, which will play a significant role in the following work.

DEFINITION 3. *Suppose $\succsim'$ is an ordering on a group $\mathscr{F}$, and $\mathscr{H}$ is a subgroup of $\mathscr{F}$.*

1. *$\mathscr{H}$ is said to be* convex *in $\mathscr{F}$ iff for each $\alpha \in \mathscr{F}$ if there exists $\beta \in \mathscr{H}$ such that both $\beta \succsim' \alpha$ and $\beta \succsim' \alpha^{-1}$, then $\alpha \in \mathscr{H}$.*
2. *$\mathscr{H}$ is said to be* normal *in $\mathscr{F}$ iff for $\alpha \in \mathscr{F}$, $\alpha\mathscr{H}\alpha^{-1} = \mathscr{H}$.*

Consider the real affine group of transformations and suppose $\alpha(x) = rx + s$, $r > 0$, and $\beta(x) = ux + v$, $u > 0$; and define

$$\alpha \succsim' \beta \quad \text{iff} \quad \text{either} \quad r > u \quad \text{or} \quad r = u \quad \text{and} \quad s \geqslant v.$$

The translations are of the form $\tau(x) = x + t$, and they form a group. We show that it is convex. Suppose $\tau \succsim' \alpha$ and $\tau \succsim' \alpha^{-1}$. Since $\alpha^{-1}(x) = (x - s)/r$, we see that $1 \geqslant r$ and $1 \geqslant 1/r$, and so $r = 1$, proving $\alpha \in \mathscr{T}$. As we shall see in Theorem 5, this convex group is important.

The issue is to devise a suitable definition of an ordering of the automorphisms in an abstract context. The following works.

DEFINITION 4. *Suppose $\mathscr{A}$ is an ordered relational structure and $\mathscr{G}$ is its automorphism group. The* asymptotic order *$\succsim'$ on $\mathscr{G}$ is defined as follows: for $\alpha, \beta \in \mathscr{G}$*

$$a \succsim' \beta \quad \textit{iff} \quad \textit{for some } b \in A, \quad \alpha(a) \succsim \beta(a) \textit{ for all } a \succ b.$$

It is easy to verify that $\succsim'$ is transitive, but it may not be connected. It is also easy to verify that when a structure has an interval-scale representation, the asymptotic order is identical to the order just described on the positive affine transformations.

THEOREM 5. *Suppose that* $\mathscr{R} = \langle \text{Re}, \geqslant, R_j \rangle_{j \in J}$ *is a numerical relational structure that is homogeneous and unique. Then the following are true.*

(i) $\mathscr{R}$ *is either* 1- *or* 2-*point unique.*

(ii) *The asymptotic order on* $\mathscr{G}$ *is a total order, and there is a subgroup* $\mathscr{H}$ *of* $\mathscr{G}$ *with the following properties*:

(a) $\mathscr{H}$ *is the minimal, nontrivial, convex subgroup of* $\mathscr{G}$;

(b) $\mathscr{H}$ *together with the asymptotic order is an Archimedean ordered group*;

(c) $\mathscr{H} = \mathscr{T}$;

(d) $\mathscr{H}$ *is normal*;

(e) $\mathscr{H}$ *is of scale type* $(1,1)$.

(iii) *If* $\mathscr{D}$ *is nontrivial, then* $\mathscr{D}$ *is homogeneous.*

(iv) $\mathscr{R}$ *is of scale type* $(1,1)$ *iff* $\mathscr{R}$ *is isomorphic to a real structure whose automorphisms are the difference group.*

(v) $\mathscr{R}$ *is of scale type* $(2,2)$ *iff* $\mathscr{R}$ *is isomorphic to a real structure whose automorphisms are the affine group.*

(vi) $\mathscr{R}$ *is of scale type* $(1,2)$ *iff* $\mathscr{R}$ *is isomorphic to a real structure whose automorphisms are a proper subgroup of the affine group that properly includes the difference group.*

From parts (iv)–(vi) we see that the set of translations $\mathscr{T}$ forms a group, and it is the subgroup $\mathscr{H}$ of part (ii). No direct proof has been given of that fact; moreover, we do not have any structural condition that is equivalent to $\mathscr{T}$ being a group, though, as we have seen in Theorem 4, there are various statements about automorphisms that are equivalent to it. Such a structural condition would probably be useful in gaining some understanding of nonhomogeneous structures.

The history of this important result is as follows. Narens (1981a) proved statement (iv), and later (1981b) he proved the more subtle part (v) and showed, in addition, that no real structure is of scale type (M, M) for $2 < M < \infty$. Alper (1985) modified Narens' proof for the $(2,2)$ case to show part (vi) and that there are no structures of type $(M, M+1)$ for $1 < M < \infty$. Later, in a work based on his 1984 B.A. thesis, Alper (1987) demonstrated that when such a real structure is both homogeneous and N-point unique, then it is 2-point unique and includes a convex $(1,1)$

subgroup. His proof draws on an unpublished result of A. M. Gleason (c1959) (see Lemma 5.6 of proof).

These results do not generalize to structures not on the continuum. For example, Cameron (1989) has shown that for all integers M and N, with $M < N$, there is a structure on the rational numbers of scale type (M, N). Thus the question of which structures can be Dedekind completed without altering the scale type is of great importance. It is not currently known what can happen on the rationals for $M = N$.

A related paper of interest is Levine (1972), in which the question considered is when a family of strictly increasing curves can be transformed isomorphically into a subfamily of the affine group. He calls such a family an *affine system*. The answer depends on the properties of a group that is called the *derived* or *commutator group*. It is formed by considering for each α and β in the given family all functions of the form $\alpha\beta\alpha^{-1}\beta^{-1}$, which are called "commutators" of the original family and everything generated from them by finite sequences of the function composition. His main result is that a group of noncommutative increasing functions is an affine system if and only if the commutator group is both noncyclic and a uniform system in the sense of Definition 6.9 in Volume I. In essence this means that the system of functions consisting of the commutator group and those functions formed from pointwise limits of sequences of functions from the commutator group is isomorphically embeddable in a system of parallel straight lines.

A general result, leading to a generalized Theorem 5, would be the affine analogue of Hölder's theorem for similarity groups, namely, conditions on an ordered group that are necessary and sufficient for it to be isomorphic to a subgroup of the affine group of real transformations.

Let us combine Theorem 5 and the Corollary to Theorem 2 to see the state of our knowledge about real relational structures. If $\mathscr{R}$ is of scale type (M, ∞), then either $M = \infty$ or $M < \text{order}(R_j)$ for some $j \in J$. We do not know much about these ∞-point unique structures; in particular, we do not have examples for each M with $0 \leqslant M < \infty$, nor are any of their properties known. For specific cases such as semiorders (Chapter 16), one can say something about M, namely, either $M = \infty$, or $M \leqslant 1$. The reason is that if P is the semiorder on Re and $\succsim$ is its induced weak order, then $\langle \text{Re}, \succsim, \text{P} \rangle$ is the relational structure; and so by the Corollary to Theorem 2, either $M = \infty$ or $M < \text{order}(\succsim) = \text{order}(\text{P}) = 2$. But even in this special case, no characterization of the automorphism group has been discovered. For the purely ordinal case of type (∞, ∞), Droste (1987) has provided a general characterization of the possible automorphism groups.

For $M > 0$ and $N < \infty$, Theorem 5 tells us that for real structures on Re^+ there are only three cases to consider: $(1,1)$, $(1,2)$, and $(2,2)$, and we know their automorphism groups. Earlier we mentioned that we had no satisfactory way to select among the infinity of alternative representations. In these cases, however, it is reasonable to restrict attention to those representations for which the automorphism group is a subgroup of either the affine or the power group. These are, of course, conventional choices, but they agree with the ones previously made in physics and economics.

For structures that form proper subsets of Re^+, we do not know what classes of transformation groups correspond to scale types $(1, 1)$, $(1, 2)$, and $(2, 2)$.

The only remaining cases are scale types $(0, N)$ with $N < \infty$. Cohen and Narens (1979) provided examples of $(0, 0)$, $(0, 1)$, and $(0, 2)$ structures. Alper (1987) has demonstrated that any $(0, N)$ case can occur, and for the Re^+ case he has given a characterization of their (complex) automorphism groups. There is little doubt that other principles of classification are needed because structures of these types range from highly regular ones with rich families of automorphisms (for example, those that are 1-point homogeneous above and below a singular point that maps into itself under all automorphisms) to structures with no automorphisms except for the identity, which simply reflects their great irregularity. Examples of structures that are homogeneous except for isolated points are (i) relativistic velocity, in which the velocity of light is special; (ii) some infinite probability structures, in which both the universal and null events are special; and (iii) many structures with an operation and an identity element that are special. The probability example makes clear that in terms of the representation the distinction between open and closed intervals is closely related to structures that are homogeneous and those that have universal fixed points and so cannot be homogeneous. New work is needed to single out classes of interesting nonhomogeneous structures.

20.2.6 Structures with Homogeneous, Archimedean Ordered Translation Groups

It is clear from Theorem 5, and especially its proof, that the fact the translations form an Archimedean ordered group is highly significant. The major reason for its importance is that by Hölder's theorem the translations can be represented in the ordered additive real numbers, that is, as translations in the usual sense. Put another way, such groups are inherently one-dimensional. Moreover, recall from Section 19.3.4 that the automorphisms of every PCS are translations and that the group is Archimedean ordered. The following result, due to Luce (1987), also suggests the importance of this property.

THEOREM 6. *Suppose* $\mathscr{A}$ *is a relational structure that is Dedekind complete and order dense. If its set* $\mathscr{T}$ *of translations is* 1*-point unique (equivalently, a group), then* $\mathscr{T}$ *is Archimedean.*

The proof of this is left as Exercise 6.

These observations make one wonder exactly what significance there is to the translations being an Archimedean ordered group. For example, is this a reasonable generalization of what we mean for a structure to be Archimedean? That concept has been defined up to now only in structures for which an operation is either given or readily defined, as in the case of difference or conjoint structures. For general relational structures one does not know how to define the Archimedean property. This may be a reasonable way to do so in general, as is argued at some length in Luce and Narens (in press). A more straightforward question is: what class of structures have Archimedean ordered groups of translations? We do not yet know the answer in general, but we do have a rather neat answer when the translations form a homogeneous group.

To formulate that result, we need two additional definitions.

DEFINITION 5. *A real relational structure* $\mathscr{R} = \langle R, \geqslant, R_j \rangle_{j \in J}$ *is said to be a* real unit structure *iff* R *is a subset of* Re^+ *and there is some subset* T *of* Re^+ *such that*

1. T *is a group under multiplication,*
2. T *maps* R *into* R, *i.e., for each* $r \in R$ *and* $t \in T$, *then* $tr \in R$,
3. T *restricted to* R *is the set of translations of* $\mathscr{R}$.

The assumptions that each $t \in T$ is an automorphism implies that for each R_j, $j \in J$, $r_i \in R$, $i = 1, \ldots, n(j)$, and $t \in T$, if $(r_1, \ldots, r_{n(j)}) \in R_j$, then $(tr_1, \ldots, tr_{n(j)}) \in R_j$. This definition, due to Luce (1986b), generalizes a definition of Cohen and Narens (1979) for real, homogeneous, concatenation structures for which $R = T = \mathrm{Re}^+$, $J = \{1\}$, and R_1 is a binary operation. Cohen and Narens used the same term, *real unit structure*, for the case of concatenation structures; we shall use the more explicit *homogeneous, real unit, concatenation structure* when needed, but in context we may just abbreviate it to *unit structure*. We shall return to that special case in Section 20.4.2.

The next pair of definitions generalize to structures more general than concatenation ones the idea of the structure distributing in a conjoint structure. Recall (Section 19.6.6) that an operation $\bigcirc$ on A is said to distribute in $\mathscr{C} = \langle A \times P, \succsim \rangle$ if whenever $(a, p) \sim (a', q)$ and $(b, p) \sim$ $cb', q)$, then $(c, p) \sim (c', q)$ where $c = a \bigcirc b$ and $c' = a' \bigcirc b'$. An alternative way of saying the same thing, one that generalizes naturally, is to say that if the three equations all hold and if $c = a \bigcirc b$, then $c' = a' \bigcirc b'$. This we generalize as follows:

DEFINITION 6. *Suppose* $\mathscr{C} = \langle A \times P, \succsim \rangle$ *is a conjoint structure,* $a_i, b_i \in A$, $i = 1, \ldots, n$, *and* S *is a relation of order* n *on* A.

1. *The ordered n-tuples* $(a_i, \ldots, a_n)$ *and* $(b_1, \ldots, b_n)$ *are said to be* similar *iff there exist* $p, q \in P$ *such that for each* $i = 1, \ldots, n$, $(a_i, p) \sim (b_i, q)$.

2. S *is said to be* distributive *in* $\mathscr{C}$ *iff for each* $a_i, b_i \in A$, $i = 1, \ldots, n$, *if* $(a_1, \ldots, a_n) \in S$ *and* $(b_1, \ldots, b_n)$ *is similar to it, then* $(b_1, \ldots, b_n) \in S$. *A structure is said to be* distributive *in* $\mathscr{C}$ *iff each of its defining relations is distributive in* $\mathscr{C}$.

This definition and the following major result of this section are due to Luce (1987):

THEOREM 7. *Suppose* $\mathscr{A} = \langle A, \succsim, S_j \rangle_{j \in J}$ *is a relational structure and* $\mathscr{T}$ *is its set of translations. Then the following are equivalent:*

(i) $\mathscr{A}$ *is isomorphic to a real unit structure for which* T *is homogeneous.*

(ii) $\mathscr{T}$ *is a homogeneous, Archimedean ordered group.*

(iii) $\mathscr{T}$ *is* 1*-point unique, and there exists an Archimedean, solvable conjoint structure* $\mathscr{C}$ *with a relational structure* $\mathscr{A}'$ *on the first component such that* $\mathscr{A}'$ *is isomorphic to* $\mathscr{A}$ *and* $\mathscr{A}'$ *is distributive in* $\mathscr{C}$.

COROLLARY 1. *The conjoint structure of part* (iii) *satisfies the Thomsen condition.*

COROLLARY 2. *Suppose* $\mathscr{A}$ *is order dense and property* (ii) *holds. Then the automorphism group of the isomorphic real unit structure is a subgroup of the power group* $(x \to tx^r, t > 0, r > 0)$ *restricted to its domain, and so* $\mathscr{A}$ *is* 2*-point unique.*

This result is significant. It gives us a reasonably clear sense of just how far the dimensional structure of physics can be generalized, though at present there are no substantive examples of such a generalization. A homogeneous real unit structure that is 1-point unique has a ratio-scale representation, and it can be embedded distributively in a multiplicative conjoint structure. Thus, as we shall discuss at some length in Sections 20.2.7 and 22.7.2, such interrelated structures can be assigned units that are products of powers of other units. What is clear is that the usual physical representation involving units in no way depends upon extensive measurement or even on having an empirical operation. The key to the representation is for the structure lying on one component of an Archimedean conjoint structure to have translations that form a homogeneous, Archimedean ordered group. Thus, if one is confronted with a particular axiom system for a relational structure, the first thing to do is to investigate its translations. If they form a homogeneous, Archimedean ordered group, then one knows from Theorem 7 that the structure can be represented numerically as a homogeneous, real unit structure.

One would like to know what happens when $\mathscr{T}$ is an Archimedean ordered but not homogeneous group. Theorem 7 is proved by mapping the relational structure onto the Archimedean ordered group of translations, which in turn is mapped into the reals using Hölder's representation of an Archimedean ordered group. Since the mapping of the structure onto the translations depends upon homogeneity, this strategy does not work in the nonhomogeneous case. Perhaps something is possible when $\mathscr{T}$ is nonhomogeneous provided it is not cyclic, but we do not know of any result.

Also, one would like to know whether, in the context of Theorem 7, the structure is necessarily 2-point unique or if not, when N-point uniqueness implies it. Corollary 2 gives a sufficient condition, but we do not know to what extent density can be dropped.

20.2.7 Representations of Dedekind Complete Distributive Triples

In Section 19.6.6 we began to study what happens when a conjoint structure $\mathscr{C} = \langle A \times P, \succsim \rangle$ has an operation $\bigcirc$ on A that is distributive in $\mathscr{C}$. We are now in a position to study this more thoroughly and more generally in the Dedekind complete case. Indeed, the results hold not just for concatenation structures but for any structure with a real unit representation. The exact result (Luce, 1987) is as follows.

THEOREM 8. *Suppose $\mathscr{C} = \langle A \times P, \succsim \rangle$ is a conjoint structure that is unrestrictedly A-solvable, restrictedly P-solvable, and Archimedean. Suppose, further, that $\mathscr{A} = \langle A, \succsim_A, S_j \rangle_{j \in J}$ is a relational structure whose translations form an Archimedean ordered group.*

(i) *If $\mathscr{A}$ is distributive in $\mathscr{C}$, then $\mathscr{A}$ is 1-point homogeneous, and $\mathscr{C}$ satisfies the Thomsen condition.*

(ii) *If, in addition, $\mathscr{A}$ is Dedekind complete, then under some mapping φ from A onto Re^+ $\mathscr{A}$ has a homogeneous unit representation and there exists a mapping ψ from P into Re^+ such that $\varphi\psi$ is a representation of $\mathscr{C}$.*

(iii) *If, further, $\mathscr{C}$ is unrestrictedly P-solvable and there is a Dedekind complete relational structure on P that, under some ψ from P onto Re^+, has a* *homogeneous unit representation, then for some real constant* ρ, $\varphi\psi^{\rho}$ *is a representation of* $\mathscr{C}$.

This result extends considerably the structures for which the classical product of powers representation of physics holds (Section 10.7). Two things are clear. First, one need not postulate that $\mathscr{C}$ has a product representation since that follows from the uniqueness of the translations of $\mathscr{A}$ and the assumption that $\mathscr{A}$ is distributive in $\mathscr{C}$. Second, the representation is not restricted to extensive or even concatenation structures; it holds for any structure isomorphic to a real unit structure. Theorem 8 will be invoked in the discussion of dimensional analysis in Section 22.7.

It is important to recognize that the solvability assumptions are crucial to this result and that we do not know what can happen when they fail. The following is an example of a (1, 1) structure that is distributive in a conjoint structure, but the conjoint structure fails to satisfy the Thomsen condition and, of course, is not solvable. Set $R = \mathrm{Re}^{+} \cup \{0\}$, and let $\mathscr{A} = \langle R, \geqslant, \bigcirc \rangle$, where $x \bigcirc y = (x + y)/2$, and $\mathscr{C} = \langle R \times R, \succsim \rangle$, where

$$(x, u) \succsim (y, v) \quad \text{iff} \quad xu + u^2 \geqslant yv + v^2.$$

It is easy to verify that $\succsim$ satisfies the monotonicity axioms but not solvability and that $\bigcirc$ is distributive, since if

$$xu + u^2 = x'v + v^2 \quad \text{and} \quad yu + u^2 = y'v + v^2,$$

then adding and dividing by 2 yields

$$\frac{1}{2}(x + y)u + u^2 = \frac{1}{2}(x' + y')v + v^2.$$

$\mathscr{C}$ does not satisfy the Thomsen condition since

$$(5,1) \sim (1,2) \quad \text{and} \quad (1,3) \sim (11,1) \quad \text{but} \quad (5,3) < (11,2).$$

20.3 PROOFS

20.3.1 Theorem 2 (p. 117)

Suppose $\mathscr{A} = \langle A, \succsim, S_j \rangle_{j \in J}$ *is a totally ordered relational structure that is M-point homogeneous. If* $\mathrm{order}(S_j) \leqslant M \leqslant |A|$, $j \in J$, *then* $\mathscr{A}$ *and* $\langle A, \succsim \rangle$ *have the same automorphism group.*

COROLLARY. *If, in addition, $\langle A, \succsim \rangle$ is isomorphic to $\langle \text{Re}, \geqslant \rangle$, either for some $j \in J$, $M <$ order(S_j) or $\mathscr{A}$ is of scale type (∞, ∞).*

PROOF. Clearly, auto$(\mathscr{A}) \subseteq$ auto$(\langle A, \succsim \rangle)$. Conversely, suppose $\alpha \in$ auto$(\langle A, \succsim \rangle)$. Let $m(j) =$ order(S_j), and suppose $a_1, a_2, \ldots, a_{m(j)} \in A$. Since $m_i \leqslant M \leqslant |A|$, there exists a set X having M distinct elements that includes $\{a_1, a_2, \ldots, a_{m(j)}\}$. Since α is one-to-one, $\alpha(X)$ has M distinct elements. So by the M-point homogeneity of $\mathscr{A}$, there exists an automorphism β of $\mathscr{A}$ such that $\beta(x) = \alpha(x)$ for $x \in X$. In particular,

$$
\begin{aligned}
[a_1, a_2, \ldots, a_{m(j)}] \in S_j \quad & \text{iff} \quad [\beta(a_1), \beta(a_2), \ldots, \beta(a_{m(j)})] \in S_j \\
& \text{iff} \quad [\alpha(a_1), \alpha(a_1), \ldots, \alpha(a_{m(j)})] \in S_j.
\end{aligned}
$$

Thus α is an automorphism of $\mathscr{A}$.

The corollary follows immediately since $\langle \text{Re}, \geqslant \rangle$ is clearly of scale type (∞, ∞). ◇

20.3.2 Theorem 3 (p. 118)

If a totally ordered relational structure $\mathscr{A}$ has the property that there is a set $\mathscr{F}$ of $\mathscr{A}$-invariant operations, including the identity function, such that the algebraic closure of each subset of N distinct elements is order dense in $\mathscr{A}$, then $\mathscr{A}$ is N-point unique.

PROOF. Suppose B is a set of N distinct fixed points of an automorphism α, and define C_k inductively as follows:

$$
C_0 = B, \qquad \text{and} \qquad C_{k+1} = \{F(c_1, \ldots, c_n) | c_k \in C_k \,\&\, F \in \mathscr{F}\}
$$

and let

$$
C = \bigcup_{k=0}^{\infty} C_k .
$$

Note that C is the algebraic closure of B under $\mathscr{F}$, and that since $\iota \in \mathscr{F}$, C_k is included in C_{k+1}. Over C_0, $\alpha = \iota$. Suppose this is true over C_k, and let $c \in C_{k+1}$, i.e., for some $F \in \mathscr{F}$ and $c_i \in C_k$, we know $c = F(c_1, \ldots, c_n)$. Then, by the invariance property,

$$
\begin{aligned}
\alpha(c) &= \alpha F(c_1, \ldots, c_n) \\
&= F[\, a(c_1), \ldots, \alpha(c_n)] \\
&= F(c_1, \ldots, c_n) \\
&= c.
\end{aligned}
$$

So, by induction, $\alpha = \iota$ on C. Suppose for some $a \in A$, $\alpha(a) \neq a$; with no loss of generality, suppose $\alpha(a) \succ a$. Since C is order dense, there is some c in C with $\alpha(a) \succsim c = \alpha(c) \succsim a$. Thus both $c \succsim a$ and $\alpha(a) \succsim \alpha(c)$, which since α is order-preserving implies $\alpha(a) = a$, contrary to choice. So $\alpha = \iota$, proving $\mathscr{A}$ is N-point unique. ◇

20.3.3 Theorem 4 (p. 118)

In a totally ordered relational structure, the following are equivalent:

(i) $\mathscr{T}$ *is a group under function composition;*

(ii) $\mathscr{T}$ *is* 1*-point unique;*

(iii) *each dilation, except for the identity, and each translation intersect.*

PROOF.

(i) implies (ii). Suppose for some $a \in A$ and $\alpha, \beta \in \mathscr{T}$, $\alpha(a) = \beta(a)$. Thus $a = \alpha^{-1}\beta(a)$, which, since $\alpha^{-1}\beta \in \mathscr{T}$, is possible only if $\alpha^{-1}\beta = \iota$, proving $\mathscr{T}$ is 1-point unique.

(ii) implies (iii). Suppose α is a dilation at a, $\alpha \neq \iota$, $\tau \in \mathscr{T}$. If they do not intersect, then $\tau^{-1}\alpha \in \mathscr{T}$. However, $\tau^{-1}\alpha(a) = \tau^{-1}(a)$, and so by the 1-point uniqueness of $\mathscr{T}$, $\tau^{-1}\alpha = \tau^{-1}$, whence $\alpha = \iota$, contrary to choice.

(iii) implies (i). Suppose $\mathscr{T}$ is not a group, which is possible under function composition only if it is not closed. So let $\alpha, \beta \in \mathscr{T}$ and $\alpha\beta$ be a dilation, $\neq \iota$ since $\alpha\beta$ is not a translation. By hypothesis, there is some $a \in A$ such that $\alpha\beta(a) = \alpha(a)$, whence $\beta(a) = a$, contradicting the assumption that $\beta \in \mathscr{T}$. ◇

COROLLARY. *If $\mathscr{A}$ is N-point unique, there is no singular point, and the automorphisms of $\mathscr{A}$ commute, then $\mathscr{A}$ is* 1*-point unique.*

PROOF. We show $\mathscr{G} = \mathscr{T}$, in which case $\mathscr{T}$ is a group and thus is 1-point unique by the theorem. Suppose for some a, $\alpha(a) = a$. Since a is not singular, for some $\beta \in \mathscr{G}$, $\beta(a) \succ a$. By commutativity, $\beta(a) = \beta\alpha(a) = \alpha\beta(a)$, and so $\beta(a)$ is a second fixed point of α. Continue inductively, and $\beta^n(a)$ is a fixed point, and so by N-point uniqueness, α is the identity, proving that there are no nontrivial dilations. ◇

20.3.4 Theorem 5 (p. 120)

Suppose that $\mathscr{R} = \langle \text{Re}, \geqslant, R_j \rangle_{j \in J}$ is a numerical relational structure that is of scale type (M, N) with $M > 0$ and $N < \infty$. Then the following are true.

(i) $1 \leqslant N \leqslant 2$.

(ii) *The asymptotic order on $\mathscr{G}$ is a total order, and there is a subgroup $\mathscr{H}$ of $\mathscr{T}$ with the following properties*:

(a) *$\mathscr{H}$ is the minimal, nontrivial, convex subgroup of $\mathscr{G}$,*

(b) *$\mathscr{H}$ is Archimedean ordered,*

(c) *$\mathscr{H} = \mathscr{T}$.*

(d) *$\mathscr{H}$ is normal, and*

(e) *$\mathscr{H}$ is of scale type* $(1, 1)$.

(iii) *If $\mathscr{D}$ is nontrivial, then $\mathscr{D}$ is homogeneous.*

(iv) *$\mathscr{R}$ is of scale type* $(1, 1)$ *iff $\mathscr{R}$ is isomorphic to a real structure whose automorphisms are the difference group.*

(v) *$\mathscr{R}$ is of scale type* $(2, 2)$ *iff $\mathscr{R}$ is isomorphic to a real structure whose automorphisms are the affine group.*

(vi) *$\mathscr{R}$ is of scale type* $(1, 2)$ *iff $\mathscr{R}$ is isomorphic to a real structure whose automorphisms are a proper subgroup of the affine group that properly includes the difference group.*

PROOF. We proceed as follows. The several parts of statement (ii) are proved first. Lemma 2 constructs the ordering on $\mathscr{G}$; Lemmas 3–7 lead to (ii, a); Lemma 8 establishes (ii, b), (ii, c), (ii, d), and that $\mathscr{H}$ is 1-point unique, which is half of (ii, e); the other half, that $\mathscr{H}$ is 1-point homogeneous, is proved in Lemmas 9–12. Part (iii) is the Corollary to Lemma 11. Using these facts, we then characterize the automorphism group and thereby establish parts (i) and (iv)–(vi). The fact that $\mathscr{H}$ is $(1, 1)$ is crucial to doing this. ◇

The basic assumption of all the lemmas is that $\mathscr{R}$ is a homogeneous and unique relational structure on Re.

LEMMA 1. *Suppose $\alpha \in \mathscr{G}$ is such that for some x, y, $\alpha(x) < x$ and $\alpha(y) > y$, then α is a dilation at some z between x and y.*

PROOF. Since homeomorphisms are strictly increasing and *onto* functions they are continuous, and so there is some z between x and y that is a fixed point. ◇

LEMMA 2. *Suppose $\mathscr{G}$ is a group of homeomorphisms such that, except for the identity, each α in $\mathscr{G}$ either has no fixed point or a maximum one. Then the asymptotic order $\succsim'$ on $\mathscr{G}$ is a total order and is monotonic relative to the composition of automorphisms.*

PROOF.

(i) $\succsim'$ is clearly transitive.

(ii) $\succsim'$ is connected. If $\alpha \succsim' \beta$ is not true, then for each x in Re there exists some $y > x$ such that $\beta(y) > \alpha(y)$. Equally, if $\beta \succsim' \alpha$ is not true, for each x in Re there is some $y > x$ such that $\alpha(y) > \beta(y)$. Consider $\gamma = \beta^{-1}\alpha$. Clearly γ is not the identity since it must repeatedly fluctuate between $\gamma(y) > y$ and $\gamma(y) < y$. So by Lemma 1, the set of fixed points must be infinite and unbounded. Since this is contrary to assumption, $\succsim'$ is connected.

(iii) $\alpha \succsim' \beta$ and $\beta \succsim' \alpha$ imply $\alpha \equiv \beta$. For some x and all $y \geqslant x$, we know $\alpha(y) \geqslant \beta(y) \geqslant \alpha(y)$. Thus $\beta^{-1}\alpha(y) = y$, which since there is no maximal fixed point means $\beta^{-1}\alpha = \iota$.

(iv) $\succsim'$ is monotonic relative to the compositions of automorphisms. Suppose $\alpha \succsim' \beta$ and $\gamma \in \mathscr{G}$. For some x and all $y > x$, $\alpha(y) \geqslant \beta(y)$. Let $z = \gamma^{-1}(x)$, then for all $y > z$, since $\gamma(y) > x$, $\alpha\gamma(y) \geqslant \beta\gamma(y)$, whence $\alpha\gamma \succsim' \beta\gamma$. And for all $y > x$, $\gamma\alpha(x) \geqslant \gamma\beta(y)$, whence $\gamma\alpha \succsim' \gamma\beta$.

LEMMA 3. *The collection of all convex subgroups is totally ordered by inclusion.*

PROOF. Suppose $\mathscr{F}$ and $\mathscr{H}$ are distinct convex subgroups. With no loss of generality, suppose $\mathscr{H} - \mathscr{F} \neq \varnothing$ and let $\alpha \in \mathscr{H} - \mathscr{F}$. Suppose $\beta \in \mathscr{F}$. Neither $\beta \succsim' \alpha, \alpha^{-1}$ nor $\beta^{-1} \succsim' \alpha, \alpha^{-1}$, for otherwise by the convexity of $\mathscr{F}$ we could conclude $\alpha \in \mathscr{F}$, contrary to choice. So $\alpha \succsim' \beta, \beta^{-1}$, and by the convexity of $\mathscr{H}$, $\beta \in \mathscr{H}$, proving $\mathscr{F} \subseteq \mathscr{H}$. ◇

In terms of the order defined in Lemma 2, we introduce two definitions: for $\alpha, \beta \in \mathscr{G}$,

$$|\alpha| = \begin{cases} \alpha & \text{if} \quad \alpha \succsim' \iota \\ \alpha^{-1} & \text{if} \quad \alpha \prec' \iota; \end{cases}$$

$$\alpha \gg \beta \quad \text{iff} \quad \text{for each integer } n, \quad |\alpha| \succsim' |\beta|^n$$

LEMMA 4. *Suppose $\mathscr{F}$ and $\mathscr{G}$ are convex and $\mathscr{F} \subset \mathscr{H}$. If $\alpha \in \mathscr{H} - \mathscr{F}$ and $\beta \in \mathscr{F}$, then $\alpha \gg \beta$.*

PROOF. Suppose not, then for some integer n, $|\beta|^n \succ' |\alpha|$. Since $\mathscr{F}$ is a group, $|\beta|^n \in \mathscr{F}$, and since $\mathscr{F}$ is convex, we have $\alpha \in \mathscr{F}$, contrary to hypothesis. ◇

LEMMA 5. *Suppose α is a homeomorphism and $x, y \in$ Re are such that $\alpha(x) > x$, $x < y$, and α has no fixed point in $[x, y]$. Then for some integer n, $\alpha^n(x) > y$.*

PROOF. Suppose, on the contrary, $\alpha^n(x) \leqslant y$ for all integers n. Let $z = \text{l.u.b}\{\alpha^n(x) | n \text{ an integer}\}$, which clearly exists and is $\leqslant y$. Moreover,

$\alpha(z) > z$ since, if $\alpha(z) \leqslant z$, then by the continuity of α there would be a fixed point in $[x, y]$. Since $z > \alpha^{-1}(z)$, by definition of the l.u.b. there exists an n such that $z \geqslant \alpha^n(x) > \alpha^{-1}(z)$. Thus $\alpha^{n+1}(x) > z$, which contradicts the choice of z. ◇

LEMMA 6. (This is Alper's modification of an unpublished result of A. M. Gleason, c. 1959). *Suppose $\mathscr{G}$ is a group of homeomorphisms that have no singular point (i.e., no common fixed point) and each nontrivial α in $\mathscr{G}$ either has no fixed point or a maximum one. If there exist $\alpha_i \in \mathscr{G}$, $i = 1, \ldots, n+1$, such that $\alpha_{n+1} \gg \alpha_n \gg \cdots \gg \alpha_1 \gg \iota$, then there exists $\beta \in \mathscr{G}$, $\beta \neq \iota$, with n distinct fixed points.*

PROOF. With no loss of generality, we can select the $\alpha_i \succ' \iota$. Define $\gamma_i = \alpha_{i+1}\alpha_{i+2} \cdots \alpha_{n+1}$, and $\gamma_{n+1} = \iota$. Observe that $\gamma_i = \alpha_{i+1}\gamma_{i+1} \succ' \alpha_{i+1}$. Let x_0 be larger than any of the fixed points of $\alpha_1, \ldots, \alpha_{n+1}$. Let $\beta_0 = \iota$, then $\gamma_0(x_0) > \alpha_1(x_0) > x_0 = \beta_0(x_0)$.

Select any $x_1 > x_0$. By Lemma 5, there exists an integer $n(1)$ such that $\alpha_1^{n(1)}(x_1) > \gamma_1(x_1) \geqslant \alpha_2(x_1)$. Let $\beta_1 = \alpha_1^{n(1)}$.

We proceed by induction. Suppose for some k we have found $x_0 < x_1 < \cdots < x_k$ and $\beta_0 = \iota, \beta_1, \ldots, \beta_k$ such that $\beta_k(x_k) \leqslant \gamma_k(x_k)$, $\beta_k^{-1}(x_{k-1}) > \gamma_k(x_{k-1})$, and so on, alternating between β_k and β_k^{-1} back through the x_i. Since $\beta_k^{-1}\alpha_{k+1} \succ' \iota$, we know that for some $x_{k+1} > x_k$ and for all $y \geqslant x_{k+1}$, $\beta_k^{-1}\alpha_{k+1}(y) > y$. So by Lemma 5, there exists an integer $n(k+1)$ such that

$$(\beta_k^{-1}\alpha_{k+1})^{n(k+1)}(x_{k+1}) > \gamma_{k+1}(x_{k+1}).$$

Let $\beta_{k+1} = (\beta_k^{-1}\alpha_{k+1})^{n(k+1)}$.

Three things must be shown to complete the induction:

(i) $\alpha_{k+2} \succ' |\beta_{k+1}|$. By the corresponding induction hypothesis, the assumption that $\alpha_{k+1} \succ' \iota$, and monotonicity we see that $\alpha_{k+1}^2 \succ' |\beta_k|\alpha_{k+1}$. Since $\alpha_{k+2} \gg \alpha_{k+1}$, monotonicity yields $\alpha_{k+2} \succ' (\alpha_{k+1})^{2n(k+1)} \succ' (\beta_k^{-1}\alpha_{k+1})^{n(k+1)} = \beta_{k+1}$.

(ii) $\beta_{k+1}(x_{k+1}) > \gamma_{k+1}(x_{k+1})$. This was established prior to the definition of β_{k+1}.

(iii) And that β_{k+1} oscillates appropriately over the $\gamma_{k+1}(x_i)$.

For $i \leqslant k$, there are two possibilities: either

$$\beta_k(x_i) > \gamma_k(x_i),$$

in which case,

$$\alpha_{k+1}^{-1}\beta_k(x_i) > \gamma_{k+1}(x_i) \geqslant x_i,$$

and so

$$\beta_{k+1}^{-1}(x_i) = (\alpha_{k+1}^{-1}\beta_k)^{n(k+1)}(x_i) > \gamma_{k+1}^{n(k+1)}(x_i) \geqslant \gamma_{k+1}(x_i);$$

or

$$\beta_k^{-1}(x_i) > \gamma_k(x_i),$$

in which case,

$$\alpha_{k+1}(x_i) > x_i > \beta_k\gamma_{k+1}(x_i),$$

and so

$$\alpha_{k+1}^{-1}\beta_k^{-1}\alpha_{k+1}(x_i) > \alpha_{k+1}^{-1}\gamma_k(x_i) = \gamma_{k+1}(x_i).$$

Applying α_{k+1} to this inequality and recognizing that $\gamma_{k+1}(x_i)$ is larger than any fixed point of α_{k+1}, we therefore have

$$\beta_k^{-1}\alpha_{k+1}(x_i) > \alpha_{k+1}\gamma_{k+1}(x_i) > \gamma_{k+1}(x_i) > x_i.$$

From this,

$$\beta_{k+1}(x_i) = (\beta_k^{-1}\alpha_{k+1})^{n(k+1)}(x_i) > \gamma_{k+1}(x_i).$$

Thus the appropriate oscillations occur.

Thus, by induction, $\beta = \beta_{n+1}$ reverses itself $n+1$ times relative to $\iota = \gamma_{n+1}$. Therefore, since it is a homeomorphism on Re and so is continuous, it has n fixed points. ◇

LEMMA 7. *If $\mathscr{R}$ is of scale type (M, N) with $M > 0$ and $N < \infty$, then the number of distinct, nontrivial, convex subgroups is $\leqslant N$, and so there is a minimal one.*

PROOF. Suppose there are at least $N+1$ distinct, nontrivial, convex subgroups $\mathscr{H}_k$, $k = 1, \ldots, N+1$, ordered so that $\mathscr{H}_k$ properly includes $\mathscr{H}_{k-1}$. Select $\alpha_k \in \mathscr{H}_k - \mathscr{H}_{k-1}$. Then by Lemma 4, $\alpha_{k+1} \gg \alpha_k$, $k = 1, \ldots, N$. By 1-point homogeneity, $\mathscr{G}$ does not have a singular point and by N-point uniqueness each nontrivial automorphism has either no fixed point or a maximal one. Thus Lemma 6 implies there is a nontrivial automorphism with N fixed points, which is impossible. Any totally ordered finite sequence has a minimal element. ◇

In the next few lemmas, $\mathscr{H}_k$, $k = 1, \ldots, K$ where $K \leqslant N$ denotes the finite chain of convex subgroups described in the proof of Lemma 7; we write $\mathscr{H}$ for the minimal subgroup $\mathscr{H}_1$.

LEMMA 8. *If $\mathscr{R}$ is of scale type (M, N) with $M > 0$ and $N < \infty$, then $\mathscr{H}$ is Archimedean, a subset of $\mathscr{T}$, 1-point unique, and normal.*

PROOF. $\mathscr{H}$ is Archimedean since if it were not, it would not be minimal. By Hölder's theorem, $\mathscr{H}$ is commutative.

To show $\mathscr{H}$ is normal, we first show $\beta\mathscr{H}\beta^{-1}$ is convex. Suppose $\gamma \in \mathscr{G}$ and for some $\alpha \in \mathscr{H}$, $|\gamma| \precsim' \beta\alpha\beta^{-1}$. Thus $\eta = \beta^{-1}|\gamma|\beta \precsim' \alpha$. Since $\mathscr{H}$ is convex, $\eta \in \mathscr{H}$, whence $|\gamma| = \beta\eta\beta^{-1}$, proving $\beta\mathscr{H}\beta^{-1}$ is convex. Suppose $\gamma = \beta\alpha\beta^{-1} \in \beta\mathscr{H}\beta^{-1}$. Since $\mathscr{H}$ is minimal, $\mathscr{H} \subseteq \beta^{-1}\mathscr{H}\beta$, and so for $\alpha \in \mathscr{H}$ there exists $\eta \in \mathscr{H}$ such that $\alpha = \beta^{-1}\eta\beta$. Thus $\eta = \beta\alpha\beta^{-1} = \gamma$, proving $\beta\mathscr{H}\beta^{-1} \subseteq \mathscr{H}$, whence $\beta\mathscr{H}\beta^{-1} = \mathscr{H}$.

Next we show that $\mathscr{H}$ has no singularity. Suppose, on the contrary, x is singular. Since $\mathscr{G}$ does not have a singularity, we can select $\beta \in \mathscr{G}$ such that $\beta^{-1}(x) \succ x$. By the normality of $\mathscr{H}$, for each $\alpha \in \mathscr{H}$, $x = \beta\alpha\beta^{-1}(x)$, and so $\beta^{-1}(x)$ is also a singular point of $\mathscr{H}$. By induction, so, for each integer n, is $\beta^{-n}(x)$, which is contrary to $N < \infty$.

To show $\mathscr{H}$ is included in $\mathscr{T}$, suppose $\alpha \in \mathscr{H}$, $\alpha \neq \iota$, has a fixed point, and let x be the largest. Since $\mathscr{H}$ is Archimedean and $\mathscr{H}$ does not have a singularity, there exists $\beta \in \mathscr{H}$ such that $\beta(x) \succ x$. Using the fact that x is a fixed point of α and $\beta(x)$ is not, we have

$$\beta\alpha(x) = \beta(x) \neq \alpha\beta(x),$$

contradicting the fact the $\mathscr{H}$ is commutative. So α has no fixed point, i.e., $\mathscr{H} \subseteq \mathscr{T}$.

By the Corollary to Theorem 4, $\mathscr{H}$ is 1-point unique. ◇

LEMMA 9. *If $\mathscr{H}_k \subseteq \mathscr{T}$, then $k = 1$.*

PROOF. Suppose $k > 1$. Then for any $\tau \in \mathscr{H}_{k-1}$, $\tau \succ' \iota$, there exists $\tau' \in \mathscr{H}_k$ such that for each integer n, $\tau' \succsim' \tau^n$. Since $\mathscr{H}_k \subseteq \mathscr{T}$, we see that for all x, $\tau'(x) \geqslant \tau^n(x)$. Since the structure is Dedekind complete, $u(x) =$ l.u.b.$\{\tau^n(x) | n \text{ an integer}\}$ exists. Clearly $\tau u(x) \geqslant u(x)$. If $\tau u(x) > u(x) \geqslant \tau^n(x)$ for all n, then $u(x) > \tau^{-1}u(x) \geqslant \tau^{n-1}(x)$ for all n, proving that $\tau^{-1}u(x)$ is a smaller bound than $u(x)$, contrary to choice. Therefore $u(x)$ is a fixed point of τ, which is impossible since $\tau \succ' \iota$ and $\tau \in \mathscr{T}$. So $k = 1$. ◇

LEMMA 10. *If $k > 1$ and $\alpha, \beta \in \mathscr{H}_k$, then $\alpha^{-1}\beta^{-1}\alpha\beta \in \mathscr{H}_{k-1}$.*

PROOF. Since there is no intermediate convex subgroup between $\mathscr{H}_k$ and $\mathscr{H}_{k-1}$, then according to Fuchs (1963, p. 50) $\mathscr{H}_k/\mathscr{H}_{k-1} = \{\alpha\mathscr{H}_{k-1} | \alpha \in \mathscr{H}_k\}$, with the group operation defined by $(\alpha\mathscr{H}_{k-1})(\beta\mathscr{H}_{k-1}) = \alpha\beta\mathscr{H}_{k-1}$, is isomorphic to a subgroup of the additive real numbers. Thus it is commutative, and so $\alpha\beta\mathscr{H}_{k-1} = \beta\alpha\mathscr{H}_{k-1}$. Therefore,

$$\begin{aligned}\mathscr{H}_{k-1} &= (\beta\alpha)^{-1}(\beta\alpha)\mathscr{H}_{k-1}\\ &= (\beta\alpha)^{-1}(\beta\alpha\mathscr{H}_{k-1})\\ &= (\beta\alpha)^{-1}(\alpha\beta\mathscr{H}_{k-1})\\ &= (\beta\alpha)^{-1}(\alpha\beta)H_{k-1}\\ &= (\alpha^{-1}\beta^{-1}\alpha\beta)\mathscr{H}_{k-1},\end{aligned}$$

and so $\alpha^{-1}\beta^{-1}\alpha\beta \in \mathscr{H}_{k-1}$. ◇

LEMMA 11. *If $\mathscr{R}$ is of scale type (M, N), $M > 0$ and $N < \infty$, and $\mathscr{H}_k$ is 1-point homogeneous and $\mathscr{H}_k - \mathscr{T} \neq \varnothing$, then $(\mathscr{H}_k - \mathscr{T}) \cup \{\iota\}$ is 1-point homogeneous.*

PROOF. Let $x, y \in \text{Re}$. For $x = y$, clearly $\iota(x) = x = y$. Once the result is proved for $x < y$, the result for $x > y$ follows by taking the inverse of the automorphism from y to x.

We first show there exists $\alpha \in \mathscr{H}_k - \mathscr{T}$ such that $\alpha(x) = x$ and for all $u > x$, $\alpha(u) > u$. By hypothesis, there exists $\delta \in \mathscr{H}_k - \mathscr{T}$. By N-point uniqueness, it has a maximal fixed point, say z. By the homogeneity of $\mathscr{H}_k$, there exists $\beta \in \mathscr{H}_k$ such that $\beta(z) = x$. Then $\alpha = \beta\delta\beta^{-1} \in \mathscr{H}_k$,

$$\alpha(x) = \beta\delta\beta^{-1}(x) = \beta\delta(z) = \beta(z) = x,$$

and for $u > x$, $\alpha(u) \neq u$ since, otherwise, $\delta\beta^{-1}(u) = \beta^{-1}(u)$ and $\beta^{-1}(u) > \beta^{-1}(x) = z$, which would violate the choice of z as the maximal fixed point of δ. By Lemma 1, either $\alpha(u) > u$ for all $u > x$ or $\alpha(u) < u$ for all $u > x$. In the latter case, α^{-1} has the asserted property.

For each $w > x$, by Lemma 5 there exists $\beta \in \mathscr{H}_k - \mathscr{T}$ such that $\beta(x) = x$ and $\beta(y) > w$.

Finally, we construct $\mu \in \mathscr{H}_k - \mathscr{T}$ such that $\mu(x) = y$. By 1-point homogeneity of $\mathscr{H}_k$, select $\delta \in \mathscr{H}_k$ with $\delta(x) = y$. If δ is a dilation, we are done. So suppose it is a translation. Since $x < y$, it follows by Lemma 1 that $\delta(u) > u$ for all u. Select $z > \delta(y)$, and let $\beta \in \mathscr{H}_k - \mathscr{T}$ be such that $\beta(x) = x$ and $\beta(y) > z$, and so $\beta^{-1}(z) < y$. Consider $\mu = \delta\beta^{-1}$. First, $\mu(x) = \delta\beta^{-1}(x) = \delta(x) = y$. Second, $\mu \in \mathscr{H}_k$. And third, since $\mu(z) = \delta\beta^{-1}(z) < \delta(y) < z$ and $\mu(x) = y > x$, Lemma 1 establishes that μ is a dilation. Thus $(\mathscr{H}_k - \mathscr{T}) \cup \{\iota\}$ is homogeneous. ◇

COROLLARY. *If $\mathscr{R}$ is of scale type (M, N), $M > 0$ and $N < \infty$, and $\mathscr{D} \neq \varnothing$, then $\mathscr{D}$ is 1-point homogeneous.*

PROOF. Since $\mathscr{G}$ is convex and $\mathscr{D} = (F - T) \cup \{\iota\}$, the result follows immediately. ◇

LEMMA 12. *If $\mathscr{R}$ is of scale type (M, N), $M > 0$ and $N < \infty$, then $\mathscr{H}$ is 1-point homogeneous.*

PROOF. By Lemma 7 the result follows from a finite induction on the property that if $k > 1$ and $\mathscr{H}_k$ is 1-point homogeneous, then $\mathscr{H}_{k-1}$ is 1-point homogeneous, which we now show. By Lemma 9, $\mathscr{H}_k - \mathscr{T} \neq \varnothing$; and so by Lemma 11, $(\mathscr{H}_k - \mathscr{T}) \cup \{\iota\}$ is 1-point homogeneous. So for each $x, y \in \mathrm{Re}$, there exists $\beta(x) = y$. Let z be a fixed point of β, then there exists $\alpha \in \mathscr{H}_k$ such that $\alpha(y) = z$. Thus,

$$\alpha^{-1}\beta^{-1}\alpha\beta(x) = \alpha^{-1}\beta^{-1}\alpha(y) = \alpha^{-1}\beta^{-1}(z) = \alpha^{-1}(z) = y,$$

which by Lemma 10 establishes that $\mathscr{H}_{k-1}$ is 1-point homogeneous. ◇

CONTINUATION OF PROOF. We turn now to the remaining parts of the theorem. We first show that $\mathscr{R}$ is isomorphic to a structure whose automorphism group is a subgroup of the affine group and that it includes the similarity group. This establishes that $N \leqslant 2$.

Since by part (ii) of Theorem 5 $\mathscr{H}$ is an Archimedean ordered group, by Hölder's theorem, there is an isomorphism f from $\mathscr{H}$ into the additive real numbers. Define the one-to-one transformation g of Re onto Re as follows. For some fixed $x_0 \in \mathrm{Re}$ and each $x \in \mathrm{Re}$, since $\mathscr{H}$ is homogeneous (Lemma 12) there exists $\tau_x \in \mathscr{H}$ such that $x = \tau_x(x_0)$. Let $g(x) = f(\tau_x)$. Let $\mathscr{R}_g$ be the isomorphic copy of $\mathscr{R}$ under g and $\mathscr{G}_g$ be its automorphism group. Thus $\tau_g \in \mathscr{H}g$ iff for some $\tau \in \mathscr{H}$, $\tau_g(x) = g\tau g^{-1}(x)$. Observe that

$$\begin{aligned}
\tau_g(x) &= g\tau g^{-1}(x) \\
&= g\tau(y), \qquad \text{where} \quad y = g^{-1}(x), \\
&= g\tau\tau_y(x_0) \\
&= f(\tau\tau_y) \\
&= f(\tau) + f(\tau_y) \\
&= f(\tau) + g\tau_y(x_0) \\
&= f(\tau) + x.
\end{aligned}$$

So, with no loss of generality, we can identify $\mathscr{R}$ with $\mathscr{R}_g$ and $\mathscr{H}$ with $\mathscr{H}_g$. Now suppose $\alpha \in \mathscr{G}$. For x, s in Re, let $\tau(x) = x + s$ and $y = \alpha(x)$.

Since, by Lemma 8, $\tau' = \alpha\tau\alpha^{-1} \in \mathscr{H}$, there is some $s' \in \mathrm{Re}$ such that $\tau'(x) = x + s'$. Thus,

$$\begin{aligned} \alpha(x) + s' &= y + s' \\ &= \tau'(y) \\ &= \alpha\tau\alpha^{-1}(y) \\ &= \alpha\tau(x) \\ &= \alpha(x + s). \end{aligned}$$

Setting $x = 0$, we see that $s' = \alpha(s) - \alpha(0)$, and so the last equation can be rewritten as

$$\alpha(x + s) - \alpha(0) = \alpha(x) - \alpha(0) + \alpha(s) - \alpha(0).$$

Since α is strictly increasing and the equation holds for all x, y, the solution is well known to be $\alpha(x) = \gamma x + t$ where $t = \alpha(0)$. Thus, $\mathscr{G}$ is a subgroup of the affine group. In the course of the proof, we have shown that $\mathscr{H}$ is isomorphic to the similarity group. Thus, $\mathscr{H} = \mathscr{T}$.

We turn to proving part (iv) of the theorem. Clearly, if $\mathscr{G}$ is the translation group, the structure is $(1, 1)$. Suppose the structure is $(1, 1)$; then, by what has just been shown, we know $\mathscr{G}$ includes the translation group. Suppose more is included, say $\alpha_{r,s}(x) = rx + s$. For any x, let $y = \alpha_{r,s}(x)$. But equally, for $s' = y - x$, then $y = x + s'$. So by 1-point uniqueness, $r = 1$ and $s = s'$, and so $\mathscr{G}$ is the translation group.

Consider part (v) of Theorem 5. Clearly, if $\mathscr{G}$ is the affine group, the structure is $(2, 2)$. Suppose the structure is $(2, 2)$ and $\mathscr{G}$ is a proper subgroup of the affine group. Let $\alpha_{r,s}(x) = rx + s$ be one of the missing transformations. Since a conjugation from the affine group into itself preserves translations, by parts (ii, e) and (iv) of the theorem, for each real s, $\tau_s = x + s$ is included. Thus $\alpha_{r,s'}$ is also missing since, otherwise,

$$\tau_s\tau_{-s'}\alpha_{r,s'}(x)[(rx + s') - s'] + s = rx + s$$

would be in $\mathscr{G}$, contrary to choice. Select any $x_1 < x_2$ and $y_1 < y_2$ such that $(y_2 - y_1)/(x_2 - x_1) = r$. By 2-point homogeneity, there exists $\alpha_{r',s'}$ such that $\alpha_{r',s'}(x_1) = y_i$, $i = 1, 2, \ldots,$ from which it follows that $y_2 - y_1 = r'(x_2 - x_1) = r(x_2 - x_1)$, and so $r' = r$, a contradiction. So no $\alpha_{r,s}$ can be missing.

In part (vi) of Theorem 5, by excluding the other two cases, the $(1, 2)$ case must correspond to a proper subgroup of the affine group that properly includes the similarity group. ◇

20.3.5 Theorem 7 (p. 124)

Suppose $\mathscr{A} = \langle A, \succsim, S_j \rangle_{j \in J}$ *is a relational structure and* $\mathscr{T}$ *is its set of translations. Then the following are equivalent:*

(i) $\mathscr{A}$ *is isomorphic to a real unit structure for which the translations are homogeneous.*

(ii) $\mathscr{T}$ *is a homogeneous, Archimedean ordered group.*

(iii) $\mathscr{T}$ *is* 1-*point unique, and there exists an Archimedean, solvable conjoint structure* $\mathscr{C}$ *with a relational structure* $\mathscr{A}'$ *on the first component such that* $\mathscr{A}'$ *is isomorphic to* $\mathscr{A}$ *and* $\mathscr{A}'$ *is distributive in* $\mathscr{C}$.

PROOF.

(i) implies (ii). If $\mathscr{A}$ is isomorphic to a homogeneous real unit structure, then $\mathscr{T}$ is isomorphic to T, and by definition, $\langle T, \geqslant, \cdot \rangle$ is a homogeneous, Archimedean ordered group.

(ii) implies (i). We first embed $\mathscr{A}$ isomorphically in $\langle \mathscr{T}, \succsim' \rangle$, where $\succsim'$ is the asymptotic ordering of $\mathscr{T}$. Let $n(j) = \text{order}(S_j)$. For $\tau_i \in \mathscr{T}$, $i = 1, \ldots, n(j)$, and for a fixed $a \in A$, define S_j' on $\mathscr{T}$ by

$$(\tau_1, \ldots, \tau_{n(j)}) \in S_j' \qquad \text{iff} \qquad (\tau_1(a), \ldots, \tau_{n(j)}(a)) \in S_j.$$

Note that the definition of S_j' is independent of the choice of a. For suppose we had chosen $b \in A$, then by the fact that $\mathscr{T}$ is homogeneous we know there exists $\sigma \in \mathscr{T}$ such that $b = \sigma(a)$. Using this, the fact that σ is an automorphism and so is invariant under the defining relations S_j, and the fact that by Hölder's theorem elements of $\mathscr{T}$ commute, we have

$$\begin{aligned}
[\tau_1(a), \ldots, \tau_{n(j)}(a)] \in S_j \qquad & \text{iff} \qquad (\sigma\tau_1(a), \ldots, \sigma\tau_{n(j)}(a)) \in S_j \\
& \text{iff} \qquad (\tau_1\sigma(a), \ldots, \tau_{n(j)}\sigma(a)) \in S_j \\
& \text{iff} \qquad (\tau_1(b), \ldots, \tau_{n(j)}(b)) \in S_j.
\end{aligned}$$

$$\text{iff} \qquad F(\tau) \succsim F(\sigma).$$

For fixed a, define the function F from T into A by for each $\tau \in \mathscr{T}$,

$$F(\tau) = \tau(a).$$

It is onto A because T is homogeneous, and it is 1:1 because $\mathscr{T}$ is 1-point unique. It is order-preserving because, by the following argument, the elements of $\mathscr{T}$ are uncrossed. Suppose that there exists $\gamma \in T$ with $\gamma(x) \succ x$ and $\gamma(y) \prec y$. By homogeneity, there exists $\sigma \in T$ such that $\sigma(x) = y$, and so using the commutativity of T,

$$\sigma\gamma(x) \succ \sigma(x) = y \succ \gamma(y) = \gamma\sigma(x) = \sigma\gamma(x),$$

which is impossible. Thus,

$$\tau \succsim' \sigma \qquad \text{iff} \qquad \tau(a) \succsim \sigma(a) \qquad \text{iff} \qquad F(\tau) \succsim F(\sigma).$$

Finally,

$$(\tau_1, \ldots, \tau_{n(j)}) \in S_j' \quad \text{iff} \quad [\tau_1(a), \ldots, \tau_{n(j)}(a)] \in S_j$$
$$\text{iff} \quad [F(\tau_1), \ldots, F(\tau_{n(j)})] \in S_j.$$

Thus F is the isomorphism asserted.

By Hölder's theorem, let φ denote the isomorphism between $\langle \mathscr{T}, \succsim', \cdot \rangle$, where $\cdot$ denotes function composition, and the real multiplicative subgroup $\langle R, \geqslant, \cdot \rangle$. Define the relation R_j of order $n(j)$ on R by, for $r_i \in R$, $i = 1, \ldots, n(j)$,

$$(r_1, \ldots, r_{n(j)}) \in R_j \quad \text{iff} \quad \left[\varphi^{-1}(r_1), \ldots, \varphi^{-1}(r_{n(j)})\right] \in S_j'.$$

It is easy to verify that $\langle \mathscr{T}, \succsim', S_j' \rangle_{j \in j}$ and $\langle R, \geqslant, R_j \rangle_{j \in J}$ are isomorphic. We show that R_j is homogeneous. For $s \in R$, let $\sigma = \varphi^{-1}(s)$. Then, since φ maps function composition onto multiplication and S_j' is invariant under automorphisms,

$$(r_1, \ldots, r_{n(j)}) \in R_j \quad \text{iff} \quad \left[\varphi^{-1}(r_1), \ldots, \varphi^{-1}(r_{n(j)})\right] \in S_j'$$
$$\text{iff} \quad \left[\sigma\varphi^{-1}(r_1), \ldots, \sigma\varphi^{-1}(r_{n(j)})\right] \in S_j'$$
$$\text{iff} \quad \left[\varphi^{-1}(sr_1), \ldots, \varphi^{-1}(sr_{n(j)})\right] \in S_j'$$
$$\text{iff} \quad [sr_1, \ldots, sr_{n(j)}] \in R_j.$$

(ii) is equivalent to (iii). To establish this, we need the following lemma, which generalizes Theorem 19.18. The concept of a right multiplication is found in Definition 19.16.

LEMMA 1. *Suppose $\mathscr{C} = \langle A \times P, \succsim \rangle$ is a conjoint structure that is solvable relative to $(x_0, p_0) \in A \times P$, that $*$ is the induced Holman operation on A, and that $\mathscr{T}(*)$ is the set of right multiplications of $*$. Suppose, further, that S is a relation of order n on A, that $\mathscr{E}$ is the set of endomorphisms, and $\mathscr{G}$ the set of automorphisms of $\langle A, \succsim_A, S \rangle$. Then, the following are true:*

1. *If S is distributive in $\mathscr{C}$, then $\mathscr{T}(*) \subseteq \mathscr{E}$.*
2. *$\mathscr{T}(*) \subseteq \mathscr{G}$ iff S is distributive in $\mathscr{C}$ and $\mathscr{C}$ is unrestrictedly A-solvable.*
3. *$\mathscr{T}(*) = \mathscr{G}$ iff $\mathscr{T}(*) \subseteq \mathscr{G}$ and $\mathscr{G}$ is 1-point unique.*

PROOF.

1. For $\tau \in \mathscr{T}(*)$, we know $\tau(a) = a * b$ for some b. So,

$$[\tau(a), p_0] \sim (a * b, p_0) \sim [a, \pi(b)].$$

Thus $[\tau(a_1),\ldots,\tau(a_n)]$ is similar to $(a_1,\ldots,a_n)$. Therefore, if $(a_1,\ldots,a_n) \in S$, then since S is distributive, $[\tau(a_1),\ldots,\tau(a_n)] \in S$. So τ is an endomorphism.

2. Suppose S is distributive in $\mathscr{C}$ and $\mathscr{C}$ is unrestrictedly A-solvable. If $\tau \in \mathscr{T}(*)$, then by unrestricted solvability, it is onto; since it is order-preserving it is 1:1; and by part (1) it is an endomorphism. So $\mathscr{T}(*)$ is a subset of $\mathscr{G}$.

Conversely, suppose $\mathscr{T}(*)$ is a subset of $\mathscr{G}$. Let d^R denote the right multiplication $\iota * d$. Observe that $(a, p) \sim (b, q)$ iff $\pi^{-1}(p)^R(a) = \pi^{-1}(q)^R(b)$. Now suppose $(a_1,\ldots,a_n) \in S$ and $(a_i, p) \sim (b_i, q)$, $i = 1,\ldots,n$, then using invariance under automorphisms we see that

$$\left[\pi^{-1}(p)^R(a_1),\ldots,\pi^{-1}(p)^R(a_n)\right] = \left[\pi^{-1}(q)^R(b_1),\ldots,\pi^{-1}(q)^R(b_n)\right] \in S,$$

and so $(b_1,\ldots,b_n) \in S$, proving that S is distributive in $\mathscr{C}$.

3. The same proof as in Theorem 19.18. ◇

We now continue the proof of the theorem.

(*iii*) *implies* (*ii*). Fix $a \in A$, and let $*$ be one of the Holman operations for $\mathscr{C}$ relative to a. By the lemma, the set of right multiplications of $*$, $\mathscr{T}(*)$, is a subset of the automorphisms of $\mathscr{A}$. We show $\mathscr{T}(*)$ is a subset of $\mathscr{T}$. Suppose it is not; then for some right multiplication, there is a fixed point c, i.e., $c * d = c$. Since all Holman induced operations are monotonic (Theorem 11, Chapter 19), $d \succsim a$ iff $c = c * d \succsim c * a = c$. Thus $d = a$, and so the right multiplication is the identity. Next we show that $\mathscr{T}(*) = \mathscr{T}$. Suppose $\tau \in \mathscr{T}$, and let $b = \tau(a)$. Then $\tau(a) = b = a * b$, whence by 1-point uniqueness $\tau = \iota * b \in \mathscr{T}(*)$.

To show $\mathscr{T}$ is homogeneous, suppose $b, c \in A$. Observe that the right multiplications of $*$, and so translations of $\mathscr{A}$, $d^R = \iota * d$, have the property $d^R(a) = a * d = d$. Since the translations are a group, $b^R(c^R)^{-1}(c) = b^R(a) = b$, proving $\mathscr{T}$ is homogenous.

To show that $\mathscr{T}$ is Archimedean, we first observe that $*$ is associative. Suppose $b, c, d \in A$. By the closure of $\mathscr{T}(*)$, there exists some u such that $d^R c^R = u^R$ and, indeed, $u = c * d$ since

$$u = a * u = u^R(a) = d^R c^R(a) = (a * c) * d = b * d.$$

Thus,

$$b * (c * d) = b * u = u^R(b) = d^R c^R(b) = (b * c) * d.$$

So, by induction, $(b^R)^n = \iota * nb$. Since by Theorem 19.11 $\langle A, \succsim_A, *, a\rangle$

is a total concatenation structure, we know that for $b, c \succ_A a$, there is some n such that $b(n) \succsim_A c$, whence $(b^R)^n \succsim' c^R$, proving that $\mathscr{T}$ is Archimedean.

(ii) implies (iii). We first embed $\mathscr{T}$ in $\mathscr{A}$. For $a \in A$, define $*_a$ as follows: for each $b, c \in A$, by 1-point homogeneity and by 1-point uniqueness (which follows from the fact that $\mathscr{T}$ is a group), there exist unique $\tau, \sigma \in \mathscr{T}$ such that $\tau(a) = b$ and $\sigma(a) = c$. Let $b *_a c = \sigma\tau(a)$. It is easy to show that $\langle A, \succsim, *_a \rangle$ is isomorphic to $\langle T, \succsim', * \rangle$, where $*$ denotes function composition. Note that this means it is (trivially) a total concatenation structure that is commutative and associative. By Theorem 19.11 (ii) we can construct a conjoint structure $\mathscr{C}$ on $A \times A$ such that the induced Holman operation relative to (a, a) is $*_a$. Map the defining relations of $\mathscr{A}$ onto $\langle A, \succsim_A, *_a \rangle$ under the isomorphism used to construct $\mathscr{C}$. By Lemma 1, these relations are distributive provided that $\mathscr{T}(*_a)$ is a subset of $\mathscr{T}$. For $\tau \in \mathscr{T}(*_a)$, there is some d such that for all $b \in A$, $\tau(b) = b *_a d$. By monotonicity, this can have a fixed point if and only if $d = a$, in which case $\tau = \iota$. So τ is a translation. ◇

COROLLARY 1. *The conjoint structure of part* (iii) *of Theorem 7 satisfies the Thomsen condition.*

PROOF. Corollary 1 of Theorem 19.11.

COROLLARY 2. *Suppose $\mathscr{A}$ is order dense and property* (ii) *holds. Then the automorphism group of the isomorphic real unit structure is a subgroup of the positive power group restricted to its domain, and so $\mathscr{A}$ is* 2*-point unique.*

PROOF. Note that by the construction of R in part (i) of the theorem, R is a subgroup of the multiplicative, positive, real numbers. If α is an automorphism and τ is a translation in $\mathscr{R}$, then $\alpha\tau\alpha^{-1}$ is a translation, for if x were a fixed point, then $\alpha^{-1}(x)$ would be a fixed point of τ, contrary to choice. Thus we can write for some $t, s \in T$, $\tau(x) = tx$ and $\alpha\tau\alpha^{-1}(x) = sx$. Let $y = \alpha(x)$,

$$\begin{aligned} s\alpha(x) &= sy \\ &= \alpha\tau\alpha^{-1}(y) \\ &= \alpha\tau(x) \\ &= \alpha(tx). \end{aligned}$$

Since R is a subgroup of the multiplicative, positive reals, $1 \in R$ and so $s = \alpha(t)/\alpha(1)$. Setting $h(x) = \alpha(x)/\alpha(1)$, we see that

$$h(sx) = h(s)h(x).$$

Since $\mathscr{A}$ is order dense, so is $\mathscr{R}$, which being multiplicative is also topologically dense. Therefore, since α is order-preserving, so is h, and hence h can be extended to a real interval so that the functional equation for h still holds. It is well known that the solution is $h(x) = x^r$, and so $\alpha(x) = tx^r$, where $t = \alpha(1)$. Obviously, under this condition, $\mathscr{A}$ is at most 2-point unique. ◇

20.3.6 Theorem 8 (p. 125)

Suppose $\mathscr{C} = \langle A \times P, \succsim \rangle$ is a conjoint structure that is unrestrictedly A-solvable, restrictedly P-solvable, and Archimedean. Suppose, further, that $\mathscr{A} = \langle A, \succsim_A, S_j \rangle_{j \in J}$ is a relational structure whose translations form an Archimedean ordered group.

(i) *If $\mathscr{A}$ is distributive in $\mathscr{C}$, then $\mathscr{A}$ is 1-point homogeneous and $\mathscr{C}$ satisfies the Thomsen condition.*

(ii) *If, in addition, $\mathscr{A}$ is Dedekind complete, then under some mapping φ from A onto Re^+ $\mathscr{A}$ has a homogeneous unit representation and there exists a mapping ψ from P into Re^+ such that $\varphi\psi$ is a representation of $\mathscr{C}$.*

(iii) *If, further, $\mathscr{C}$ is unrestrictedly P-solvable and there is a Dedekind complete relational structure on P that, under some ψ from P onto Re^+, has a homogeneous unit representation, then for some real constant ρ, $\varphi\psi^\rho$ is a representation of $\mathscr{C}$.*

PROOF.

(i) Since the set $\mathscr{T}$ of translations are assumed to be a group, they are 1-point unique; and so, by Theorem 7, $\mathscr{T}$ is homogeneous and, by the corollary, $\mathscr{C}$ satisfies the Thomsen condition.

(ii) By Theorem 6.2, $\mathscr{C}$ has a multiplicative representation, say $\varphi'\psi'$, and so φ' is a real representation of the total concatenation structure induced by $\mathscr{C}$ on A. By part (i) and the proof of Theorem 7, we know that $\mathscr{T}(*) = \mathscr{T}$. Thus, for $\tau \in \mathscr{T}$, there exists an $a \in A$ such that for all $b \in A$, $\tau(b) = b * a$. Thus,

$$\varphi'[\tau(b)] = \varphi'(b * a) = \varphi'(b)\varphi'(a).$$

Since φ is a unit representation of $\mathscr{A}$, for some real function r on $\mathscr{G}$,

$$\varphi[\tau(b)] = r(\tau)\varphi(b).$$

Since φ and φ' both preserve the order $\succsim_A$, there is a strictly increasing function g such that $\varphi' = g(\varphi)$. Thus,

$$\begin{aligned} g[\varphi(b)]g[\varphi(a)] &= \varphi'(b)\varphi'(a) \\ &= \varphi'[\tau(b)] \\ &= g(\varphi[\tau(b)]) \\ &= g[r(\tau)\varphi(b)]. \end{aligned}$$

Since $\mathscr{A}$ is Dedekind complete, φ is onto Re$^+$. So for some b, $\varphi(b) = 1$, whence $g(1)g[\varphi(a)] = g[r(\tau)]$, and so

$$g[r(\tau)]g[\varphi(b)] = g[r(\tau)\varphi(b)]g(1).$$

Since $\mathscr{A}$ is homogeneous, r is onto Re$^+$, and so by a well-known result (Aczél, 1966, p. 41), for some $\beta > 0$,

$$g(x) = g(1)x^{\beta}.$$

Thus,

$$[\varphi'\psi']^{1/\beta} = [g(1)\varphi^{\beta}\psi']^{1/\beta} = \varphi\psi,$$

where $\psi = [g(1)\psi']^{1/\beta}$ is a representation of $\mathscr{C}$.

(iii) This follows immediately from a proof parallel to that of (ii) and the fact that multiplicative representations of $\mathscr{C}$ form a log-interval scale. ◇

20.4 HOMOGENEOUS CONCATENATION STRUCTURES

20.4.1 Nature of Homogeneous Concatenation Structures

We now consider the special, but important, case of $\mathscr{A} = \langle A, \succsim, \bigcirc \rangle$, where $\succsim$ is a total order and $\bigcirc$ is a partial operation. The following results of Luce and Narens (1983, 1985) give some insight into the homogeneous case.

THEOREM 9. *Suppose $\mathscr{A}$ is a concatenation structure that is homogeneous.*

(i) *The partial operation $\bigcirc$ is an operation.*

(ii) *Either $\bigcirc$ is idempotent (for all $a \in A$, $a \bigcirc a \sim a$), weakly positive (for all $a \in A$, $a \bigcirc a \succ a$), or weakly negative (for all $a \in A$, $a \bigcirc a \prec a$).*

(iii) *In the idempotent cases the structure is intern and dense; in the weakly positive and negative cases it is unbounded.*

(iv) *If $\mathscr{A}$ is weakly positive or negative, then* $M = 1$; *if it is idempotent, then* $M \leqslant 2$.

(v) *If $\mathscr{A}$ is unique, then it is of scale type* $(1,1)$, $(1,2)$, *or* $(2,2)$, *and in the last two cases it is necessarily idempotent.*

Note that part (v) of this theorem is the same conclusion as parts (i) and (iii) of Theorem 5, but it holds for any homogeneous concatenation structure, not just Dedekind complete ones. The major difficulty in trying to use this result is to know when homogeneity and/or uniqueness hold. We return to homogeneity in Section 20.4.3.

We know (Corollary to Theorem 19.2) that any PCS without a maximum element is 1-point unique and that any concatenation structure that is closed, idempotent, solvable, and Archimedean in differences is 2-point unique (Theorem 19.21). In the homogeneous case, Theorem 9 makes it clear that no other cases are of great interest. For the nonhomogeneous case, other possibilities can arise. A sufficient condition for a closed concatenation structure on the real numbers to be 2-point unique is for the operation to be continuous (Luce and Narens, 1985). Since the proof of this is lengthy, and it gains little beyond the cases just mentioned, we do not prove it here.

20.4.2 Real Unit Concatenation Structures

The basic idea of this section is to use what we know about the scale types of homogeneous concatenation structures to place limits on the possible real representations. By Theorem 9 we know that for unique structures (e.g., those with a real representation with a continuous operation) the only possible homogeneous scale types are $(1,1)$, $(1,2)$, and $(2,2)$. We begin by first examining the nature of concatenation structures that are homogeneous real unit structures (Definition 5). The first result provides a reformulation of that property into the form first isolated by Cohen and Narens (1979) and that they called a "real unit structure."

THEOREM 10. *Suppose* $\mathscr{R} = \langle \mathrm{Re}^+, \geqslant, \bigcirc \rangle$ *is a closed concatenation structure. Then,* $\mathscr{R}$ *is a homogeneous, real unit structure iff there exists a function* $f: \mathrm{Re}^+ \to$ *(onto)*Re^+ *such that*

(i) *f is strictly increasing,*

(ii) *f/ι, where ι is the identity, is strictly decreasing,*[2]

[2] f/ι is the function defined by $(f/\iota)(x) = f(x)/\iota(x) = f(x)/x$ for all $x > 0$.

(iii) *for all* $x, y \in \mathrm{Re}^+$,

$$x \bigcirc y = yf(x/y).$$

The proof of this result is left as Exercise 10.

When discussing concatenation structures that are homogeneous, real unit structures, we shall make f explicit in the notation as $\langle \mathrm{Re}^+, \succsim, \bigcirc, f \rangle$.

THEOREM 11. *Suppose* $\mathscr{R} = \langle \mathrm{Re}^+, \succsim, \bigcirc \rangle$ *is a real concatenation structure. Then* $\mathscr{R}$ *is unique and homogeneous iff it is isomorphic to a real unit structure* $\langle \mathrm{Re}^+, \succsim, \bigcirc', f \rangle$. *Moreover,*

(i) $\mathscr{R}$ *is of scale type* $(1,1)$ *iff the equation*

$$f(x^\rho) = f(x)^\rho, \qquad x \in \mathrm{Re}^+, \tag{1}$$

is satisfied only for $\rho = 1$;

(ii) $\mathscr{R}$ *is of scale type* $(2,2)$ *iff Equation* (1) *holds for all* $\rho > 0$;

(iii) $\mathscr{R}$ *is of scale type* $(1,2)$ *iff Equation* (1) *is satisfied for* $\rho = k^n$, *where* $k > 0$ *is fixed and* n *is any integer.*

Uniqueness: *If* $\langle \varphi, f \rangle$ *and* $\langle \psi, g \rangle$ *are two unit representations of* $\mathscr{R}$ *with isomorphisms* φ *and* ψ, *then there exist constants* $s > 0$ *and* $t > 0$ *such that for all* r *in* Re^+,

$$\varphi = s\psi^{1/t} \qquad \text{and} \qquad f(r) = g(r^t)^{1/t}.$$

The unit structure result is due to Cohen and Narens (1979), and the characterization of scale type is from Luce and Narens (1985).

The following results provide some amplification of the meaning of this result. First, we show various properties of f for any unit structure. Second, we examine idempotent structures with doubling functions. Third, we mention a characterization of f for the $(1,2)$ case. And finally, we reformulate the characterization of the $(2,2)$ case.

THEOREM 12. *If* $\mathscr{R} = \langle \mathrm{Re}^+, \succsim, \bigcirc, f \rangle$ *is a real unit structure, then the following hold.*

(i) $k = \lim_{x \to \infty} f(x)/x \leqslant f(1)$.

(ii) $\mathscr{R}$ *has half-elements.*

(iii) *If* $\mathscr{R}$ *is weakly positive, then* $f(1) > 1$.

(iv) *If* $\mathscr{R}$ *is positive, then* $f(x) > x$ *and* $k \geqslant 1$.

(v) *If $\mathscr{R}$ is positive, then $\mathscr{R}$ is Archimedean in standard sequences.*

(vi) *If $\mathscr{R}$ is positive and restrictedly solvable, then $k = 1$.*

(vii) *If $\mathscr{R}$ is weakly negative, then $f(1) < 1$ and $\mathscr{R}$ cannot be restrictedly solvable.*

(viii) *If $\mathscr{R}$ is negative, then $f(x) < x$ and $k \leqslant 1$.*

(ix) *If $\mathscr{R}$ is idempotent, then $f(1) = 1$; for $x < 1$, $x < f(x) < 1$; and for $x > 1$, $x > f(x) > 1$.*

The proof is left as Exercise 16.

Recall that in Section 19.6.5, we introduced the notion of an idempotent structure having a doubling function δ, and we showed it is closely related to a PCS having half-element function δ^{-1} and the same automorphism group. Presumably in the homogeneous case we can say more about such structures. With $k = \lim_{x \to \infty} f(x)/x$ and $f_n(k)$ defined inductively by $f_1(k) = 1$ and $f_n(k) = f[f_{n-1}(k)/k]$, we have the following theorem.

THEOREM 13. *Suppose $R = \langle \mathrm{Re}^+, \geqslant, *, f \rangle$ is a real unit structure that is idempotent. Necessary and sufficient conditions for $\mathscr{R}$ to have a doubling function are*

(i) $\lim_{x \to 0} f(x) \geqslant k$, *and*

(ii) $\lim_{n \to \infty} f_n(k) = \infty$.

The doubling function is $\delta = \iota/k$.

The proof is left as Exercise 13.

This means that if $\langle \mathrm{Re}^+, \geqslant, \bigcirc, g \rangle$ is the induced PCS, then $g = f/k$ because

$$yg\left(\frac{x}{y}\right) = x \bigcirc y = \delta(x * y) = \frac{(x * y)}{k} = yf\left(\frac{x}{y}\right)\bigg/k.$$

In Luce and Narens (1985, Theorem 3.12), the $(1,2)$ cases for which f is differentiable are fully characterized. Because we do not make any use of the result and it is not especially edifying, we do not reproduce it.

Finally, consider the $(2,2)$ case, which is characterized by, for all $x > 0$ and $\rho > 0$,

$$f(x^\rho) = f(x)^\rho. \tag{2}$$

Since by Theorem 9(iv) this case is idempotent and by Theorem 12(ix) $f(1) = 1$, for $x < 1$, $x < f(x) < 1$, and for $x > 1$, $1 < f(x) < x$, we see that two distinct equations are involved. For $x > 1$, $x^\rho > 1$, and so we have one equation that is well known to have a solution of the form $f(x) = x^c$. Since $1 < f(x) < x$, we see that $0 < c < 1$. Similarly, for $x < 1$, the equation is of the form $f(x) = x^d$, and since $x < f(x) < 1$, then $0 < d < 1$. Thus the general solution is

$$f(x) = \begin{cases} x^c & \text{if} \quad x > 1 \\ 1 & \text{if} \quad x = 1 \\ x^d & \text{if} \quad x < 1 \end{cases}$$

or, putting it back into the definition of a unit structure,

$$x \bigcirc y = \begin{cases} x^c y^{1-c} & \text{if} \quad x > y \\ x & \text{if} \quad x = y \\ x^d y^{1-d} & \text{if} \quad x < y. \end{cases} \tag{3}$$

For some purposes, especially outside of physics, it is appropriate to transform this by a logarithm into a representation on Re instead of Re^+. Letting u and v denote the resulting variables, we obtain

$$u \bigcirc v = \begin{cases} cu + (1-c)v & \text{if} \quad u > v \\ u & \text{if} \quad u = v \\ du + (1-d)v & \text{if} \quad y < v. \end{cases} \tag{4}$$

We refer to the representation of Equation (4) [or equally, Equation (3)] as *dual bilinear*. It is easy to verify that this is invariant under affine transformations, and Theorem 11 says that it is the most general form for such invariance. The class of interval-scale representations is very limited indeed. Recall that when $c = d$, Equation (4) is the idempotent, bisymmetric case that was axiomatized in Section 6.9. That raises the question about axiomatizing the various unit structures. As we shall see, our understanding of the situation is not fully satisfactory. Following that, we explain the use of Equation (4) in the problem of the utility of gambles.

20.4.3 Characterization of Homogeneity: PCS

When we turn to the question of the conditions under which a concatenation structure is homogeneous, distinct results exist for PCSs and for solvable idempotent structures. For the PCS case, recall that in defining the standard concatenation sequence, we introduced the concept of $a(n)$ by the inductive definition: $a(1) = a$, $a(n) = a(n-1) \bigcirc a$. For n fixed, we speak of this as the *n-copy operator* defined over A.

THEOREM 14. *Suppose $\mathscr{A}$ is a closed* PCS *and $\mathscr{G}$ is its automorphism group. Then the following are true*:

(i) *If $\mathscr{A}$ is homogeneous, then each n-copy operator is an automorphism.*

(ii) *If each n-copy operator is an automorphism, then $\mathscr{G}$ is order dense.*

(iii) *If $\mathscr{A}$ is Dedekind complete and $\mathscr{G}$ is order dense, then $\mathscr{A}$ is homogeneous.*

This is due to Cohen and Narens (1979). Exercise 17 shows that one cannot simply omit Dedekind completeness in part (iii).

The following theorem answers satisfactorily in the homogeneous case one of the questions raised following Theorem 19.10 about embedding a PCS $\mathscr{A}$ into a Dedekind complete PCS **A**. Recall that we did not know if the automorphisms of $\mathscr{A}$ extend naturally to those of **A** or if those of **A** are necessarily extensions of those of $\mathscr{A}$. The following theorem concerns the first question in the homogeneous case.

THEOREM 15. *Suppose $\mathscr{A}$ is a closed* PCS *whose automorphism group is dense. Then the following are true*:

(i) *$\mathscr{A}$ is densely embeddable in a* PCS **A** *that has a unit representation, and*

(ii) *each automorphism of $\mathscr{A}$ extends to one of* **A**.

There are three versions of this result in the literature. The first, due to Cohen and Narens (1979), assumed that every n-copy operator is an automorphism, which by Theorem 14 is presumably a weaker result than Theorem 15. Narens (1981a) introduced the following concept: $\mathscr{A}$ satisfies *automorphism density* (not to be confused with $\mathscr{G}$ being order dense) if for each $a, b, c \in A$ with $a > b$, there exists $\alpha \in \mathscr{G}$ such that $a > \alpha(c) > b$. Note that if $\mathscr{A}$ satisfies automorphism density, then $\mathscr{G}$ is order dense (Exercise 18). He then claimed to show (Theorem 5.3 of 1981a) that automorphism density implies that each n-copy operation is an automorphism, whence the conclusion by Cohen and Narens' result. The proof of Theorem 5.3 appears to be faulty. Theorem 15 was first stated in Narens (1985): Theorems 13.2 (p. 153), 14.1 (p. 158), and 14.2 (p. 159). The proof we give (Section 20.5.6) is somewhat different from that of Narens.

Some of the relations among concepts are not yet understood. Among the unresolved issues for PCSs are the following. If $\mathscr{G}$ is dense, then is each n-copy operator an automorphism? If $\mathscr{G}$ is dense, then does $\mathscr{A}$ satisfy automorphism density? And if each n-copy operator is an automorphism, then is $\mathscr{A}$ homogeneous? In each case, a proof or counterexample is needed.

20.4.4 Characterizations of Homogeneity: Solvable, Idempotent Structures

For idempotent structures, we do not have nearly so satisfactory a criterion as the n-copy operator being an automorphism. At present the following results are known (Luce, 1986b).

THEOREM 16. *Suppose $\mathscr{A}$ is a concatenation structure that is closed, Dedekind complete, unique, idempotent, and solvable relative to some point. $\mathscr{A}$ is homogeneous iff $\mathscr{A}$ is solvable and all induced total concatenation structures are isomorphic.*

A somewhat more explicit result can be formulated; it is analogous to the one for PCSs.

THEOREM 17. *Suppose $\mathscr{A}$ is a concatenation structure that is closed, idempotent, solvable, and Dedekind complete. For $a, b \in A$ and n an integer, define*

$$\theta(a, b, n) = \begin{cases} a \bigcirc b & \text{if } n = 1 \\ \theta(a, b, n-1) \bigcirc b & \text{if } n > 1. \end{cases}$$

Then $\mathscr{A}$ is homogeneous iff there exists some $\tau \in \mathscr{T}$, $\tau \neq \iota$, such that for each integer n, $\theta(\tau, \iota, n)$, $\theta(\tau^{-1}, \iota, n) \in \mathscr{T}$.

To use this result, one must first find a nontrivial translation τ of $\mathscr{A}$ and then verify that, for each integer n, $\theta(\tau, \iota, n)$ is a translation. It is easy to see by Theorem 11 that if the condition is satisfied for some τ, then it is satisfied for all τ. So it does not matter which translation one begins with. Further, the condition is actually a countable set of conditions that is closely similar to showing in a PCS that each n-copy operator is an automorphism (= translation). In fact, the result for PCSs, Theorem 14, could be stated as: $\iota \bigcirc \iota$ is an automorphism, and for each n, $\theta(\iota \bigcirc \iota, \iota, n)$ is also an automorphism. The major difference between the two results is that in the idempotent case we do not have a specified automorphism with which to begin. So it is not really as satisfactory an axiomatization of homogeneity as Theorems 14 and 15.

For the (2, 2) case, however, we do have a partially satisfactory axiomatization. The difficulty with it is that the conditions are stated in terms of defined quantities that entail finding somewhat complex solutions to equations. The basic idea is this. In the dual bilinear representation, Equation (4), were we to define a new numerical operation $*$ by, for all real u and v,

$$u * v = cu + (1 - c)v,$$

then $*$ agrees with $\bigcirc$ for $u > v$, and $*$ is bisymmetric. Similarly, we can extend $\bigcirc$ for $u < v$ to $*'$ defined in terms of the parameter d. It is easy to verify that $*$ and $*'$ interlock through the following generalized bisymmetry property:

$$\begin{aligned}(x * y) *' (u * v) &= d(x * y) + (1 - d)(u * v) \\ &= dcx + d(1 - c)y + (1 - d)cu + (1 - d)(1 - c)v \\ &= (x *' y) * (u *' v).\end{aligned}$$

This observation suggests that an axiomatization might be possible if we are able to devise suitable qualitative extensions of a general concatenation operation and to impose an axiom to the effect that these extensions satisfy generalized bisymmetry. The problem is to extend an idempotent operator in an appropriate way.

DEFINITION 7. *Suppose* $\mathscr{A} = \langle A, \succsim, \bigcirc \rangle$ *is a concatenation structure that is closed and solvable.* $\mathscr{A}$ *is said to be* reflectable *iff for each* $a, b \in A$:

1. *If* $a \prec b$ *and if there exist* $u, v, w \in A$ *such that*

$$u \succ a, \quad v \succ b, \quad u \bigcirc a \succ v \bigcirc b, \quad u \bigcirc v \succ w, \quad a \prec w \prec b, \tag{5a}$$

and

$$(u \bigcirc a) \bigcirc (v \bigcirc b) = (u \bigcirc v) \bigcirc w, \tag{5b}$$

then Equation (5b) *holds with this* w *for all* $u, v \in A$ *satisfying Equation* (5a).

2. *If* $a \succ b$ *and there exist* $p, q, z \in A$ *such that*

$$p \prec a, \quad q \prec b, \quad p \bigcirc a \prec q \bigcirc b, \quad p \bigcirc q \prec z, \quad a \succ z \succ b, \tag{6a}$$

and

$$(p \bigcirc a) \bigcirc (q \bigcirc b) = (p \bigcirc q) \bigcirc z, \tag{6b}$$

then Equation (6b) *holds with this* z *for all* $p, q \in A$ *satisfying Equation* (6a).

If $\mathscr{A}$ *is idempotent and reflectable, define* $*$ *and* $*'$ *by*

$$a * b = \begin{cases} a \bigcirc b & \text{if } a \succ b, \\ a & \text{if } a = b, \\ w & \text{if } a \prec b, \end{cases}$$

where w is defined in Equations (5a, b), *and*

$$a *' b = \begin{cases} z & \text{if } a \succ b, \\ a & \text{if } a = b, \\ a \bigcirc b & \text{if } a \prec b, \end{cases}$$

where z is defined in Equations (6a, b).

In the course of proving the main theorem, we show that $*$ and $*'$ are well defined, closed, intern, monotonic, and solvable.

THEOREM 18. *Suppose* $\mathscr{A} = \langle A, \succsim, \bigcirc \rangle$ *is a totally ordered concatenation structure that is closed, idempotent, solvable, and Dedekind complete. Then* $\mathscr{A}$ *has a dual bilinear representation* [*equivalently, is of scale type* (2, 2)] *iff* $\mathscr{A}$ *is reflectable,* $*$ *and* $*'$ *are both right autodistributive, and they satisfy generalized bisymmetry: for all* $a, b, p, q \in A$,

$$(a * b) *' (p * q) = (a *' p) * (b *' q). \tag{7}$$

20.4.5 Mixture Spaces of Gambles

Let X denote a set of outcomes among which a person has preferences and $\mathscr{E}$ a set of chance events arising from an "experiment" in the sense used in statistics. We assume that X includes binary gambles generated from $x, y \in X$ and $A \in \mathscr{E}$ where the outcome is x if A occurs and y if A fails to occur. This is sometimes called a *mixture* of x and y, and we denote it by $x \bigcirc_A y$. Successive mixtures are assumed to be generated by independent experiments. Thus, for example, $(x \bigcirc_A y) \bigcirc_B z$ means that two independent experiments are run and that x is the outcome if B occurs in the first and A in the second, y if B and not A, and z if not B. Observe that B may equal A, in which case one writes $(x \bigcirc_A y) \bigcirc_A z$ and means that event A occurs in the two independent experiments. Care must be exercised in interpreting the symbol A, which has a double meaning. On one hand, it is the name of an event, and on the other hand, A in the symbol $\bigcirc_A$ denotes the equivalence class of independent realizations of the particular event also called A. Thus repeated uses of A subscripting $\bigcirc$ mean not the same but independent realizations. This is not really different from the ambiguity in the notation $(a \bigcirc a) \bigcirc a$ in concatenation structures, where it really stands for $(a_1 \bigcirc a_2) \bigcirc a_3$ and $a_1 \sim a$, $l = 1, 2, 3$.

DEFINITION 8. *Suppose* X *and* $\mathscr{E}$ *are nonempty sets,* $\succsim$ *is a binary relation on* X, *and for each* $A \in \mathscr{E}$, $\bigcirc_A$ *is a partial binary operation on* X. *If*

$\succsim$ *is a weak order, then the relational structure* $\mathscr{M} = \langle X, \succsim, \bigcirc_A \rangle_{A \in \mathscr{E}}$ *is said to be a* mixture space (*of gambles*). $\mathscr{M}$ *is said to be*:

1. Regular *iff each* $\mathscr{X}_A = \langle X, \succsim, \bigcirc_A \rangle$ *is a weakly ordered concatenation structure that is closed, idempotent, solvable, and Dedekind complete, and there is* $\Omega \in \mathscr{E}$ *such that for every* $x, y \in X$,

$$x \bigcirc_\Omega y \sim x. \tag{8}$$

2. Closed *iff there exists a function* $\chi: \mathscr{E} \times \mathscr{E} \to \mathscr{E}$ *such that for every* $A, B, C \in \mathscr{E}$ *and* $x, y \in X$, *then for* $C = \chi(A, B)$,

$$(x \bigcirc_A y) \bigcirc_B y \sim x \bigcirc_C y. \tag{9}$$

3. Restrictedly solvable in $\mathscr{E}$ *iff for each* $x, y, u, v \in X$ *and* $A, B', B'' \in \mathscr{E}$ *such that*

$$u \bigcirc_{B''} v \succsim x \bigcirc_A y \succsim u \bigcirc_{B'} v,$$

there exists $B \in \mathscr{E}$ *such that* $u \bigcirc_B v \sim x \bigcirc_A y$.

4. Commutative *iff for all* $x, y \in X$ *and* $A, B \in \mathscr{E}$,

$$(x \bigcirc_A y) \bigcirc_B y \sim (x \bigcirc_B y) \bigcirc_A y. \tag{10}$$

5. Strongly outcome independent *iff for all* $x, y, u, v \in X$, *with* $x \succ y$ *and* $u \succ v$, *and* $A, B \in \mathscr{E}$,

$$x \bigcirc_A y \succsim x \bigcirc_B y \quad \textit{iff} \quad u \bigcirc_A v \succsim u \bigcirc_B v. \tag{11}$$

6. Complemented *iff* $\mathscr{E}$ *is closed under complementation relative to* Ω *and for all* $x, y \in X$ *and* $A \in \mathscr{E}$,

$$x \bigcirc_A y \sim y \bigcirc_{\bar{A}} x. \tag{12}$$

We comment on the various properties using the numbering in the definition.

In Chapter 8 we discussed modeling preferences as a weak order, and that discussion is applicable here.

1. The closure of $\bigcirc_A$ is structural and plausible in a potential sense: given any pair of outcomes and any event, one can imagine forming a gamble from them. Idempotence is substantive and is the first of several axioms that can be classed as "rational accounting postulates." They have

the common feature that if you examine the conditions under which each outcome occurs, they are identical except for order on both sides of an equivalence. Thus $x \bigcirc_A x$ means x is the outcome if A occurs and x is also the outcome if A does not occur, which logically is just x. We assert that in addition to being logically equivalent, they are also judged to be indifferent. Finally, solvability and Dedekind completeness insure an *onto*, continuous representation of outcomes. This seems plausible if the gambles are based on money. The event Ω is simply the sure event.

2 and 3. Closure and restricted solvability of events guarantees considerable richness in the event space. Both are, of course, structural conditions.

4. Commutativity is a rationality condition since on both sides x is the outcome if both A and B occur and y otherwise, the only difference being their order.

5. Strong outcome independence basically says one can infer a probability ordering from preferences: if $x \succ y$ and $x \bigcirc_A y \succsim x \bigcirc_B y$, then one infers that A is judged at least as likely as B, and the axiom says it does not matter what x and y are used so long as $x \succ y$.

6. Complementation is another simple accounting postulate.

Classical subjective expected-utility (SEU) theory (see Chapter 8 and Fishburn, 1981a) provided axioms on a mixture space sufficient to show the existence of an interval scale U on X and an absolute scale P over $\mathscr{E}$ such that U is order preserving and

$$U(x \bigcirc_A y) = U(x)P(A) + U(y)[1 - P(A)].$$

Moreover, $\mathscr{E}$ is usually assumed to be an algebra of events, and P is finitely additive. Since numerous experiments and observations (for a summary, see Kahneman and Tversky, 1979) show SEU not to be descriptive, although it seems normatively rather compelling, the question to be raised is what generalizations of it are possible. Our strategy, as in much of this chapter, is to map out the territory of possibilities without necessarily being able to axiomatize those possibilities.

20.4.6 The Dual Bilinear Utility Model

In attempting to generalize SEU, we make two nonaxiomatized assumptions, both of which are true in SEU. First, we shall assume that each substructure $\mathscr{X}_A = \langle X, \succsim, \bigcirc_A \rangle$ is homogeneous. And second, we shall assume that there is a common utility function U on X that is order preserving and that enters into a unit representation for each $\mathscr{X}_A$. Thus, for each A in $\mathscr{E}$, there exists an f_A such that for x, y in X,

$$U(x \bigcirc_A y) = U(y) f_A\left[\frac{U(x)}{U(y)}\right]. \tag{13}$$

THEOREM 19. *Suppose $\mathscr{X}_A = \langle X, \succsim, \bigcirc_A \rangle_{A \in \mathscr{E}}$ is a mixture space satisfying conditions* 1–5 *of Definition* 8, *that each $\mathscr{X}_A$ is homogeneous, and there are unit representations as in Equation* (13) *with a common utility function U. Then*:

(i) *There exists a strictly increasing function h from* Re^+ *onto* Re *and functions S^+, S^- from $\mathscr{E}$ onto* $[0,1]$ *such that for all $A \in \mathscr{E}$ and $r \in$* Re,

(a) $h(1) = 0$ *and* $S^1(\Omega) = 1$, $1 = +, -$;

(b)

$$f_A(r) = \begin{cases} h^{-1}[h(r)S^+(A)] & \text{if } r > 1, \\ 1 & \text{if } r = 1, \\ h^{-1}[h(r)S^-(A)] & \text{if } r < 1. \end{cases} \tag{14}$$

(ii) *If, in addition, $\mathscr{M}$ is complemented, h is continuously differentiable on* Re^+ *except possibly at* 1, *and there exist positive constants μ and δ such that*

$$\lim_{x \downarrow 1} \frac{d[h(x)^{\mu}]}{dx} \quad \text{and} \quad \lim_{x \uparrow 1} \frac{d[h(x)^{\delta}]}{dx}$$

exist and are nonzero, then for all $A \in \mathscr{E}$,

$$S^+(A) + S^-(\bar{A}) = 1, \tag{15}$$

$$U(x \bigcirc_A y) = \begin{cases} U(x)S^+(A) + U(y)[1 - S^+(A)] & \text{if } U(x) > U(y) \\ U(x) & \text{if } U(x) = U(y) \\ U(x)S^-(A) + U(y)[1 - S^-(A)] & \text{if } U(x) < U(y), \end{cases} \tag{16}$$

and U is an interval scale.

This result is proved in Luce and Narens (1985) and is due, in part, to M. A. Cohen.

The significance of part (ii) is this: under the conditions of Definition 9, rather weak smoothness conditions force one into the interval-scale case, Equation (16), which is called *dual bilinear utility*. The two really substantive assumptions involved are commutativity and complementation, both of which are accounting equations in the sense that, except for order of events, the outcomes occur under exactly the same circumstances on both sides of the equation.[3] Since complementation seems highly likely to hold empirically, the only real possibilities are to accept the dual bilinear model or to reject commutativity. A very minor amount of empirical evidence plus plausibility arguments suggest that, despite the fact that Equation (10) is an accounting equation, some subjects exhibit preferences for the order of the events. In particular, if $x \succ y$ and A is more likely than B, then some prefer $(x \bigcirc_B y) \bigcirc_A y$ apparently because it increases the chance that the second gamble will in fact be run. As yet, no theory has been developed in which commutativity is violated.

Turning to the dual bilinear utility model, observe that it differs from SEU in two respects. First, S^+ need not be equal to S^-. Second, neither S^+ nor S^- need be finitely additive. These two changes appear to be adequate to bypass the difficulties with SEU; however, not enough time has passed for penetrating critiques of the dual bilinear model to have been mounted.

It is interesting to investigate which of the properties postulated are entailed by Equations (15) and (16), and then what properties drive the dual bilinear model toward SEU.

THEOREM 20. *Suppose Equation* (16) *holds on a mixture space, then $\mathscr{M}$ is regular, commutative, and strongly outcome independent. If, in addition, $\mathscr{E}$ is closed under complementation and Equation* (15) *holds, then it is complemented.*

We are ready now to introduce some properties not unlike those already given in Definition 8, and so we keep the same numbering.

DEFINITION 8 (continued). *Suppose $\mathscr{M} = \langle X, \succsim, \bigcirc_A \rangle_{A \in \mathscr{E}}$ is a mixture space. Then $\mathscr{M}$ is said to be:*

7. Self distributive *iff $x \bigcirc_A w \sim u \bigcirc_A z$ and $y \bigcirc_A w \sim v \bigcirc_A z$ imply*

$$(x \bigcirc_A y) \bigcirc_A w \sim (u \bigcirc_A v) \bigcirc_A z.$$

8. Right autodistributive *iff*

$$(x \bigcirc_A y) \bigcirc_A z \sim (x \bigcirc_A z) \bigcirc_A (y \bigcirc_A z).$$

[3] Much the same notion has been discussed under other names, e.g., extensionality (Arrow, 1982) and invariance (Tversky and Kahneman, 1986).

9. Bisymmetric *iff*

$$(x \bigcirc_A y) \bigcirc_A (y \bigcirc_A v) \sim (x \bigcirc_A u) \bigcirc_A (y \bigcirc_A v).$$

10. Monotonic in events *iff for all* $A, B, C \in \mathscr{E}$ *such that* $A \cup C$, $B \cup C \in \mathscr{E}$, *and* $A \cap C = B \cap C = \varnothing$, *then*

$$x \bigcirc_A y \succsim x \bigcirc_B y \quad \text{iff} \quad x \bigcirc_{A \cup C} y \succsim x \bigcirc_{B \cup C} y.$$

11. Consistent with event inclusion *iff when* $B \subseteq A$,

$$x \succsim y \quad \text{iff} \quad x \bigcirc_A y \succsim x \bigcirc_B y.$$

Properties 7–9 are each an accounting property in the sense, defined above, that bookkeeping on each side yields the same result. They are a bit more complicated than the earlier ones in that they involve three or more outcomes, not just two. This makes them more difficult to see through at an intuitive level. Monotonicity in events and consistency with event inclusion are somewhat more subtle rationality conditions, much in the spirit of strong outcome independence. None of these is entailed by Equations (15) and (16). Indeed,

THEOREM 21. *Under the conditions of Theorem* 19(ii),

(i) *the following are equivalent*:
  (a) $S^+ = S^-$,
  (b) *self distribution*,
  (c) *right autodistribution*,
  (d) *bisymmetry*.

(ii) *For* A, B, C, $A \cup C$, $B \cup C \in \mathscr{E}$, $A \cap C = B \cap C = \varnothing$, *monotonicity of events holds iff for* $i = +, -$,

$$S^i(A) \geqslant S^i(B) \quad \text{iff} \quad S^i(A \cup C) \geqslant S^i(B \cup C).$$

(iii) *Consistency with event inclusion holds iff*

$$A \subseteq B \quad \text{implies} \quad S^i(A) \geqslant S^i(B), \qquad i = +, -.$$

Theorems 20 and 21 are adaptations of results in Luce and Narens (1985). The proofs are left as Exercises 20 and 21. Note that the equivalence of parts (b), (c), (d) of Theorem 21(i) is established in Theorem 24, Chapter 19. (Proved in Section 20.5.3.)

It is to be noted that a particular class of empirical failures of monotonicity of events is known as the Ellsberg paradox. Several other empirical failures of properties that are entailed by SEU and not by dual bilinear utility are the reflection effect and the isolation effect (Kahneman and Tversky, 1979). The reflection effect is that $x \bigcirc_A 0 \succ y \bigcirc_B 0$ does not entail $-x \bigcirc_A 0 \prec -y \bigcirc_B 0$. In the dual bilinear model, this can arise as much from $S^+ \neq S^-$ as from special properties of U. The isolation effect is based on violations of $x \bigcirc_A 0 \succsim y \bigcirc_B 0$ iff $x \bigcirc_C 0 \succsim y$, where C is such that $P(C)P(B) = P(A)$. In the dual bilinear model, the equivalence holds if $A = \chi(C, B)$, where χ is given in part 2 of Definition 9, and $S^i(A) = S^i(B)S^i(C)$. Since the weights S^i are not necessarily probabilities, violations may occur.

The dual bilinear model, with its weighted-average form and the dependence of the weights on the ranking of outcomes, has a close relation to models that have arisen during the 1980s in economics. The major papers are Chew (1983), Gilboa (1987), Quiggin (1982), Segal (1987), and Yaari (1987). In all of this work, axiom systems lead to weighted-average expressions for utility, with the weights dependent upon the event involved and the preference ordering of the outcomes. The major differences between these models and the dual bilinear model is the number of alternatives and a crucial assumption about the domains. In the economic work, any number of alternatives are possible provided they can be thought of as random variables. In the present model, of course, only two alternatives are admitted. However, the use of random variables precludes, it turns out, any effective treatment of the ubiquitous framing effects whereas the formulation as mixtures does not force the kinds of accounting equations that deny framing effects. The extension of the dual bilinear model to gambles with any finite number of outcomes has been worked out in Luce (1988).

20.5 PROOFS

20.5.1 Theorem 9 (p. 142)

Suppose $\mathscr{A}$ is a concatenation structure that is homogeneous. Then,

(i) $\bigcirc$ *is an operation;*

(ii) $\bigcirc$ *is either idempotent, weakly positive, or weakly negative;*

(iii) *in the idempotent case,* $\bigcirc$ *is intern and* $\mathscr{A}$ *is dense; in the weakly positive or negative case,* $\mathscr{A}$ *is unbounded;*

(iv) *if* $\bigcirc$ *is weakly positive or negative, then* $M = 1$; *if* $\bigcirc$ *is idempotent, then* $M \leqslant 2$;

(v) *if* $\mathscr{A}$ *is unique, then* $N \leqslant 2$; *and when* $N = 2$, *then* $\bigcirc$ *is idempotent.*

PROOF. With no loss of generality, assume $\mathscr{A}$ is totally ordered.

(i) Let $\mathscr{G}$ be the automorphism group of A. Since there exist u, v such that $u \bigcirc v$ is defined, by local definability $w \bigcirc w$ is defined for $w = \min(u, v)$. Consider any a, b in A, and let $c = \max(a, b)$. By homogeneity, there exists $\alpha \in \mathscr{G}$ such that $\alpha(w) = c$, and so $c \bigcirc c$ is defined. By local definability, $a \bigcirc b$ is defined, proving $\bigcirc$ is an operation.

(ii) Suppose for some a, $a \bigcirc a = a$. Then for each b, there exists $\alpha \in \mathscr{G}$ such that $\alpha(a) = b$. By monotonicity,

$$b \bigcirc b = \alpha(a) \bigcirc \alpha(a) = \alpha(a \bigcirc a) = \alpha(a) = b,$$

proving that $\bigcirc$ is idempotent. The proofs in the other two cases are similar.

(iii) If $\bigcirc$ is idempotent, then for $a \succ b$, $a = a \bigcirc a \succ a \bigcirc b$, $b \bigcirc a \succ b \bigcirc b = b$, proving that $\bigcirc$ is both intern and dense. If $\bigcirc$ is weakly positive and u were a bound, then by part (i) $u \bigcirc u \succ u$ exists, contradicting that u is an upper bound. The weakly negative case is similar.

(iv) Suppose $\mathscr{A}$ is weakly positive and $M \geqslant 2$. By part (iii), the structure is unbounded, and so by Theorem 1(i), $\mathscr{A}$ is 2-point homogeneous. Select any a, b, c in $\mathscr{A}$ such that $b \prec c \neq b \bigcirc b$. Since $a \prec a \bigcirc a$, there exists an automorphism α with $\alpha(a) = b$ and $\alpha(a \bigcirc a) = c$. Thus,

$$\alpha(a \bigcirc a) = c \neq b \bigcirc b = \alpha(a) \bigcirc \alpha(a) = \alpha(a \bigcirc a),$$

which is impossible. So $M = 1$. The weakly negative case is similar.

Suppose $\mathscr{A}$ is idempotent and $M \geqslant 3$. By part (iii) and Theorem 1(i), $\mathscr{A}$ is 3-point homogeneous. Select a, b, c, d, and e such that $a \prec b$ and $c \prec e \prec d$, $e \neq c \bigcirc d$. Since $a \prec a \bigcirc b \prec b$, there exists an automorphism α such that $\alpha(a) = c$, $\alpha(b) = d$, and $\alpha(a \bigcirc b) = e$. But

$$\alpha(a \bigcirc b) = e \neq c \bigcirc d = \alpha(a) \bigcirc \alpha(b) = \alpha(a \bigcirc b),$$

which is impossible. So $M \leqslant 2$.

(v) Suppose $\mathscr{A}$ is N-point unique. Since $\mathscr{A}$ is 1-point homogeneous and either dense or unbounded, by Theorem 1(iii), $N \geqslant 1$. Suppose $\mathscr{A}$ is not idempotent, then by part (ii) it is either weakly positive or negative.

Without loss of generality, suppose the former. Note that by weak positivity,

$$\begin{aligned} a &\prec a \bigcirc a \\ &\prec (a \bigcirc a) \bigcirc (a \bigcirc a) \\ &\prec [(a \bigcirc a) \bigcirc (a \bigcirc a)] \bigcirc [(a \bigcirc a) \bigcirc (a \bigcirc a)] \\ &\vdots \end{aligned}$$

Now, suppose $\alpha, \beta \in \mathscr{G}$ and agree at a. Then, by induction on the fact that

$$\alpha(a \bigcirc a) = \alpha(a) \bigcirc \alpha(a) = \beta(a) \bigcirc \beta(a) = \beta(a \bigcirc a),$$

they agree at all points in the above sequence of inequalities. Thus, by N-point uniqueness, $\alpha \equiv \beta$, and so $N = 1$.

Next, suppose $\mathscr{A}$ is idempotent and so is intern. Let $a, b \in A$ and $\alpha, \beta \in \mathscr{G}$ be such that $a \prec b$, $\alpha(a) = \beta(a)$, and $\alpha(b) = \beta(b)$. Since $a \prec a \bigcirc b \prec b$ and

$$\alpha(a \bigcirc b) = \alpha(a) \bigcirc \alpha(b) = \beta(a) \bigcirc \beta(b) = \beta(a \bigcirc b),$$

by induction it follows that α and β agree at N distinct points and so are identical, thus proving $N \leqslant 2$. ◇

20.5.2 Theorem 11 (p. 144)

A real concatenation structure $\mathscr{R} = \langle \mathrm{Re}^+, \geqslant, \bigcirc \rangle$ *is unique and homogeneous iff* $\mathscr{R}$ *is isomorphic to a real unit structure* $\langle \mathrm{Re}^+, \geqslant, \bigcirc', f \rangle$. *Moreover,*

(i) $\mathscr{R}$ *is of scale type* $(1,1)$ *iff*

$$f(x^\rho) = f(x)^\rho, \qquad x > 0,$$

holds only for $\rho = 1$;

(ii) $\mathscr{R}$ *is of scale type* $(2,2)$ *iff Equation* (1) *holds for all* $\rho > 0$;

(iii) $\mathscr{R}$ *is of scale type* $(1,2)$ *iff there exists some* $k > 0$ *such that Equation* (1) *holds just for* $\rho = k^n$, *n integer.*

If $\langle \varphi, f \rangle$ *and* $\langle \psi, g \rangle$ *are two unit representations, then for some* $s > 0$ *and* $t > 0$, *and all* $r > 0$, $\varphi = s\psi^{1/t}$ *and* $f(r) = g(r^t)^{1/t}$.

PROOF. Since $\mathscr{R}$ is unique and homogeneous, by Theorem 5 $\mathscr{R}$ is isomorphic to a real structure with the similarity group included in its automorphism group; and so we can without loss of generality assume $\mathscr{R}$ is that structure. Thus, by Theorem 7, $\mathscr{R}$ is a homogeneous, real unit structure that, by Theorem 10, is of the form $\langle \mathrm{Re}^+, \geqslant, \bigcirc, f \rangle$.

(i) Now suppose that $\mathscr{R}$ is of scale type (1, 1) and that, for some $\rho \neq 1$, $f(x^\rho) = f(x)^\rho$. We show that this implies that $\alpha(x) = x^\rho$ is an automorphism, which is impossible since it is not a similarity. Obviously, α is increasing and onto, and

$$\begin{aligned}\alpha(x \bigcirc y) &= y^\rho f\left(\frac{x}{y}\right)^\rho \\ &= y^\rho f\left(\frac{x^\rho}{y^\rho}\right) \\ &= \alpha(y) f\left[\frac{\alpha(x)}{\alpha(y)}\right] \\ &= \alpha(x) \bigcirc \alpha(y),\end{aligned}$$

and so α is indeed an automorphism, which is impossible. So $\rho \neq 1$ cannot exist.

Conversely, suppose $\rho = 1$ is the only solution. Since a unit structure operation is continuous, by Luce and Narens (1985), the structure is 2-point unique. Thus, by Theorem 5(vi), $\mathscr{G}$ is a subgroup of the affine group. If $\alpha(x) = \sigma x^\rho$ is in $\mathscr{G}$, then by the above argument $f(x^\rho) = f(x)^\rho$, and so $\rho = 1$, proving 1-point uniqueness. Thus the unit structure is of scale type (1, 1).

(ii) Suppose $\mathscr{R}$ is of scale type (2, 2), and so, by Theorem 5, the automorphism group is the affine group. That is, for $r > 0$ and $\rho > 0$,

$$r(x \bigcirc y)^\rho = rx^\rho \bigcirc ry^\rho.$$

Substituting the unit representation yields

$$r\left[yf\left(\frac{x}{y}\right)\right]^\rho = ry^\rho f\left[\left(\frac{x}{y}\right)^\rho\right],$$

whence Equation (1) for all $\rho > 0$. The converse retraces the steps.

(iii) Suppose R is of scale type (1, 2). By Theorem 5(vi), the automorphism group $\mathscr{G}$ of this representation is a subgroup of the affine group and includes the multiplicative positive reals. Thus they are of the form σx^ρ, where $\sigma > 0$ and ρ is in a subgroup $\mathscr{H}$ of the positive reals. $\mathscr{H}$ cannot be the identity since, by part (i), that would mean $\mathscr{R}$ is of type (1, 1), not (1, 2).

If $\mathscr{H}$ is dense in the reals, then by the continuity of f (it is *onto* and strictly increasing) it would follow that Equation (1) is true for all $\rho > 0$, in which case by part (ii) $\mathscr{R}$ would be (2, 2), not (1, 2). So $\mathscr{H}$ must be discrete, in which case it is generated by some $k > 0$, and so the elements of $\mathscr{H}$ are of the form k^n, n an integer.

Again, the converse is immediate.

The uniqueness is left as Exercise 15. 

20.5.3 Theorem 24, Chapter 19 (p. 103)

Suppose $\mathscr{A} = \langle A, \succsim, \bigcirc \rangle$ is a closed concatenation structure that is idempotent, solvable, and Dedekind complete. Then the following are equivalent:

(i) *$\mathscr{A}$ is bisymmetric,*

(ii) *$\mathscr{A}$ is right autodistributive,*

(iii) *$\mathscr{A}$ is self distributive.*

The proof will require an intermediate result that is of no great interest in and of itself but because it is self-contained we state it formally as:

LEMMA. *Suppose A is a nonempty set, $\succsim$ is a binary relation on A, $\bigcirc$ is a binary operation on A, and $\succsim'$ and $\succsim''$ are binary relations on $A \times A$ that satisfy the following conditions:*

(i) *$\mathscr{A} = \langle A, \succsim, \bigcirc \rangle$ is a closed, totally ordered, Dedekind complete, concatenation structure that is homogeneous and unique.*

(ii) *$\mathscr{C}' = \langle A \times A, \succsim' \rangle$ and $\mathscr{C}'' = \langle A \times A, \succsim'' \rangle$ are both unrestrictedly solvable conjoint structures.*

(iii) *$\mathscr{C}'$ is Archimedean and satisfies the Thomsen condition.*

(iv) *The induced orderings of $\succsim'$ and $\succsim''$ are both equal to $\succsim$.*

(v) *The operation $\bigcirc$ is distributive in both $\mathscr{C}'$ and $\mathscr{C}''$.*

Let $*'$ and $*''$ denote the operations induced by $\succsim'$ and $\succsim''$, respectively, relative to a_0 in A. Then $*' = *''$, and so $\mathscr{C}''$ also satisfies the Thomsen condition and is Archimedean.

PROOF OF LEMMA. Let $\mathscr{G}$ denote the group of automorphisms of $\mathscr{A}$, and let $\mathscr{T}'$ and $\mathscr{T}''$ denote the sets of right multiplications of $*'$ and $*''$, respectively. First, we show that $\mathscr{T}'$ and $\mathscr{T}''$ are 1-point homogeneous. Let $a, b \in A$. Then, by the solvability of $\mathscr{C}'$, there exists p such that $bp_0 \sim ap \sim *'\pi^{-1}(p)p_0$, whence by independence $a *' \pi^{-1}(p) = b$. The proof for $\mathscr{T}''$ is similar. Next, we show that $\mathscr{T}'$ and $\mathscr{T}''$ are 1-point unique. Suppose that $c *' a = c *' b$, then by the monotonicity of $*'$, $a = b$, proving $\mathscr{T}'$ is 1-point unique. The proof for $\mathscr{T}''$ is similar. By Theorem 19.18, both $\mathscr{T}'$ and $\mathscr{T}''$ are subsets of $\mathscr{G}$.

Next, we establish that $\mathscr{T}'$ is a group. From assumption (iii) and Corollary 1 to Theorem 19.11, $*'$ is associative. Thus, if $a^R, b^R \in \mathscr{T}'$,

$$a^R b^R(c) = (c *'b) *'a = c *'(b *'a) = (b *'a)^R(c).$$

By 1-point uniqueness, $a^R b^R = (b *'a)^R$. By 1-point homogeneity, for τ_a there is some d such that $d^R a^R(c) = c = \iota(c)$, and so by 1-point uniqueness $d^R = (a^R)^{-1}$.

Since $\mathscr{A}$ is homogeneous and unique, by Theorem 5(vi), $\mathscr{G}$ is isomorphic to a subgroup of the affine group. Since the elements of $\mathscr{T}'$ and $\mathscr{T}''$ are not dilations, both must be isomorphic to the subgroup of translations of the affine group. And since $\mathscr{T}'$ is a (1,1) group, it is isomorphic to the translations, and so $\mathscr{T}''$ is included in $\mathscr{T}'$. Thus, for each a in A and $(b^R)''$ in $\mathscr{T}''$ there exists a c in A such that $(c^R)'$ is in $\mathscr{T}'$ and $(c^R)' = (b^R)''$. Thus,

$$c = a_0 *'c = (c^R)'(a_0) = (b^R)''(a_0) = c_0 *''b = b,$$

and so for all $a, b \in A$, $a *'b = a *''b$. Since $\mathscr{C}'$ is Archimedean, so is the induced structure $*'$, and therefore so is $*''$, whence $\mathscr{C}''$ is also Archimedean. By Corollary 1 to Theorem 11, Chapter 19, $\mathscr{C}''$ satisfies the Thomsen condition because $*'' = *'$ is associative. ◇

PROOF OF THEOREM. By Theorem 19.23(iv) and (v), (i) → (ii) → (iii). So we show that (iii) → (i). By Theorem 19.21, $\mathscr{A}$ is at most 2-point unique and has a numerical representation. Let $\mathscr{C}'' = \langle A \times A, \succsim'' \rangle$ be the conjoint structure induced by $\bigcirc$. Since $\bigcirc$ is self-distributive, it is distributive in $\mathscr{C}''$. So by Theorem 19.18, the set $\mathscr{T}$ of right multiplications is included in the automorphism group $\mathscr{G}$ of $\mathscr{A}$. We show that $\mathscr{T}$, and so $\mathscr{G}$, is 1-point homogeneous. Suppose $a, b \in A$. Since $\mathscr{A}$ is solvable, there exists p such that $ba \sim ap \sim a *_{\pi^{-1}(p)} a$, whence $\tau_{\pi^{-1}(p)}(a) = b$. Thus, by Theorem 10, $\mathscr{A}$ has a unit representation $\langle \varphi, f \rangle$, and so by Theorem 7 there exists a conjoint structure $\mathscr{C}' = \langle A \times A, \succsim' \rangle$ fulfilling the properties of $\mathscr{C}'$ in the lemma. $\mathscr{C}''$ does also, and $\succsim$ is the order induced on A by both $\succsim'$ and $\succsim''$. So, by the lemma, $\mathscr{C}''$ satisfies the Thomsen condition. Thus there exist ψ_1, ψ_2 such that $\psi_1\psi_2$ represents $\succsim$. Note that φf represents $\succsim'$ because $\langle \varphi, f \rangle$ is a unit representation.

We show that it is possible to select $\psi_1 = \varphi$. Because $\bigcirc$ is distributive in $\mathscr{C}''$, we have that

$$\frac{\psi_1(x)}{\psi_1(y)} = \frac{\psi_1(u)}{\psi_1(v)}$$

implies

$$\frac{\psi_1(x \bigcirc y)}{\psi_1(y)} = \frac{\psi_1(u \bigcirc v)}{\psi_1(v)},$$

and so, as in Theorem 10, we can define a function g such that ψ_1 and g form a unit representation. By the uniqueness of unit representations (Theorem 10), for some $\rho, \sigma > 0$, $\psi_1 = \sigma\varphi^\rho$. So, setting $\psi = \psi_2^{1/\rho}$, we see that $\varphi\psi$ represents $\succsim''$. Moreover, by considering $u \bigcirc a \succsim u \bigcirc b$, we see that there is a strictly increasing function h such that $\psi = h(\varphi)$. Thus we have two representations of $\bigcirc$,

$$\begin{aligned} b \succsim p \bigcirc q \quad & \text{iff} \quad \varphi(a)h[\varphi(b)] \geqslant \varphi(p)h[\varphi(q)] \\ & \text{iff} \quad \varphi(b)f\left[\frac{\varphi(a)}{\varphi(b)}\right] \geqslant \varphi(q)f\left[\frac{\varphi(p)}{\varphi(q)}\right], \end{aligned}$$

and so, for some strictly increasing F,

$$F(\varphi(a)h[\varphi(b)]) = \varphi(b)f\left[\frac{\varphi(a)}{\varphi(b)}\right]. \tag{17}$$

If we select $b = e$ such that $h[\varphi(e)] = 1$, and set $E = \varphi(e)$, then

$$F[\varphi(a)] = Ef\left[\frac{\varphi(a)}{E}\right]. \tag{18}$$

Now, substituting Equation (18) into Equation (17) and setting $W = \varphi(a)/E$, $Y = \varphi(b)/E$, and $H(c) = h(cE)$, we obtain

$$f[WH(Y)] = Yf\left(\frac{W}{Y}\right). \tag{19}$$

Observe that by setting $W = Y$ in Equation (19) and noting that $f(1) = 1$ because the structure is idempotent,

$$H(W) = \frac{f^{-1}(W)}{W}. \tag{20}$$

If we set $U = W/Y$ and $V = f^{-1}(Y)$ and substitute Equation (20) into

Equation (19), then f is characterized by

$$f(UV) = f(U)f(V),$$

which, since f is strictly increasing, is well known to have as its unique solution,

$$f(U) = U^c, \qquad c > 0.$$

Since f is a unit representation, f/ι is decreasing, and so $c < 1$. This defines the (2, 2) bilinear representation with $c = d$, and so the structure is bisymmetric. ◇

20.5.4 Theorem 14 (p. 147)

In a PCS $\mathscr{A}$, *the following are true.*

(i) *Homogeneity of* $\mathscr{A}$ *implies every n-copy operator is an automorphism.*

(ii) *The latter implies that the automorphism group* $\mathscr{G}$ *is dense.*

(iii) $\mathscr{A}$ *is Dedekind complete and* $\mathscr{G}$ *is dense implies* $\mathscr{A}$ *is homogeneous.*

PROOF.

(i) Let θ_n denote the n-copy operator. $\theta_1 = \iota$, and so is in $\mathscr{G}$. For any integer $m > 1$ and a, b in $\mathscr{A}$, by 1-point homogeneity there exist automorphisms λ and η such that $\lambda(a) = \theta_{m-1}(a) \bigcirc a$ and $\eta(a) = b$. By the commutativity of automorphisms of a PCS (Theorem 19),

$$\begin{aligned}
\lambda(b) &= \lambda\eta(a) \\
&= \eta\lambda(a) \\
&= \eta[\theta_{m-1}(a) \bigcirc a] \\
&= \eta\theta_{m-1}(a) \bigcirc \eta(a) \\
&= \theta_{m-1}\eta(a) \bigcirc b \\
&= \theta_{m-1}(b) \bigcirc b \\
&= \theta_m(b),
\end{aligned}$$

so $\theta_m \equiv \lambda$ is an automorphism.

(ii) We know $\mathscr{G}$ is nontrivial since by hypothesis θ_n is in $\mathscr{G}$. Suppose, therefore, it is discrete, and let η be the smallest positive automorphism. Thus, for each positive integer n, there exists an integer $m(n)$ such that $\eta^{m(n)} = \theta_n$. Since $\theta_{n+1}(a) = a(n+1) \succ a(n) = \theta_n(a)$, it follows that $\eta^{m(n+1)} \succ \eta^{m(n)}$. Since η is positive and minimal, $\eta^{m(n+1)} \succsim \eta\eta^{m(n)} = \eta^{m(n)+1}$. Since $\mathscr{A}$ is Archimedean, $\eta^{m(n)}(a)$ can be made arbitrarily large, and so by restricted solvability there exists some m such that $\eta(a) \succ$

$a \bigcirc \eta^{-m(n)}(a)$. Thus,

$$\begin{aligned}\eta^{m(n)+1}(a) &= \eta^{m(n)}[\eta(a)] \\ &\succ \eta^{m(n)}[a \bigcirc \eta^{-m(n)}(a)] \\ &= \eta^{m(n)}(a) \bigcirc a \\ &= \theta_n(a) \bigcirc a \\ &= \theta_{n+1}(a) \\ &= \eta^{m(n+1)}(a) \\ &\succsim \eta^{m(n)+1}(a),\end{aligned}$$

which is a contradiction.

(iii) We first show that $\mathscr{G}$ is Dedekind complete. Let Γ be a nonempty set of automorphisms bounded from above by the automorphism ξ. For each a in A, $\{\alpha(a) \mid \alpha \in \Gamma\}$ is a nonempty subset of A bounded by $\xi(a)$. Since $\mathscr{A}$ is Dedekind complete, $\eta(a) = l.u.b.\{\gamma(a) \mid \gamma \in \Gamma\}$ is defined. We prove η is an automorphism.

First, suppose $a, b \in A$. We prove $\eta(a \bigcirc b) = \eta(a) \bigcirc \eta(b)$. Suppose not. If $\eta(a \bigcirc b) \succ \eta(a) \bigcirc \eta(b)$, then because $\mathscr{A}$ is a Dedekind complete PCS, and $\mathscr{G}$ is dense (see Exercise 18), there is some θ in $\mathscr{G}$ such that $\eta(a \bigcirc b) \succ \theta(a \bigcirc b) \succ \eta(a) \bigcirc \eta(b)$. For any $\gamma \in \Gamma$, $\eta(a) \succsim \gamma(a)$, $\eta(b) \succ \gamma(b)$, and so

$$\eta(a) \bigcirc \eta(b) \succsim \gamma(a) \bigcirc \gamma(b) = \gamma(a \bigcirc b).$$

Thus $\theta \succ \gamma$, and so $\theta \succsim \eta$, contrary to choice. Next, suppose $\eta(a \bigcirc b) \prec \eta(a) \bigcirc \eta(b)$. As above, there is a θ in $\mathscr{G}$ such that

$$\eta(a \bigcirc b) \prec \theta(a \bigcirc b) \prec \eta(a) \bigcirc \eta(b).$$

But $\eta(a \bigcirc b) \succsim \gamma(a \bigcirc b)$ for $\gamma \in \Gamma$, and so $\theta \succ \gamma$. Therefore, since η is an upper bound, $\theta(a \bigcirc b) = \theta(b) \bigcirc \theta(b) \succ \eta(a) \bigcirc \eta(b)$. This is a contradiction; thus $\eta(a \bigcirc b) = \eta(a) \bigcirc \eta(b)$.

Next, we must show that $a \succsim b$ iff $\eta(a) \succsim \eta(b)$. It suffices to assume $a \succ b$ and to prove $\eta(a) \succ \eta(b)$. By the construction used in Theorem 2, there exist c in A and a positive integer m such that $a \succ c(m+1) \succ c(m) \succ b$. From the definition of η and the property just shown,

$$\eta(a) \succsim \eta[c(m+1)] = (m+1)\eta(c) \succ m\eta(c) = \eta(c(m)) \succsim \eta(b),$$

which was to be shown.

We now show that η is *onto*. Let a be any element of $\mathscr{A}$, and consider $B = \{\gamma^{-1}(a) \mid \gamma \in \Gamma\}$. Since B is bounded from below by $\eta^{-1}(a)$, and since $\mathscr{A}$ is Dedekind complete, there is a g.l.b. b. From $\eta^{-1}(a) \succsim b$, we have $a \succsim \eta(b)$. Suppose $a \succ \eta(b)$, then there exists $\theta \in \mathscr{A}$ such that $a \succ \theta[\eta(b)] \succsim \theta[\gamma(b)]$ since $\eta \succsim' \gamma$. But composition of automorphisms of a PCS is commutative; so $a \succ \gamma[\theta(b)]$, whence $\gamma^{-2}(a) \succ \theta(b) \succ b$, contradicting that b is g.l.b. of B. So $a = \eta(b)$, and η is onto A.

Last, we show that η is a l.u.b. of Γ. If θ is another bound, then for each a in A and γ in Γ, $\theta(a) \succsim \gamma(a)$, and so $\theta(a) \succsim \eta(a)$, whence $\theta \succsim' \eta$. Now, for any $a, b \in A$, define

$$B = \{\xi \mid \xi \in \mathscr{G} \,\&\, \xi(a) \precsim b\}$$
$$C = \{\xi \mid \xi \in \mathscr{G} \,\&\, \xi(a) \succ b\}.$$

Because $\mathscr{G}$ is dense, B and $\mathscr{C}$ are nonempty. Since $\mathscr{G}$ is Dedekind complete, let η be the cut element and suppose $\eta(a) \neq b$. If $\eta(a) \prec b$, then let γ in $\mathscr{G}$ be such that $\eta(a) \prec \gamma(\eta(a)) \prec b$. Since $\gamma(\eta)$ is in B and $\eta \prec \gamma(\eta)$, this contradicts that η is the cut element. A similar argument shows that $\eta(a) \succ b$ is impossible. So $\eta(a) = b$. ◇

20.5.5 Theorem 15 (p. 147)

Suppose $\mathscr{A}$ *is a closed* PCS *for which the automorphism group is dense. Then,*

(i) $\mathscr{A}$ *is densely embeddable in a* PCS $\mathscr{A}$ *having a unit representation, and*

(ii) *each automorphism of* $\mathscr{A}$ *extends to one of* $\mathscr{A}$.

PROOF. We use the density of $\mathscr{G}$ to establish that $\mathscr{A}$ satisfies upper and lower semicontinuity. Toward that end, we show that if $a \succ b$, there exists $\alpha, \beta \in \mathscr{G}$ such that $a \succ \alpha(a)$, $\beta(b) \succ b$. We show the latter first. Suppose on the contrary that, for all $\beta \succ' \iota$, $\beta(b) \succ a$. By restricted solvability, let c be such that $a \succ b \bigcirc c$. Consider any $\gamma \succ' \iota$, and we show for each n, $\gamma^n(b) \succ c(n)$. For $n = 1$, $\gamma(b) \succsim a \succ b \bigcirc c \succ c$. By the induction hypothesis and $\gamma(b) \succsim a \succ b \bigcirc c$, we have

$$\begin{aligned}\gamma^n(b) &= \gamma^{n-1}\gamma(b) \succ \gamma^{n-1}(b \bigcirc c)\\ &= \gamma^{n-1}(b) \bigcirc \gamma^{n-1}(c) \succ c(n-1) \bigcirc c = c(n).\end{aligned}$$

Since $\mathscr{A}$ is Archimedean, there exists an m such that $c(m) \succ \beta(b)$. Since $\mathscr{G}$

is dense, there exists a $\gamma \succ' \iota$ with $\beta \succ' \gamma^m$. So

$$\beta(b) \succ \gamma^m(b) \succ c(m) \succ \beta(b),$$

which is a contradiction. Let $\alpha = \beta^{-1}$. Since $\beta(b) \succ b$, it follows that $b = \alpha\beta(b) \succ \alpha(b)$, and so $\alpha \prec' \iota$. Thus $a \succ \alpha(a) \succ \alpha\beta(b) = b$. To show lower semicontinuity, suppose $a \bigcirc b \succ c$, and let α in $\mathscr{G}$ be such that $a \bigcirc b \succ \alpha(a \bigcirc b) = \alpha(a) \bigcirc \alpha(b) \succ c$. Since $\alpha \prec' \iota$, $a \succ \alpha(a)$, $b \succ \alpha(b)$, and lower semicontinuity holds. The proof for upper semicontinuity is similar.

By Theorems 19.4 and 19.6, there is no loss of generality in supposing that $\mathscr{A}$ is real and its automorphism group is multiplication by numbers from a dense subgroup $\mathscr{R}$ of Re^+. By Corollary 1 of Theorem 19.9 and Exercise 18 of Chapter 19, $\mathscr{A}$ can be densely embedded in a Dedekind-complete real $\mathscr{A}$ for which both upper and lower semicontinuity hold. Thus, by Theorem 19.6, $\bigcirc$ is continuous in each variable. Let A^* be the image of A, which is dense in Re^+. For any r, x, y in Re^+, select increasing sequences r_i, x_i, and y_i such that $r_i \in \mathscr{R}$, x_i, $y_i \in A^*$ and they converge respectively to r, x, and y. Since by choice of the representation of $\mathscr{A}$, $r_i(x_i \bigcirc y_i) = r_i x_i \bigcirc r_i y_i$ and, by the continuity of $\bigcirc$, $x_i \bigcirc y_i$ converges to $x \bigcirc y$, $r_i(x_i \bigcirc y_i)$ to $r(x \bigcirc y)$, $r_i x_i$ to rx, and $r_i y_i$ to ry, it follows that $r(x \bigcirc y) = rx \bigcirc ry$. This establishes both (i) that $\mathscr{A}$ is homogeneous and therefore a unit structure and (ii) that the automorphisms of $\mathscr{A}$ extend to ones of $\mathscr{A}$. ◇

20.5.6 Theorem 16 (p. 148)

Suppose $\mathscr{A}$ is a concatenation structure that is closed, idempotent, Dedekind complete, unique, and solvable relative to some $a \in \mathscr{A}$. Then $\mathscr{A}$ is homogeneous iff $\mathscr{A}$ is solvable relative to all $a \in A$, and all induced total concatenation structures are isomorphic.

PROOF. Suppose, first, that $\mathscr{A}$ is homogeneous, and let b be in A. Let α be an automorphism such that $\alpha(a) = b$. For any $y \in A$, let x solve $x \bigcirc a = \alpha^{-1}(y)$, then

$$y = \alpha\alpha^{-1}(y) = \alpha(x \bigcirc a) = \alpha(x) \bigcirc \alpha(b) = \alpha(x) \bigcirc b,$$

in which case $\alpha(x)$ is the solution relative to b. By Theorem 5(iii), the translations of A are homogeneous, and so by Theorem 19.12 all of the induced total concatenation structures are isomorphic.

The converse is immediate by Theorem 19.12. ◇

20.5.7 Theorem 17 (p. 148)

Suppose $\mathscr{A}$ is a closed, idempotent, solvable, Dedekind complete concatenation structure. Then $\mathscr{A}$ is homogeneous iff for some $\tau \in \mathscr{T}$, $\tau \neq \iota$, and for each integer n, $\theta(\tau, \iota, n) \in \mathscr{T}$ and $\theta(\tau^{-1}, \iota, n) \in \mathscr{T}$.

PROOF OF NECESSITY. By Theorems 19.1 and 19.21, $\mathscr{A}$ is isomorphic to a real representation, and by Dedekind completeness it is onto Re^+. By Theorem 19.21, it is 2-point unique. Since $\mathscr{A}$ is homogeneous, Theorem 10 implies it has a unit representation $\mathscr{R} = \langle \mathrm{Re}^+, \geqslant, \bigcirc, f \rangle$. Note that $\mathscr{T}$ is a group. With no loss of generality, we assume $\mathscr{A} = \mathscr{R}$. Suppose $\tau \in \mathscr{T}$, $\tau \succ' \iota$, and consider $\tau' = \tau \bigcirc \iota$. Thus, for some $c > 1$, $\tau(x) = cx$. Observe that

$$\tau'(x) = \tau(x) \bigcirc x = xf\left[\frac{\tau(x)}{x}\right] = xf\left(\frac{cx}{x}\right) = xf(c) = c'x,$$

and so τ' is also a similarity (translation) with $c' = f(c) > f(1) = 1$. By induction, $\theta(\tau, \iota, n) \in \mathscr{T}$.

PROOF OF SUFFICIENCY. We establish this more difficult part of the proof of Theorem 17 as a series of six lemmas.

LEMMA 1. *Suppose $\mathscr{A}$ is a concatenation structure that is closed, solvable, order dense, and Dedekind complete. Then upper and lower semicontinuity hold.*

PROOF. Because the two halves are similar, we show only lower semicontinuity. Suppose $c \prec a \bigcirc b$. By order density, there exists a u such that $c \prec u \prec a \bigcirc b$. By order density and Dedekind completeness, we can find an increasing sequence $\{a_i\}$ such that for every i, $a_i \prec a$ and l.u.b. $\{a_i\} = a$. For each i, let b_i solve $u = a_i \bigcirc b_i$, and let b^* solve $u = a \bigcirc b^*$. Since $u \prec a \bigcirc b$, $b^* \prec b$. Observe that $\{b_i\}$ is necessarily a decreasing sequence with g.l.b. b^*. So, for sufficiently large i, $b^* \prec b_i \prec b$. Thus $a_i \bigcirc b_i$ fulfills the condition since by construction and choice, $a_i \prec a$, $b_i \prec b$, and $c \prec u = a_i \bigcirc b_i \prec a \bigcirc b$. ◇

LEMMA 2. *Suppose $\mathscr{A}$ is a concatenation structure that is closed, intern, Dedekind complete, and upper (lower) semicontinuous. Then for each $a, b, c \in A$ with $a \succ c \succ b$, there exists an integer n $[m]$ such that $\theta(a, b, n) \prec c$ $[\theta(b, a, m) \succ c]$.*

PROOF. Observe that $\{\theta(a, b, n) \mid n \text{ an integer}\}$ is bounded from below by b; and so, by Dedekind completeness, there is a g.l.b. d. If $d = b$, we are done; so suppose that $d \succ b$. Since $\mathscr{A}$ is intern, $d \succ d \bigcirc b \succ b$. By upper semicontinuity, there exists $d' \succ d$ such that $d \succ d' \bigcirc b$. By the choice of d, there exists an n such that $\theta(a, b, n) \prec d'$, in which case

$$\theta(a, b, n+1) = \theta(a, b, n) \bigcirc b \prec d' \bigcirc b \prec d,$$

which is contrary to the choice of d. So $d = b$. The other case is similar. $\diamondsuit$

Let $\mathscr{H}$ be a subset of the automorphism group $\mathscr{G}$. We shall say $\mathscr{H}$ is *dense in* $\mathscr{A}$ iff for each $a, b \in A$ with $a \succ b$, there exist $\alpha, \beta \in \mathscr{H}$ such that $a \succ \alpha(a)$, $\beta(b) \succ b$.

LEMMA 3. *Under the conditions of Theorem* 17, *if* $\mathscr{T}$ *satisfies the condition stated, then it is dense in* $\mathscr{A}$.

PROOF. By assumption, there exists $\tau \in \mathscr{T}$, $\tau \succ' \iota$ such that $\theta[\tau(b), b, n] \in \mathscr{T}$. Suppose $a \succ b$. If $a \succ \tau(b) \succ b$, we are done; so suppose that $\tau(b) \succsim a \succ b$. By Lemmas 1 and 2, there exists an n such that $a \succ \theta[\tau(b), b, n] \succ b$, which yields half of $\mathscr{T}$ being dense in $\mathscr{A}$. The other half, which uses $\theta[\tau^{-1}(a), a, n]$, is similar. $\diamondsuit$

LEMMA 4. *Suppose* $\mathscr{A}$ *is a Dedekind complete relational structure, and* $\mathscr{H}$ *is a group included in* $\mathscr{T}$. *If for* $\alpha, \beta \in \mathscr{H}$ *and* $a \in A$, $\alpha(a) \succ \beta(a)$, *then for all* $b \in A$, $\alpha(b) \succ \beta(b)$.

PROOF. Suppose not, then for some b, we have both $b \prec \alpha^{-1}\beta(b)$ and $a \succ \alpha^{-1}\beta(a)$. By Lemma 1 of Theorem 5, $\alpha^{-1}\beta$ has a fixed point, which is impossible since $\alpha^{-1}\beta$ is not the identity and is in $\mathscr{T}$ since it is in $\mathscr{H}$. $\diamondsuit$

Define $\succsim'$ on $\mathscr{H}$ by $\alpha \succsim' \beta$ iff $\alpha(a) \succsim \beta(a)$ for all $a \in A$.

LEMMA 5. *Suppose* $\mathscr{A}$ *is a concatenation structure that is closed, solvable, lower semicontinuous, and Dedekind complete. Suppose* $\mathscr{H}$ *is a maximal subgroup of* $\mathscr{T}$, *then* $\langle \mathscr{H}, \succsim' \rangle$ *is Dedekind complete.*

PROOF. Let Γ be a bounded subset of $\mathscr{H}$. Since $\succsim'$ is a total ordering, the bound of Γ yields a bound of $\{\alpha(a) \mid \alpha \in \Gamma\}$ for each $a \in A$. So, $\delta(a) = \text{l.u.b.}\ \{\alpha(a) \mid \alpha \in \Gamma\}$ is defined. We show that δ is in $\mathscr{H}$.

1. $\delta(a \bigcirc b) = \delta(a) \bigcirc \delta(b)$. Suppose not. If $\delta(a \bigcirc b) \succ \delta(a) \bigcirc \delta(b)$, then since δ is a l.u.b., there exists $\alpha \in \Gamma$ such that

$$\delta(a \bigcirc b) \succsim \alpha(a \bigcirc b) \succ \delta(a) \bigcirc \delta(b) \succsim \alpha(a) \bigcirc \alpha(b) = \alpha(a \bigcirc b),$$

which is impossible. Suppose the other inequality; then by lower semicontinuity, there exist $a', b' \in A$ such that $a' \prec \delta(a)$, $b' \prec \delta(b)$, and $\delta(a \bigcirc b) \prec a' \bigcirc b'$. Thus there exist $\alpha, \beta \in \Gamma$ with $a' \prec \alpha(a) \prec \delta(a)$ and $b' \prec \beta(b) \prec \delta(b)$. But $\mathscr{T}$ is ordered, so select the larger of α and β, say α, then we have

$$\delta(a) \bigcirc \delta(b) \succsim \alpha(a) \bigcirc \alpha(b) = \alpha(a \bigcirc b') \succ a' \bigcirc b' \succ \delta(a \bigcirc b),$$

which is impossible since $\delta(a \bigcirc b) \succsim \alpha(a \bigcirc b)$.

2. δ is order preserving. Suppose, on the contrary that $a \succ b$ and $\delta(a) = \delta(b)$. By solvability, there exists a nontrivial standard difference sequence $\{x_i\}$ such that $x_i \bigcirc b = x_{i+1} \bigcirc a$. By part 1 and monotonicity, $\delta(x_i) = \delta(x_{i+1})$. Let u denote this common value. By the choice of Γ there exists some $\tau \in \mathscr{H}$ such that $\tau \succsim' \alpha$ for all $\alpha \in \Gamma$, i.e., $\tau(a) \succsim \alpha(a)$ is true independent of a. Thus $\tau(x_i) \succsim \delta(x_i) = u$. But since $\mathscr{A}$ is Archimedean in differences (Theorem 19.1) and τ is an automorphism, it follows that $\tau(a) \succsim u$ for all $a \in A$. Because $\mathscr{A}$ is solvable there is no minimal element, which means τ is not *onto*, contradicting the assumption it is in $\mathscr{H}$. So δ must be order preserving.

3. δ is onto A. Suppose $b \in A$ and $\alpha \in \Gamma$. Because α is *onto* there exists an $a_\alpha \in A$ such that $\alpha(a_\alpha) = b$. Since Γ is included in $\mathscr{T}$, the elements of Γ are ordered, and moreover, they are uncrossed. Thus, if $\alpha \succ' \beta$, then since $\beta(a_\beta) = b = \alpha(a_\alpha) \succ \beta(a_\alpha)$, we see by the fact that β is order preserving, $a_\beta \succ a_\alpha$. Since Γ is bounded, let γ be an upper bound, and so $\{a_\alpha \mid \alpha \in \Gamma\}$ is bounded from below by a_γ. Let a be the g.l.b. Since $a_\alpha \succsim a$, we see that $b = \alpha(a_\alpha) \succsim \alpha(a)$. Thus b is an upper bound of $\{\alpha(a) \mid \alpha \in \Gamma\}$, and so $b \succsim \delta(a)$. Suppose $b \succ \delta(a)$. By idempotence and monotonicity, the structure is intern, and so $b \bigcirc \delta(a) \succ \delta(a)$. By lower semicontinuity (Lemma 1 and the fact a closed intern structure is order dense), there exist $u, v \in A$ such that

$$u \prec b,\ v \prec \delta(a) \qquad \text{and} \qquad u \bigcirc v \succ \delta(a).$$

So we can select $\alpha \in \Gamma$ such that

$$u \prec \alpha(a) \prec b \qquad \text{and} \qquad v \prec \alpha(a) \prec \delta(a),$$

whence,

$$\alpha(a) = \alpha(a) \bigcirc \alpha(a) \succ u \bigcirc v \succ \delta(a),$$

which is a contradiction. So $\delta(a) = b$, proving δ is *onto*.

Thus δ is an automorphism. We next show that it is in $\mathscr{T}$. Suppose, for some $a \in A$, $\delta(a) = a$. If $\delta \neq \iota$, then for some $b \neq a$, $\delta(b) \neq b$. Suppose

$\delta(b) \succ b$, then for some $\alpha \in \Gamma$, $\alpha(b) \succ b$. By the fact that $\mathscr{T}$ is uncrossed, $\alpha(a) \succ a$, and so $\delta(a) \succsim \alpha(a) \succ a$, contrary to assumption. So $\delta(b) \prec b$. Since $\mathscr{A}$ is solvable, there exists $c \in A$ such that $a = b \bigcirc c$ and $\delta(c) \precsim c$. So, by the fact that δ is an automorphism and by use of monotonicity,

$$a = \delta(a) = \delta(b \bigcirc c) = \delta(b) \bigcirc \delta(c) \prec d \bigcirc c = a,$$

which is a contradiction. So $\delta = \iota \in \mathscr{T}$.

Suppose δ, δ' are, respectively, completions of Γ, Γ', both subsets of $\mathscr{H}$. Clearly,

$$\underset{\alpha\alpha' \in \Gamma \times \Gamma'}{\text{l.u.b.}} \{\alpha\alpha'(a)\} \succsim \underset{\alpha \in \Gamma}{\text{l.u.b.}} \left\{ \alpha \left[\underset{\alpha' \in \Gamma'}{\text{l.u.b.}} \{\alpha'(a)\} \right] \right\}.$$

Suppose the inequality holds, then for some $\beta \in \Gamma$ and $\beta' \in \Gamma'$,

$$\begin{aligned} \beta\beta'(a) &\succ \underset{\alpha \in \Gamma}{\text{l.u.b.}} \left\{ \alpha \left[\underset{\alpha' \in \Gamma'}{\text{l.u.b.}} \{\alpha'(a)\} \right] \right\} \\ &\succsim \underset{\alpha \in \Gamma}{\text{l.u.b.}} \{\alpha\beta'(a)\} \\ &\succsim \beta\beta'(a), \end{aligned}$$

which is a contradiction. Thus $\delta\delta'$ is the l.u.b. of $\Gamma \times \Gamma'$, and so it is in $\mathscr{H}'$, as was to be shown.

Next, we show that if δ is the completion of a bounded subset Γ of $\mathscr{G}$, then there is a bounded subset Γ' of $\mathscr{G}$ such that δ^{-1} is the completion of Γ'. Let

$$\Gamma' = \{\beta \mid \beta \in \mathscr{G} \text{ and, for all } \alpha \in \Gamma,\ \beta \prec' \alpha^{-1}\}.$$

Since Γ has an upper bound γ, then $\gamma^{-1} \in \Gamma'$. Moreover, Γ' is bounded by construction. Let δ' be the completion of Γ'. Suppose $\delta^{-1} \succ' \delta'$, then for each $\alpha \in \Gamma$, $\alpha^{-1} \succ' \delta^{-1} \succ' \delta'$, and so $\delta^{-1} \in \Gamma'$, which contradicts that δ' is its l.u.b. Suppose $\delta^{-1} \prec' \delta'$, then $\delta \succ' \delta'^{-1}$, and so there exists an $\alpha \in \Gamma$ such that $\delta \succ' \alpha \succ' \delta'^{-1}$. Thus $\delta' \succ' \alpha^{-1}$, contrary to choice.

So we have shown that the set of all completions forms a group lying between $\mathscr{H}$ and $\mathscr{T}$. Since $\mathscr{H}$ is a maximal group included in $\mathscr{T}$, it follows that $\langle \mathscr{H}, \succsim' \rangle$ is Dedekind complete. ◇

Given Lemmas 3, 4, and 5, the following lemma completes the proof that $\mathscr{A}$ is homogeneous.

LEMMA 6. *Suppose $\mathscr{A}$ is dense, Dedekind complete, and unbounded, and $\mathscr{H}$ is a group that is a subset of $\mathscr{T}$. If $\langle \mathscr{H}, \succsim' \rangle$ is Dedekind complete and $\mathscr{H}$ is dense in $\mathscr{A}$, then $\mathscr{A}$ is homogeneous.*

PROOF. We first show that $\mathscr{H}$ is uncrossed. Suppose on the contrary, for some $a, b \in A$ and $\alpha \in \mathscr{H}$ is such that $\alpha(a) \succ a$ and $\alpha(b) \prec b$. Since $\mathscr{C} = \{c \mid \alpha(c) \succ c \ \&\ c \prec b\}$ is nonempty and bounded by b, the hypothesis that $\mathscr{A}$ is Dedekind complete implies that $\mathscr{C}$ has a l.u.b. u. Suppose $\alpha(u) \succ u$, then $u \prec b$ since $\alpha(b) \prec b$. By order density, there exists a v such that $u \prec v \prec \min[b, \alpha(u)]$. By definition of l.u.b., $\alpha(v) \prec v$. But since α is order preserving, $\alpha(v) \succ \alpha(u) \succ v$, which is a contradiction. Next, suppose that $\alpha(u) \prec u$. Since u is a l.u.b., we know there exists a w with $\alpha(w) \succ w$ and $\alpha(u) \prec w \prec u$. Thus $\alpha(w) \succ \alpha(u)$ and so $w \succ u$, which is a contradiction. So u is a fixed point of α. But that is impossible since $\mathscr{H}$ is included in $\mathscr{T}$.

Thus $\succsim'$ is a total order on $\mathscr{H}$. Suppose $a, b \in A$. Define

$$\mathscr{B} = \{\tau \mid \tau \in \mathscr{H} \text{ and } \tau(a) \succsim b\}$$
$$\mathscr{B}' = \{\tau \mid \tau \in \mathscr{H} \text{ and } \tau(a) \prec b\}.$$

If $a \prec b$, then by the density of $\mathscr{H}$ in $\mathscr{A}$, there exists $\tau' \in \mathscr{H}$ such that $a \prec \tau'(a) \prec b$. So $\iota \prec' \tau'$. For some integer n, $b \prec \tau'^n(a)$, for otherwise, by Dedekind completeness, τ' has a fixed point, contrary to the assumption that $\mathscr{H}$ is included in $\mathscr{T}$. So $\mathscr{B}$ and $\mathscr{B}'$ are both nonempty. The argument is similar for $a \succ b$. Since $\langle \mathscr{H}, \succsim' \rangle$ is Dedekind complete, there is a cut element δ. Suppose $\delta(a) \succ b$, then by the density of $\mathscr{H}$ in $\mathscr{A}$, there is an $\alpha \in \mathscr{G}$ such that $\delta(a) \succ \alpha[\delta(a)] \succ b$. Since $\succsim'$ is a total order, $\delta \succ' a\delta$, which together with $\alpha\delta \in \mathscr{H}$ contradicts the choice of δ as the cut element. A similar argument shows $\delta(a) \prec b$ is impossible. So $\delta(a) = b$, proving that $\mathscr{A}$ is homogeneous. ◇

20.5.8 Theorem 18 (p. 150)

Suppose $\mathscr{A}$ is a totally ordered concatenation structure that is closed, idempotent, solvable, and Dedekind complete. Then $\mathscr{A}$ has a dual bilinear representation iff $\mathscr{A}$ is reflectable, $*$ and $*'$ are right autodistributive, and they satisfy generalized bisymmetry.

PROOF. By Theorems 19.1 and 19.21 and Dedekind completeness, we are in the situation of Theorem 11. Suppose $\mathscr{A}$ has a dual bilinear representation. It suffices to show that $*$ (Definition 7) coincides with $\oplus$ defined by $x \oplus y = cx + (1 - c)y$ and that $*'$ coincides with $\oplus'$ defined using d instead of c. For $x \geqslant y$, the first two agree by definition. For $x < y$, $x * y = r$, where for $x > x$, $v > y$,

$$(u \bigcirc x) \bigcirc (v \bigcirc y) = (u \bigcirc v) \bigcirc r,$$

which is equivalent to

$$\begin{aligned} c^2u + c(1-c)x + c(1-c)v + (1-c)^2 y \\ = c^2u + c(1-c)v + (1-c)r, \end{aligned}$$

whence $r = cx + (1-c)y = x \oplus y$ and $*$ is clearly reflectable. Since $\oplus$ is bisymmetric, it is right autodistributive, and earlier we showed that $\oplus$ and $\oplus'$ satisfy generalized bisymmetry.

To show the converse, we need first to establish some properties of $*$ and $*'$.

LEMMA 1. *Suppose $\mathscr{A}$ is a concatenation structure that is idempotent and solvable. For each $a, b \in A$,*

(i) *if $a \prec b$, then there exist $u, v, w \in A$ such that*

$$u \succsim a, \qquad v \succsim b, \qquad u \bigcirc a \succsim v \bigcirc b, \qquad u \bigcirc v \succsim w, \tag{21}$$

and

$$(u \bigcirc x) \bigcirc (v \bigcirc y) = (u \bigcirc v) \bigcirc w; \tag{22}$$

(ii) *if $a \succ b$, then there exist $p, q, z \in A$ such that*

$$p \precsim a, \qquad q \precsim b, \qquad p \bigcirc a \precsim q \bigcirc b, \qquad p \bigcirc q \precsim z, \tag{23}$$

and

$$(p \bigcirc x) \bigcirc (q \bigcirc y) = (p \bigcirc q) \bigcirc z. \tag{24}$$

PROOF. Since the two proofs are similar, only (i) is presented. By the nontrivialness of the structure and solvability, select u and v so that $u \succsim a$, $v \succsim b$, and $u \bigcirc a \succsim v \bigcirc b$. Define w as the solution to Equation (22). It is now sufficient to show the last inequality of Equation (21). Suppose, on the contrary, $u \bigcirc v \prec w$; then using the idempotency and monotonicity of $\bigcirc$, we obtain

$$\begin{aligned} u \bigcirc v = (u \bigcirc v) \bigcirc (u \bigcirc v) \succsim (u \bigcirc a) \bigcirc (v \bigcirc b) \\ = (u \bigcirc v) \bigcirc w \succ (u \bigcirc v) \bigcirc (u \bigcirc v) = u \bigcirc v, \end{aligned}$$

a contradiction. ◇

COROLLARY. *If Equation* (21) *holds, then $u \succsim v$.*

PROOF. Since $b \succ a$ and $u \bigcirc a \succsim v \bigcirc b$, monotonicity yields $u \bigcirc a \succsim v \bigcirc b$, and so $u \succsim v$. $\diamond$

LEMMA 2. *Suppose $\mathscr{A} = \langle A, \succsim, \bigcirc \rangle$ is a concatenation structure that is closed, idempotent, solvable, reflectable. Then $*$ and $*'$ are well defined, closed, intern, monotonic, and solvable.*

PROOF. Because the two cases are symmetric, it suffices to prove the result for $*$. By reflectableness, $*$ is a well-defined operation that is intern, and by construction, it is idempotent.

Next we show that $*$ is monotonic.

(i) If $a, b \succsim c$, then $a * c = a \bigcirc c$ and $b * c = b \bigcirc c$, and the monotonicity of $*$ follows from that of $\bigcirc$.

(ii) If $a \succ c \succ b$, then from the fact that $*$ is intern $a * c = a \bigcirc c \succ c \succ b * c$. The case $a \prec c \prec b$ is similar.

(iii) If $c \precsim a, b$, then select $v \in A$ such that $v \succ c$, and let $u, u' \in A$ solve $u \bigcirc a = v \bigcirc c = u' \bigcirc b$. Since $c \precsim a, b$, it follows by monotonicity of $\bigcirc$ that $u, u' \prec c \prec c \precsim a, b$, and by interness of $*$, $u \bigcirc v \prec a * c$ and $u' \bigcirc v \prec b * c$. Thus, by reflectability,

$$(u \bigcirc a) \bigcirc (v \bigcirc c) = (u \bigcirc v) \bigcirc (a * c) \qquad \text{and}$$
$$(u' \bigcirc b) \bigcirc (v \bigcirc c) = (u' \bigcirc v) \bigcirc (b * c).$$

Since $u \bigcirc a = u' \bigcirc b$, monotonicity yields

$$(u \bigcirc v) \bigcirc (a * c) = (u' \bigcirc v) \bigcirc (b * c).$$

Using the monotonicity of $\bigcirc$ and this relation, we see that

$$\begin{aligned} a \succsim b \quad & \text{iff} \quad u \precsim u' \qquad (\text{since } u \bigcirc a = u' \bigcirc b) \\ & \text{iff} \quad u \bigcirc v \precsim u' \bigcirc v \\ & \text{iff} \quad a * c \succsim b * c. \end{aligned}$$

The other side is similar.

Finally, we show that $*$ is solvable. Suppose a and c are given, and we search for b such that $c = a * b$. (The existence of the solution b to $c = b * a$ is shown similarly.) If $a \succsim c$, then let b solve $c = a \bigcirc b$. Since by monotonicity $a \succsim b$, we see that $a \bigcirc b = a * b$. So, suppose that $a \prec c$. Select u to be any element for which $u \succ a$ and $u \bigcirc a \succ c$. Select p so that $u \bigcirc a \succ p \succ c$. Let w solve $(u \bigcirc a) \bigcirc p = w \bigcirc c$, v solve $u \bigcirc v = w$, and y solve $v \bigcirc b = p$. Observe, by monotonicity and transitivity,

$$(u \bigcirc a) \bigcirc (v \bigcirc b) = (u \bigcirc a) \bigcirc p = w \bigcirc c = (u \bigcirc v) \bigcirc c,$$

and so by reflectability, $c = a * b$ provided the requisite inequalities hold. We have $u \succ a$ and $u \bigcirc v \succ p = v \bigcirc b$ by choice. Since $w \bigcirc c = (u \bigcirc a) \bigcirc p \succ c \bigcirc c = c$, we see $u \bigcirc v = w \succ c$. Finally, we show $v \succsim b$. From $v \bigcirc b = p$, $v \succsim b$ iff $v \succsim p$. Suppose $v \prec p$, then

$$w = u \bigcirc v \prec u \bigcirc p \prec (u \bigcirc a) \bigcirc p = w \bigcirc c,$$

whence $w \prec c$, contrary to what was previously shown. ◇

LEMMA 3. *Suppose* $\langle A, \succsim, * \rangle$ *and* $\langle A, \succsim, *' \rangle$ *are closed, idempotent, solvable, concatenation structures that are Dedekind complete and right autodistributive, and* $*$ *and* $*'$ *satisfy generalized bisymmetry. Then there exists a common bisymmetric representation for* $*$ *and* $*'$.

PROOF. By Theorem 19.24, $*$ and $*'$ are both bisymmetric, and by Theorems 19.1 and 6.10 they each have bisymmetric representations, say φ and φ'. So there is a strictly increasing function g relating them, and with no loss of generality we can select affine transformations so that $g(0) = 0$, $g(1) = 1$. In a minor abuse of notation, let $*$ and $*'$ denote the real bisymmetric operations so that they can be written, for x, y in Re,

$$x * y = cx + (1 - c)y, \tag{25}$$

$$x *' y = g^{-1}[dg(x) + (1 - d)g(y)]. \tag{26}$$

If we write $G(x, y) = x * y$ and $F(x, y) = x *' y$, then the assumed generalized bisymmetry becomes the functional equation

$$F[G(x, y), G(u, v)] = G[F(x, u), F(y, v)]. \tag{27}$$

By Aczél (1966, p. 317), the general solution to Equation (27) is

$$F(x, y) = k^{-1}[Ah(x) + Bh(y) + C] \tag{28}$$

$$G(x, y) = m[Ak(x) + Bk(y) + C], \tag{29}$$

where h, k, and m are strictly increasing functions. Since $*$ and $*'$ are idempotent,

$$m^{-1}(x) = (A + B)k(x) + C, \tag{30}$$

$$k(r) = (A + B)h(x) + C. \tag{31}$$

From Equations (25), (29), and (31),

$$\begin{aligned} Ah(x) + Bh(y) + C &= kF(x, y) \\ &= k[cx + (1 - c)y] \\ &= (A + B)h[cx + (1 - c)y] + C. \end{aligned}$$

By Aczél (1966, p. 67), h is linear; thus, by Equations (30) and (31), so are k and m. From Equations (26) and (28) and this linearity,

$$\begin{aligned} cg(x) + (1-c)g(y) &= g[G(x, y)] \\ &= g\{m[Ak(x) + Bk(y) + C]\}, \end{aligned}$$

whence g is linear. Since $g(0) = 0$ and $g(1) = 1$, g is the identity, and $\varphi' = \varphi$. ◇

The result stated in Theorem 18 follows immediately from the hypothesis and the three lemmas just proved. ◇

20.5.9 Theorem 19 (p. 153)

Suppose $\mathscr{M} = \langle X, \succsim, \bigcirc_A \rangle_{A \in \mathscr{E}}$ is a mixture space satisfying conditions 1–5 *of Definition* 8, *that each $\mathscr{X}_A = \langle X, \succsim, \bigcirc_A \rangle$ is homogeneous, and there are unit representations $\langle U, f_A \rangle$ with a common utility function U. Then:*

(i) *There exists a strictly increasing function h from Re^+ onto Re and functions s^+, s^- from $\mathscr{E}$ onto $[0,1]$ such that for all $A \in \mathscr{E}$ and $r \in \mathrm{Re}$,*

(a) $$h(1) = 0 \text{ and } s^i(\Omega) = 1,\ i = +, -;$$

(b) $$f_A(r) = \begin{cases} h^{-1}[h(r)s^+(A)], & \text{if } r > 1 \\ 1, & \text{if } r = 1 \\ h^{-1}[h(r)s^-(A)], & \text{if } r < 1. \end{cases} \tag{14}$$

(ii) *If, in addition, $\mathscr{M}$ is complemented, h is continuously differentiable on Re^+ except possibly at* 1, *and there exist positive constants μ and δ such that*

$$\lim_{x \downarrow 1} \frac{d[h(x)^{\mu}]}{dx} \quad \text{and} \quad \lim_{x \uparrow 1} \frac{d[h(x)^{\delta}]}{dx}$$

exist and are nonzero, then for $S^+ = (s^+)^{\mu}$, $S^- = (s^-)^{\delta}$, and all $A \in \mathscr{E}$,

$$S^+(A) + S^-(\bar{A}) = 1, \tag{15}$$

$$U(x \bigcirc_A y) = \begin{cases} U(x)S^+(A) + U(y)[1 - S^+(A)] & \text{if } U(x) > U(y) \\ U(x) & \text{if } U(x) = U(y) \\ U(x)S^-(A) + U(y)[1 - S^-(A)] & \text{if } U(x) < U(y), \end{cases} \tag{16}$$

and U is an interval scale.

PROOF.

(i) For $x, y \in (1, \infty)$ and $A, B \in \mathscr{E}$, define $\succsim'$ by

$$(x, A) \succsim' (y, B) \quad \text{iff} \quad f_A(x) \geqslant f_B(y).$$

We show that $\langle (1, \infty) \times \mathscr{E}, \succsim' \rangle$ is an additive conjoint structure (Definition 7, Chapter 6). The relation $\succsim'$ satisfies independence because, for each A, f_A is strictly monotonic, and for each $x > 1$, monotonicity holds over $\mathscr{M}$ by outcome independence. To show double cancellation, suppose that

$$(x, B) \succsim' (y, C) \quad \text{and} \quad (y, A) \succsim' (z, B),$$

i.e.,

$$f_B(x) \geqslant f_C(y) \quad \text{and} \quad f_A(y) \geqslant f_B(z).$$

Apply f_A to the first inequality and f_C to the second, and use the commutativity assumption,

$$f_B f_A(x) = f_A f_B(x) \geqslant f_A f_C(y) = f_C f_A(y) \geqslant f_C f_B(z) = f_B f_C(z).$$

So, by the monotonicity of f_B, $(x, A) \succsim' (z, C)$.

Restricted solvability holds because f_A is *onto* and by the assumption that it satisfies restricted solvability for events.

Each component is essential by the strict monotonicity of the fs and by the existence of events other than Ω.

A nontrivial standard sequence $\{x_n\}$ corresponds to some A, B in $\mathscr{E}$, $B \neq \Omega$, such that f_A and f_B are distinct and $(x_n, A) \sim' (x_{n+1}, B)$, which in turn corresponds to $x_{n+1} = f_B^{-1} f_A(x_n)$.

Thus, by assumption, the Archimedean property holds. So by the representation theorem for such structures (Theorem 6.2), there exist functions h, s^+, and k such that

$$k[f_A(x)] = h(x) s^+(A).$$

Since for all $x, y \in X$, $x \bigcirc_\Omega y \sim x$, we have

$$U(y) f_\Omega\left[\frac{U(x)}{U(y)}\right] = U(x),$$

whence,

$$k(z) = k[f_\Omega(z)] = h(z) s^+(\Omega).$$

If we choose s^+ so that $s^+(\Omega) = 1$, which by the uniqueness of additive conjoint structures is possible, then $k = h$. Setting $z = 1$ and noting that, for each A in $\mathscr{E}$, $f_A(1) = 1$, we see that

$$h(1) = h[f_A(1)] = h(1)s^+(A),$$

and so $h(1) = 0$.

In a completely analogous fashion, we define an additive conjoint structure for x, y in $(0,1)$, which permits us to extend h onto that interval and introduces s^- on $\mathscr{E}$. Thus Equation (14) holds.

Note that because f_A is onto and solvability holds for the events, the functions s^i map onto a continuum, which by choice of Ω is $[0,1]$.

(ii) Define θ^i by $\theta^i[s^i(\bar{A})] = s^i(A)$, $i = +, -$. Observe, for $U(x)$, $U(y) > 1$,

$$x \bigcirc_A y \sim y \bigcirc_{\bar{A}} \quad \text{iff} \quad U(y) f_A\left[\frac{U(x)}{U(y)}\right] = U(x) f_{\bar{A}}\left[\frac{U(y)}{U(x)}\right]$$

$$\text{iff} \quad \text{for all } z > 0,\ h^{-1}[h(z)s^+(A)] = zh^{-1}\left[h\left(\frac{1}{z}\right)s^+(\bar{A})\right].$$

Setting $s^+(\bar{A}) = a$, we see that h must satisfy the upper part of

$$h\left\{xh^{-1}\left[h\left(\frac{1}{x}\right)a\right]\right\} = \begin{cases} \theta^+(a), & \text{if } x > 1 \\ h(x), & \text{if } x = 1 \\ \theta^-(a), & \text{if } x < 1. \end{cases} \tag{32}$$

The proof is similar for the bottom part. Define the function H on Re^+ by

$$H(x) = \begin{cases} h^{\mu}(x), & \text{if } x > 1 \\ 0, & \text{if } x = 1 \\ h^{\delta}(x), & \text{if } x < 1. \end{cases}$$

Observe from Equation (32) that H must satisfy the functional equation:

$$\begin{aligned} \frac{H\{xH^{-1}[H(1/x)a^{\mu}]\}}{H(x)} &= \theta^+(a)^{\delta}, \quad \text{if } x > 1 \\ \frac{H\{xH^{-1}[H(1/x)a^{\delta}]\}}{H(x)} &= \theta^-(a)^{\mu}, \quad \text{if } x < 1. \end{aligned} \tag{33}$$

Note that if H satisfies Equation (33), then for $p, q > 0$,

$$H_{p,q}(x) = \begin{cases} pH(x), & \text{if } x > 1 \\ 0, & \text{if } x = 1 \\ qH(x), & \text{if } x < 1 \end{cases}$$

also satisfies Equation (33). Therefore, by an appropriate choice of p and q, there is no loss of generality in assuming that H has equal right and left derivatives at $x = 1$ and that they equal 1. Hence, H is continuously differentiable everywhere.

In Equation (33), take the limits as $x \to 1$, which we can do by using l'Hospital's rule, and it yields

$$\theta^+(a)^\delta = 1 - a^\mu,$$
$$\theta^-(a)^\mu = 1 - a^\delta.$$

Substituting this into Equation (33) yields

$$H(x) = \begin{cases} H\dfrac{\{xH^{-1}[H(1/x)a^\mu]\}}{1 - a^\mu}, & \text{if } x > 1 \\ 0, & \text{if } x = 1 \\ H\dfrac{\{xH^{-1}[H(1/x)a^\delta]\}}{1 - a^\delta}, & \text{if } x < 1. \end{cases}$$

Now, for $x = 1$, take the limit as $a \to 1$, again using l'Hospital's rule, yielding

$$H(x) = -\frac{xH(1/x)}{H'(1/x)}, \qquad x \neq 1.$$

Set $y = 1/x$, and solve for H'/H,

$$\frac{H'(y)}{H(y)} = -\frac{1}{yH(1/y)}, \qquad y \neq 1. \tag{34}$$

Since H is continuously differential, we can take the derivative of Equation

(34) to obtain

$$\frac{H''(y)}{H'(y)} - \frac{H'(y)}{H(y)}\frac{H'(y)}{H(y)} = \frac{H(1/y) + yH'(1/y)(-1/y^2)}{y^2H(1/y)^2}$$
$$= \frac{1}{yH(1/y)}\frac{1}{y} - \frac{H'(1/y)}{y^2H(1/y)}$$
$$= -\frac{H'(y)[1 + 1/H(y)]}{yH(y)}.$$

Dividing out H'/H and setting $H(y) = G(\log y)$, we see that G satisfies the differential equation

$$\frac{G''}{G'} - \frac{G'}{G} = -\frac{1}{G}.$$

It is easy to verify that this is equivalent to

$$\frac{d}{dy}\log[G'(y) - 1] - \frac{d}{dy}\log G(y) = 0,$$

i.e.,

$$\frac{G'(y) - 1}{G(y)} = \beta,$$

whence,

$$G(y) = \sigma e^{\beta y} - \frac{1}{\beta}.$$

Since $H = \log G$, H is continuous at 1, and $H(1) = 0$, we see that $\sigma = 1/\beta$, whence,

$$h(y) = \begin{cases} [(y^\beta - 1)/\beta]^{1/\delta}, & \text{if} \quad y > 1 \\ 0, & \text{if} \quad y = 1 \\ [(y^\beta - 1)/\beta]^{1/\mu}, & \text{if} \quad y < 1. \end{cases}$$

Substituting into the unit expression for $\bigcirc_A$,

$$x \bigcirc_A y = y f_A\left(\frac{x}{y}\right)$$

$$= y\begin{cases} h^{-1}\left[h\left(\dfrac{x}{y}\right)s^{+}(A)\right], & \text{if } x > y \\ 1, & \text{if } x = y \\ h^{-1}\left[h\left(\dfrac{x}{y}\right)s^{-}(A)\right], & \text{if } x < y \end{cases}$$

$$= \begin{cases} \{x^{\beta}S^{+}(A) + y^{\beta}[1 - S^{+}(A)]\}^{1/\beta}, & \text{if } x > y \\ y, & \text{if } x = y \\ \{x^{\beta}S^{-}(A) + y^{\beta}[1 - S^{-}(A)]\}^{1/\beta}, & \text{if } x < y, \end{cases}$$

where $S^{+} = (s^{+})^{\mu}$ and $S^{-} = (s^{-})^{\delta}$. Thus, in the transformed variable x^{β}, we see that the structure has a dual bilinear representation and so is, necessarily, of scale type (2, 2). ◇

20.6 HOMOGENEOUS CONJOINT STRUCTURES

To a rough approximation, this section shows that there are two broad types of two-factor conjoint structures: those with a singular point, which have representations closely related to the unit representations of the (1, 1) case, and those without such a point, which tend to be additive. The theory, which is due to Luce and Cohen (1983), is somewhat less complete and satisfactory than that for concatenation structures.

20.6.1 Component Homogeneity and Uniqueness

Although the general concepts of M-point homogeneity and N-point uniqueness apply to conjoint structures, they are not really sufficiently refined to be of much use. The lack of refinement stems from the fact that general order automorphisms do not necessarily take into account the factorial structure. If we are to make use of what we know about how conjoint structures induce a total concatenation structure on one component, then it is necessary to confine our attention to factorizable automorphisms (Section 19.6.3) and to be concerned with exactly what the factors of the automorphisms do to the components. It is reasonably clear how to define homogeneity and uniqueness on a single component, but it is rather less clear what we want to do with the other factor of the factorizable automorphism. There does not seem to be an especially good reason for the choice except that it seems to work.

DEFINITION 9. *Suppose* $\mathscr{C} = \langle A \times P, \succsim \rangle$ *is a conjoint structure, and* $\mathscr{F}$ *is its group of factorizable automorphisms.*

1. $\mathscr{F}$ *is said to satisfy* component M-point homogeneity *iff for strictly increasing sequences of* M *elements* $\{a_1\}$, $\{b_1\}$, $\{p_1\}$, *and* $\{q_1\}$, *i.e.,* $a_1 \prec_A a_2 \prec_A \cdots \prec_A a_M$, *etc., there are factorizable automorphisms* $\langle \theta, \eta \rangle$ *and* $\langle \theta', \eta' \rangle$ *such that*

$$\theta(a_i) = b_i, \qquad i = 1, \ldots, M$$
$$\eta(p_M) = q_M$$

and

$$\theta'(a_M) = b_M$$
$$\eta'(p_i) = q_i, \qquad i = 1, \ldots, M.$$

2. $\mathscr{F}$ *is said to satisfy* component N-point uniqueness *iff for all sequences* $\{a_i\}$ *and* $\{p_i\}$ *such that* $\{a_i p_i\}$ *is strictly increasing and for all* $\langle \theta, \eta \rangle$, $\langle \theta', \eta' \rangle$ *in* $\mathscr{F}$ *such that* $\theta(a_i) = \theta'(a_i)$ *and* $\eta(p_i) = \eta'(p_i)$,
   (a) *if* $\{a_i\}$ *is strictly increasing, then* $\theta = \theta'$,
   (b) *if* $\{p_i\}$ *is strictly increasing, then* $\eta = \eta'$.

It is left as Exercise 22 to show that if $\mathscr{C}$ is component M-point homogeneous, then $\mathscr{C}$ is M-point homogeneous, and that if $\mathscr{C}$ is N-point unique, then it is component N-point unique.

THEOREM 22. *Suppose* $\mathscr{C}$ *is a conjoint structure that is unrestrictedly solvable and Archimedean, then its group of factorizable automorphisms is either component* 1- *or* 2-*point unique.*

Solvability here is playing a role something like continuity in a concatenation operation to get uniqueness down to $N \leqslant 2$. The remaining work is then to understand what happens as various degrees of homogeneity are added to $\mathscr{C}$.

20.6.2 Singular Points in Conjoint Structures

The concept of a singular point for a general relational structure is a point that remains fixed under all automorphisms of the structure. To see what this means for a conjoint structure, one must recast it as a relational structure as outlined in Section 19.6.3. When done, we get the following criterion.

Let $\mathscr{F}$ denote the group of factorizable automorphisms, let $\mathscr{F}_A$ denote the set of mappings that arises as a first factor of a factorizable automorphism, and let $\mathscr{F}_P$ denote the set that arises as a second factor. Then a_0 in A is a singular point iff for each θ in $\mathscr{F}_A$, $\theta(a_0) = a_0$; and p_0 in P is a singular point iff for each η in $\mathscr{F}_P$, $\eta(p_0) = p_0$. In a slight abuse of terminology, we refer to a pair $a_0 p_0$ of singular points as a singular point of $\mathscr{C}$.

With $a_0 p_0$ fixed, we introduce the following notation: $A^+ = \{a \mid a \in A$ & $a \succ_A a_0\}$, $\mathscr{I}_{A^+}$ is the PCS on A^+ induced by $\mathscr{C}$ via $a_0 p_0$, $\mathscr{F}_{A^+}$ is the restriction of $\mathscr{F}_A$ to A^+, $\mathscr{F}_{A^+ \times A^+}$ is the restriction of $\mathscr{F}$ to $A^+ \times A^+$, etc.

THEOREM 23. *Suppose $\mathscr{C}$ is a conjoint structure that is unrestrictedly solvable and Archimedean, $\mathscr{F}$ is its group of factorizable automorphisms, and $a_0 p_0 \in A \times P$ is a singular point. Then $\mathscr{F}$ is component 2-point unique. If, in addition, $\mathscr{F}_{A^+ \times P^+}$ is component 1-point homogeneous on $A^+ \times P^+$, then*

(i) *$\mathscr{F}_{A^+}$ is the automorphism group of $\mathscr{I}_{A^+}$ and it is of scale type* (1,1).

(ii) *$\mathscr{F}_{A^-}$ is the automorphism group of $\mathscr{I}_{A^-}$, and it is of scale type* (1,1).

(iii) *Analogous statements hold for P^+ and P^-.*

We turn now to candidate representations with singularities. As may be suggested by the last theorem, we find that unit representations are involved. Two levels of distinction must be made. The first stems from the domain $A^i \times P^j$, $i, j = +, -$, and is distinguished by the subscripts on the unit representation f_{ij}. The second stems from the fact that in the mixed cases, $i = +$, $j = -$ and $i = -$, $j = +$, the situation differs depending on whether the element is above or below the singularity; these we distinguish by a superscript + or −, respectively, f_{ij}^k, $i, j, k = +, -$.

THEOREM 24. *Suppose $\mathscr{C} = \langle \mathrm{Re} \times \mathrm{Re}, \succsim \rangle$ is a real conjoint structure that is unrestrictedly solvable and Archimedean, $\mathscr{F}$ is its group of factorizable automorphisms, and $x_0^{(1)} x_0^{(2)} \in \mathrm{Re} \times \mathrm{Re}$ is singular. If $\mathscr{F}$ is 2-point unique and both $\mathscr{F}_{A^+ \times P^+}$ and $\mathscr{F}_{A^+ \times P^-}$ are 1-point homogeneous, then there exist functions φ_1: Re → (onto)Re with $\varphi_i(x_0^{(i)}) = 0$, $i = 1, 2$, and a real operation $\bigcirc$ such that $\varphi_1 \bigcirc \varphi_2$ represents $\mathscr{C}$ and $\bigcirc$ is of the following form: there exist positive constants $\eta(+)$, $\eta(-)$, $\sigma(+)$, $\sigma(-)$, ρ, and unit representations*

f_{++}:	$\mathrm{Re}^+ \to [1, \infty)$	
f_{+-}^+:	$[1, \infty) \to [0, \infty)$,	$f_{+-}^+(1) = 0$
f_{-+}^+:	$[1, \infty) \to [0, \infty)$,	$f_{-+}^+(1) = 0$
f_{+-}^-:	$(0, 1] \to (-\infty, 0]$,	$f_{+-}^-(1) = 0$
f_{-+}^-:	$(0, 1] \to (-\infty, 0]$,	$f_{-+}^-(1) = 0$
f_{--}:	$\mathrm{Re}^+ \to (-\infty, 1]$	

such that

$$x \bigcirc y = \begin{cases} x & \text{if } y = 0 \\ y & \text{if } x = 0 \\ yf_{++}(x/y) & \text{if } x > 0, y > 0 \\ |y|^{\eta(+)} f^{+}_{+-}(x/|y|^{\rho}) & \text{if } x > 0, y < 0, x \geqslant |y|^{\rho} \\ |y|^{\eta(-)} f^{-}_{+-}(x/|y|^{\rho}) & \text{if } x > 0, y < 0, x < |y|^{\rho} \\ y^{\sigma(+)} f^{+}_{-+}(|x|^{\rho}/y) & \text{if } x < 0, y > 0, |x| \geqslant y^{1/\rho} \\ y^{\sigma(-)} f^{-}_{-+}(|x|^{\rho}/y) & \text{if } x < 0, y > 0, |x| < y^{1/\rho} \\ |y|\, f_{--}(|x|/|y|) & \text{if } x < 0, y < 0. \end{cases}$$

This result corrects Theorem 13 of Luce and Cohen (1983), which is correct only under more restrictive assumptions.

20.6.3 Forcing the Thomsen Condition

In the preceding case we explored 2-point uniqueness with a kind of restricted 2-point homogeneity in which one degree of freedom is always taken up by the singularity. We now turn to the case in which there is no singular point, and so 2-point homogeneity has more force. Nonetheless, we must assume something more than that to have some restrictions on the components; so we use component 2-point homogeneity. As we see, this forces the Thomsen condition. After stating this result, which is comparatively easy to prove, we discuss several related results, which we do not prove.

THEOREM 25. *Suppose* $\mathscr{C} = \langle \mathrm{Re} \times \mathrm{Re}, \succsim \rangle$ *is a real conjoint structure that satisfies unrestricted solvability, and* $\mathscr{F}$ *is its group of factorizable automorphisms. If* $\mathscr{F}$ *satisfies* 2-*point uniqueness and component* 2-*point homogeneity, and* $\mathscr{F}_i$ *on* $\langle \mathrm{Re}, \succsim_i \rangle$, $i = 1, 2$, *both satisfy* 2-*point uniqueness, then* $\mathscr{C}$ *satisfies the Thomsen condition.*

A number of other combinations of assumptions lead to the same conclusion or a slightly weaker one. Luce and Cohen (1983) explored some, but by no means all, of these. They established that there is some tradeoff between assumptions about the differentiability of $\mathscr{C}$ and assumptions about homogeneity and uniqueness. For example, we quote (but do not prove) the most extreme case of smoothness assumptions.

THEOREM 26. *Suppose* $\mathscr{C}$ *is a conjoint structure with a real representation that is continuously differentiable of order* 4, *and* $\mathscr{F}$ *satisfies component* 1-*point homogeneity. Then* $\mathscr{C}$ *satisfies the Thomsen condition.*

In the course of proving this, they developed a criterion for additivity of a representation that is related to Scheffé's criterion (Section 6.5.3) but in some ways is simpler to use.

THEOREM 27. *Suppose $\mathscr{C}$ is a conjoint structure having a numerical representation F that is continuously differentiable in each variable. Then F is additive, i.e., there exist strictly increasing functions f, f_1, and f_2 such that*

$$F(x, y) = f[f_1(x) + f_2(x)]$$

iff there exist functions ψ_1 and ψ_2 such that $\exp \psi_1$ *and* $\exp \psi_2$ *are integrable on any bounded domain and*

$$\Psi(x, y) = \log \frac{F_x(x, y)}{F_y(x, y)} = \psi_1(x) + \psi_2(x).$$

COROLLARY. *Suppose F is continuously differentiable of order 3 and $F_x \neq 0$ and $F_y \neq 0$. Then F is additive iff $\Psi_{xy} \equiv 0$.*

Luce and Cohen (1983) gave explicit expressions for the additive representation in terms of ψ_1 and ψ_2.

We can illustrate this with the example that arose in Section 20.2.7, namely, $F(x, y) = (x + y)y$, $x, y \geqslant 0$. Observe that $F_x(x, y) = y$, $F_y(x, y) = x + 2y$, $F_{xy}(x, y) = 1$. So Scheffé's criterion is that $F_{xy}/F_xF_y = 1/y(x + 2y)$ is a function of $F = (x + y)y$. To show that it is not, it is necessary to find two (x, y) pairs with the same value of F and a different value for F_{xy}/F_xF_y. To apply the present criterion, one looks at $\Psi(x, y) = \log y - \log(x + 2y)$. It is not obvious whether or not this can be written as $\psi_1(x) + \psi_2(y)$, but it is obvious that $\Psi_{xy} = 2/(x + 2y)^2 \neq 0$.

Cohen has proved but not yet published a result much like Theorem 25 but with component 2-point homogeneity replaced by the weaker 2-point homogeneity. The result is not an additive representation but a bilinear one of the sort we have seen with (2, 2) concatenation structures. The proof is very complex, and we do not know if a simple one exists.

20.7 PROOFS

20.7.1 Theorem 22 (p. 181)

If $\mathscr{C}$ is a conjoint structure that is unrestrictedly solvable and Archimedean, then its group $\mathscr{F}$ of factorizable automorphisms is either component 1- or 2-point unique.

PROOF. Suppose $\langle \theta, \eta \rangle$ and $\langle \theta', \eta' \rangle$ are in $\mathscr{F}$ and agree at ap and bq, $a \prec_A b$. Let $\alpha = \theta^{-1}\theta'$ and $\lambda = \eta^{-1}\eta'$. Observe that $\alpha(a) = a$ and $\lambda(p) = p$. By Theorem 19.12, α is an automorphism of $\mathscr{I}_{A^+} = \langle A^+, \succ_A, *_{ap} \rangle$. Since $\alpha(b) = b$, we know by Theorems 19.2 and 19.11 that α is the identity map, and so $\theta = \theta'$ over A^+. For c in A^-, by solvability there exists c^+ in A^+ such that $c^+ *_{ap} c \sim a$. Thus,

$$\begin{aligned} \theta(c^+) *_{ap} \theta(c) &\sim \theta(c^+ *_{ap} c) \\ &\sim \theta(a) \\ &\sim \theta'(a) \\ &\sim \theta'(c^+ *_{ap} c) \\ &\sim \theta'(c^+) *_{ap} \theta'(c) \\ &\sim \theta(c^+) *_{ap} \theta'(c). \end{aligned}$$

By monotonicity, $\theta(c) = \theta'(c)$. By Theorem 19.12, $\eta = \eta'$. Thus component 2-point uniqueness holds. $\diamond$

20.7.2 Theorem 23 (p. 182)

Suppose $\mathscr{C}$ is a conjoint structure that is unrestrictedly solvable and Archimedean, $\mathscr{F}$ is its group of factorizable automorphisms, and $a_0 p_0 \in A \times P$ is singular. Then $\mathscr{F}$ is component 2-point unique. If, in addition, $\mathscr{F}_{A^+ \times P^+}$ is component 1-point homogeneous on $A^+ \times P^+$, then

(i) *$\mathscr{F}_{A^+}$ is the automorphism group of $\mathscr{I}_{A^+}$, and it is of scale type* (1, 1);

(ii) *$\mathscr{F}_{A^-}$ is the automorphism group of $\mathscr{I}_{A^-}$, and it is of scale type* (1, 1);

(iii) *analogous statements hold for P^+ and P^-.*

PROOF. With the singular point taking up one degree of freedom, it is obvious from Theorem 22 that $\mathscr{F}$ is component 2-point unique.

(i) By Theorem 19.12 it is immediate that $\mathscr{F}_A$ restricted to A^+ is contained in the automorphism group of $\mathscr{I}_{A^+}$. Suppose $a, b \in A^+$. Since $ap_0, bp_0 \succ a_0 p_0$, by hypothesis there exists $\langle \theta, \eta \rangle \in \mathscr{F}$ such that

$$bp_0 = \langle \theta, \eta \rangle (ap_0) = \theta(a)\eta(p_0) = \theta(a) p_0,$$

and so $\theta(a) = b$, establishing 1-point homogeneity over $\mathscr{I}_{A^+}$. By Theorem 19.2, the automorphism group of $\mathscr{I}_{A^+}$ satisfies 1-point uniqueness, and so the inclusion is not proper.

(ii) For each $a \in A^-$, define a^+ as the solution to $a^+ \pi(a) \sim a_0 p_0$. By monotonicity, $a^+ \in A^+$. By Theorem 19.12, we know that the automorphism groups of $\mathscr{I}_{A^+}$ and $\mathscr{I}_{A^-}$ are the restrictions of $\mathscr{F}_A$ to A^+ and A^-; and so, by 1-point homogeneity of $\mathscr{I}_{A^+}$, there exists for each $b \in A^-$ some $\theta \in \mathscr{F}_A$ such that $\theta(a^+) = b^+$. Since $\theta(a_0) = a_0$, Theorem 19.12 asserts that $\langle \theta, \pi\theta\pi^{-1} \rangle$ is a factorizable automorphism, from which we see that

$$b^+\pi(b) \sim a_0 p_0 \sim \theta(a^+) \qquad \text{and} \qquad \pi\theta\pi_{-1}\pi(a) \sim b^+\pi\theta(a).$$

By monotonicity and the 1:1 property of π, $b = \theta(a)$; and 1-point uniqueness of both structures follows from Theorem 19.2. ◇

20.7.3 Theorem 24 (p. 182)

Suppose $\mathscr{C} = \langle \mathrm{Re} \times \mathrm{Re}, \succsim \rangle$ *is a real conjoint structure that is unrestrictedly solvable and Archimedean,* $\mathscr{F}$ *is its group of factorizable automorphisms, and* $x_0^{(1)}x_0^{(2)}$ *in* Re × Re *is singular. If* $\mathscr{F}$ *is* 2*-point unique and* $\mathscr{F}_{A^+ \times P^+}$ *and* $\mathscr{F}_{A^+ \times P^-}$ *are both* 1*-point homogeneous, then there exist functions* $\varphi_1: \mathrm{Re} \to$ *(onto)*Re *with* $\varphi_1(x_0^{(i)}) = 0$, $i = 1, 2$, *and a real operation* $\bigcirc$ *such that* $\varphi_1 \bigcirc \varphi_2$ *represents* $\mathscr{C}$, *and* $\bigcirc$ *is of the following form: there exist positive constants* $\eta(+), \eta(-), \sigma(+), \sigma(-)$ *and unit representations*

$$\begin{array}{lll}
f_{++}: & \mathrm{Re}^+ \to [1, \infty) & \\
f_{+-}^+: & [1, \infty) \to [0, \infty), & f_{+-}^+(1) = 0 \\
f_{-+}^+: & [1, \infty) \to [0, \infty), & f_{-+}^+(1) = 0 \\
f_{+-}^-: & (0, 1] \to (-\infty, 0], & f_{+-}^-(1) = 0 \\
f_{-+}^-: & (0, 1] \to (-\infty, 0], & f_{-+}^-(1) = 0 \\
f_{--}: & \mathrm{Re}^+ \to (-\infty, 1] &
\end{array}$$

such that

$$x \bigcirc y = \begin{cases}
x & \text{if } y = 0 \\
y & \text{if } x = 0 \\
y f_{++}(x/y) & \text{if } x > 0, \quad y > 0 \\
|y|^{\eta(+)} f_{++}^+(x/|y|^\rho) & \text{if } x > 0, \quad y < 0, \quad x \geqslant |y|^\rho \\
|y|^{\eta(-)} f_{+-}^-(x/|y|^\rho) & \text{if } x > 0, \quad y < 0, \quad x < |y|^\rho \\
y^{\sigma(+)} f_{-+}^+(|x|^\rho/y) & \text{if } x < 0, \quad y > 0, \quad |x| \geqslant y^{1/\rho} \\
y^{\sigma(-)} f_{-+}^-(|x|^\rho/y) & \text{if } x < 0, \quad y > 0, \quad |x| < y^{1/\rho} \\
|y|\, f_{--}(|x|/|y|) & \text{if } x < 0, \quad y < 0.
\end{cases}$$

PROOF. Because $\mathscr{F}$ is 2-point unique and $x_0^{(1)}x_0^{(2)}$ is singular, $\mathscr{F}$ restricted to each of the four quadrants is 1-point unique, and by hypothesis each is 1-point homogeneous. So each has a representation $\mathscr{G}_{ij}$, $i, j = +, -$, in which the automorphisms are multiplication by the positive reals. Select as a representation $\mathscr{G}$ for $\mathscr{C}$ the one that agrees with the multiplicative one on $A^+ \times P^+$ and on $A^- \times P^-$. For $A^+ \times P^-$, there are indifferences with pairs from $A^+ \times P^+$ and from $A^- \times P^-$, and so there are monotonic functions F^+_{+-} and F^-_{+-} that transform the multiplication $\mathscr{G}_{+-}$ to agree with $\mathscr{G}_{++}$ and $\mathscr{G}_{-+}$. Similar functions exist for $A^- \times P^+$.

By the preceding theorem, the several induced subcomponents are also (1, 1), and they too have representations in which their automorphisms $\mathscr{F}_{A^+}$, etc., are represented by multiplication. Select these, and let φ_1 and φ_2 be the transformations that lead to this form and $\varphi_1(0) = 0$ and $\varphi_2(0) = 0$. With no loss of generality, suppose that $\mathscr{G}$ is the representation in these scales.

Recall from Theorem 19.12, any factorizable automorphism must be of the form $(\theta, \pi\theta\pi^{-1})$, and by our choice it is of the form (r, r') where $r, r' > 0$. Thus, for all $y > 0$ and so for $\pi(y) > 0$, $r'y = \pi[r\pi^{-1}(y)]$. Setting $y = 1$, $r' = \pi[r, \pi^{-1}(1)]$, whence $r = \pi^{-1}(r')/\pi^{-1}(1)$; and so

$$\pi^{-1}(r'y) = \frac{\pi^{-1}(r')\pi^{-1}(y)}{\pi^{-1}(1)}.$$

Thus $\pi^{-1}(y) = \pi^{-1}(1)y^{1/\beta}$ and $r' = \pi[r\pi^{-1}(1)] = r^\beta$ for some $\beta > 0$. A similar argument applies for $y < 0$, ending with a power γ. So the general automorphism is of form

$$\left.\begin{matrix} r, r^\beta \\ r, s^\gamma \\ s, r^\beta \\ s, s^\gamma \end{matrix}\right\} \quad \text{for} \quad \begin{cases} x > 0, & y > 0 \\ x > 0, & y < 0 \\ x < 0, & y > 0 \\ x < 0, & y < 0. \end{cases}$$

So we make a further change of scale, namely, the $1/\beta$ root of φ_2 for the positive part and $-|\varphi_2|^{1/\gamma}$ for the negative part, and change $\mathscr{G}$ accordingly. In these terms, the automorphism becomes

$$\begin{matrix} r, r \\ r, s \\ s, r \\ s, s, \end{matrix}$$

where, of course, $s = s(r)$. By this change, the π function becomes the identity.

For $x > 0$, $y > 0$ we have by construction for some strictly increasing function Π,

$$G(rx, ry) = \Pi(r)G(x, y).$$

As in the (1, 1) case before, this leads to

$$G(x, y) = yf_{++}\left(\frac{x}{y}\right),$$

$f_{++}(z) = G(z, 1)$. In like manner, for $x < 0$ and $y < 0$,

$$G(x, y) = |y|\, f_{--}\left(\frac{|x|}{|y|}\right).$$

For $x > 0$, $y < 0$, $G(x, y) = f_{+-}^{j}[G_{+-}(x, y)]$, where the automorphisms of G_{+-} are multiplicative. Let us work first with G_{+-}. It must satisfy

$$G_{+-}[rx, s(r)y] = \Pi(r)G_{+-}(x, y).$$

We must determine the form of $s(r)$.

For $x > 0$, let $g(x) < 0$ be the solution to $0, 0 = g(x)$, $\pi(x) = g(x)$, x. Since this is invariant under the automorphism $[s(r), r]$, we have

$$\begin{aligned} 0, 0 &= s(r)g(x), rx \\ &= g(rx), rx \qquad \text{(by definition of } g). \end{aligned}$$

So by monotonicity, $g(rx) = s(r)g(x)$. In the usual way, it follows that for some $\rho > 0$, $g(x) = g(1)x^{1/\rho}$ and $s(r) = r^{1/\rho}$. So the functional equation is

$$G_{+-}(rx, r^{1/\rho}y) = \Pi(r)G_{+-}(x, y)$$

Set $x = 0$:

$$G_{+-}(0, r^{1/\rho}y) = \Pi(r)G_{+-}(0, y)$$

Set $x = 1$, $y = 1$:

$$G_{+-}(0, r^{1/\rho}) = \Pi(r)G_{+-}(0, 1).$$

Thus $G_{+-}(0, y) = G_{+-}(0, 1)y^{\eta}$ and $\Pi(r) = r^{\eta/\rho}$. So the equation becomes

$$G_{+-}(rx, r^{1/\rho}y) = r^{\eta/\rho}G_{+-}(x, y).$$

Set $r = 1/|y|^\rho$ and

$$G_{+-}\left(\frac{x}{|y|^\rho}, 1\right) = \frac{G_{+-}(x, y)}{|y|^\eta},$$

or

$$G_{+-}(x, y) = |y|^\eta g_{+-}\left(\frac{x}{|y|^\rho}\right),$$

where $g_{+-}(z) = G_{+-}(z, 1)$. Note that $x = |y|^\rho$ is the locus of points equivalent to 0, 0, and so $g_{+-}(1) = 0$.

Recall that $G(x, y) = F^j_{+-}[G_{+-}(x, y)]$, where j depends upon the sign of G_{+-}. Consider the case $j = +$. We must consider equivalence of the form $x, y \sim x', y'$, where $x > 0$, $y < 0$, $x' > 0$, $y' > 0$, i.e., in terms of the forms derived,

$$y' f_{++}\left(\frac{x'}{y'}\right) = f^+_{+-}\left[|y|^\eta\, g_{+-}\left(\frac{x}{|y|^\rho}\right)\right].$$

Consider the automorphism $(\theta, \pi\theta\pi^{-1}) = (\theta, \theta)$ on this pair, i.e.,

$$rx', ry' \sim rx, r^{1/\rho} y,$$

and so

$$y' f_{++}\left(\frac{x'}{y'}\right) = f^+_{+-}\left[r^{\eta/\rho}\, |y|^\eta\, g_{+-}\left(\frac{x}{|y|^\rho}\right)\right].$$

Setting

$$z = |y|^\eta\, g_{+-}\left(\frac{x}{|y|^\rho}\right),$$

we see that

$$r f^+_{+-}(z) = f^+_{+-}[r^{\eta/\rho} z].$$

Proceeding in the usual way, we get $f^+_{+-}(z) = z^\theta$ for some $\theta > 0$. Thus,

$$G(x, y) = y^{\theta\eta} g^-_{+-}\left(\frac{x}{|y|^\rho}\right)^\theta.$$

So, if we set $\eta^+ = \theta\eta$ and $f^+_{+-} = g_{+-}(x/|y|^\rho)^\theta$, we get the form stated.

Similar arguments hold for $j = -$ and for $x < 0$, $y > 0$. ◇

20.7.4 Theorem 25 (p. 183)

Suppose $\mathscr{C} = \langle \mathrm{Re} \times \mathrm{Re}, \succsim \rangle$ *is a real conjoint structure that satisfies unrestricted solvability, and* $\mathscr{F}$ *is its group of factorizable automorphisms. If* $\mathscr{F}$ *satisfies 2-point uniqueness and component 2-point homogeneity, and* $\mathscr{F}_i$ *on* $\langle \mathrm{Re}, \succsim_i \rangle$, $i = 1, 2$, *both satisfy 2-point uniqueness, then* $\mathscr{C}$ *satisfies the Thomsen condition.*

PROOF. Since $\mathscr{F}$ satisfies 2-point uniqueness and component 2-point homogeneity, it is of scale type (2, 2); and so by Theorem 5(v), there is a representation of $\mathscr{C}$ so that the factorizable automorphisms are affine transformations. Note that by definition of component 2-point uniqueness and homogeneity, the $\mathscr{F}_i$ on $\langle \mathrm{Re}, \succsim_i \rangle$, $i = 1, 2$, are both of scale type (2, 2); thus a further representation of the components leads to $\mathscr{F}_i$ being affine transformations. Thus there is no loss of generality in assuming that $\succsim$ is represented by F: $\mathrm{Re} \times \mathrm{Re} \to \mathrm{Re}$ and that there are functions ρ, σ: $\mathrm{Re}^+ \times \mathrm{Re} \times \mathrm{Re}^+ \times \mathrm{Re} \to (\text{onto})\mathrm{Re}^+$ such that for some $r, r' \in R^+$ and some $s, s' \in \mathrm{Re}$, and all $x, y \in \mathrm{Re}$,

$$F(rx + x, r'x + s') = \rho(r, s, r', s')F(x, y) + \sigma(r, x, r', s'). \quad (35)$$

For any real x_0, x_0', y_0, y_0', component 1-point homogeneity implies there is a factorizable automorphism $(x, y) \to (rx + s, r'y + s')$ such that

$$rx_0 + s = x_0' \qquad \text{and} \qquad r'y_0 + s' = y_0'. \quad (36)$$

Let π and π' be the solutions defined relative to $x_0 y_0$ and $x_0' y_0'$, respectively. By Theorem 19.12, we know this automorphism must satisfy

$$r'y + s' = \pi'[r\pi^{-1}(y) + s]. \quad (37)$$

Substituting Equation (36) into Equation (37) yields

$$\pi'^{-1}[r'(y - y_0) + y_0'] = r[\pi^{-1}(y) - x_0] + x_0'. \quad (38)$$

Define

$$f(x) = \pi^{-1}(x + y_0) - x_0 \qquad \text{and} \qquad f(y)' = \pi'^{-1}(y + y_0') - x_0',$$

and so Equation (38) becomes

$$f(r'x)' = rf(x). \quad (39)$$

Set $x = 1$, solve for r, and substitute:

$$f(r'x)' = \frac{f(r')'f(x)}{f(1)}. \tag{40}$$

Note that by Equation (39), with $x = 1$, the value $r = f(1)'/f(1)$ corresponds to $r' = 1$. In Equation (40), set $r' = 1$, solve for $f(x)$, and substitute back into Equation (40) to obtain

$$f(r'x)' = \frac{f(r')'f(x)'}{f(1)'}.$$

Thus, for some β,

$$\frac{f(x)}{f(1)} = \frac{f(x)'}{f(1)'} = \begin{cases} x^{1/\beta}, & x \geqslant 0, \\ -|x|^{1/\beta}, & x < 0. \end{cases} \tag{41}$$

Observe that β appears to depend upon x_0, x_0', y_0, y_0'. However, since any π can be paired with any π', and so any r with any r', it follows that all the β's must be the same.

We next show that for any real s, $(x, y) \to (x + s, y)$ is a factorizable automorphism. Fix x_0, y_0, and let $x_0' = x_0 + s$ and $y_0' = y_0$. The question is whether the following condition from Theorem 19.12, is satisfied, namely, that for all y,

$$y = \pi'[\pi^{-1}(y) + s]$$

is satisfied. From Equation (40) with $r = r' = 1$ and $x = 1$, we see that $f(1)' = f(1)$. Using this and Equations (39) and (41), we see that

$$\pi^{-1}(y) = x_0 + f(1)\begin{cases} (y - y_0)^{1/\beta}, & y \geqslant y_0, \\ -(y_0 - y)^{1/\beta}, & y < y_0. \end{cases}$$

$$\pi'(x) = y_0 + f(1)^{-\beta}\begin{cases} (x - x_0 - s)^{\beta}, & x \geqslant x_0 + s, \\ -(x_0 + s - x)^{\beta}, & x < x_0 + s. \end{cases}$$

Thus,

$$\pi'[\pi^{-1}(y) + s] = y_0 + f(1)^{-\beta} \begin{cases} [\pi^{-1}(y) + s - x_0 - s]^\beta, \\ \qquad y \geqslant \pi(x_0) = y_0, \\ -[x_0 + s - \pi^{-1}(y) - s]^\beta, \\ \qquad y < y_0, \end{cases}$$

$$= y_0 + f(1)^{-\beta} \begin{cases} \left[f(1)(y - y_0)^{1/\beta}\right]^\beta, & y \geqslant y_0, \\ -\left[f(1)(y_0 - y)^{1/\beta}\right]^\beta, & y < y_0, \end{cases}$$

$$= y.$$

From this and Equation (33), we have

$$F(x + s, y) = \rho(s)F(x, y) + \sigma(s).$$

With no loss of generality, assume 0 is an identity of F. If we set $y = 0$,

$$s + x = F(x + s, 0) = \rho(s)F(x, 0) + \pi(s) = \rho(s)x + \sigma(s).$$

Next, set $x = 0$, and use the above relation to obtain

$$F(x, y) = \rho(s)F(0, y) + \sigma(s) = \rho(s)y + \sigma(s) = s + y,$$

thus proving F is additive. ◇

EXERCISES

1. Prove Theorem 1. (20.2.4)

2. For an ordered relational structure, prove that the set of all dilations with a common fixed point forms a group. Give a counter example to show that the set of all dilations do not form a group. (20.2.4)

3. For an ordered relational structure, suppose $\mathscr{D}$, $\mathscr{T}$, and $\mathscr{F}$ are, respectively, the set of dilations, the set of translations, and a subgroup of the automorphism group. Let $\succsim'$ be the asymptotic ordering of automorphisms defined in Lemma 2 (Section 20.3.4). If the structure is N-point unique and $\langle \mathscr{F}, \succsim' \rangle$ is an Archimedean ordered group, show that (i) either $\mathscr{F} \subseteq \mathscr{D}$ or $\mathscr{F} \subseteq \mathscr{T}$, and that (ii) if in addition $\mathscr{F}$ is homogeneous, then $\mathscr{F} \subseteq \mathscr{T}$. (20.2.5)

4. Under the assumptions of Exercise 3, for $\gamma \in \mathscr{F}$, define

$$\mathscr{H} = \{\alpha \mid \alpha \in \mathscr{G} \text{ and for some integer } n > 0,\ \gamma^n \succ' \alpha,\ \alpha^{-1}\}.$$

Prove that (i) $\mathscr{H}$ is independent of the choice of γ; (ii) $\mathscr{H}$ is a convex group; and (iii) $\mathscr{F} \subseteq \mathscr{H}$. (20.2.5)

5. Suppose that an ordered relational structure is N-point unique and that its group (Exercise 2) of dilations at a, $\mathscr{D}_a$, is Archimedean ordered under $\succsim'$ (see Definition 4) and that for each $b \in A$ with $b \neq a$, $\mathscr{D}_a \not\subseteq \mathscr{D}_b$. Show that $\mathscr{D}_a$ is 2-point unique. (20.2.5)

6. Prove Theorem 6. (20.2.6)

7. Suppose $\mathscr{C} = \langle A \times P, \succsim \rangle$ is an unrestrictedly solvable, Archimedean, conjoint structure that has a multiplicative representation in Re^+. For each $p, q \in P$, define the mapping $\beta(\cdot; p, q)$: $A \to A$ by $xp \sim \beta(x; p, q)q$. Let $\mathscr{B}$ denote the set of all such maps. Show for $\beta \in \mathscr{B}$ that (i) β is onto A; (ii) either $\beta = \iota$ or β has no fixed point; and (iii) β can be ordered in such a way as to form an Archimedean ordered group. (20.2.6)

8. Suppose $\mathscr{A} = \langle A, \succsim_A, S_j \rangle_{j \in J}$ is a structure that is distributive in the conjoint $\mathscr{C}$ of Exercise 7. If $\mathscr{T}$ is the set of translations of $\mathscr{A}$, show that $\mathscr{T} = \mathscr{B}$. (20.2.6)

9. Show that the real unit structures of Theorem 10 satisfy the conditions of Definition 5 of Section 20.2.6. (20.4.2)

10. Prove Theorem 10. (20.4.2)

11. Show that $\langle \mathrm{Re}^+, \geqslant, \otimes \rangle$, where for $x, y \in \mathrm{Re}^+$,

$$x \otimes y = x + y + \frac{axy}{x + y}, \qquad a \geqslant 0,$$

is a commutative, real unit structure. (20.4.2)

12. Suppose $\langle \varphi, f \rangle$ is the unit representation of a real unit structure, where $x \otimes y = yf(x/y)$. Develop an explicit formula for the n-copy operator $x(n)$, and use it to show that for each positive integer n, $x(n)$ is an automorphism of the structure. (20.4.2)

13. Prove Theorem 13. (19.4.2)

14. Suppose that H: $\mathrm{Re} \times \mathrm{Re} \to \mathrm{Re}$ is nontrivial and monotonic in each variable and it satisfies, for all $x, y, s \in \mathrm{Re}$ and $r \in \mathrm{Re}^+$,

$$rH(x, y) + s = H(rx + s, ry + s).$$

Prove that for some $c, d \in (0,1)$,

$$H(x, y) = \begin{cases} cx \quad (1-c)y, & x \geqslant y, \\ dx + (1-d)y, & x < y. \end{cases}$$

(Hint: establish that $H(x, y) = y + f(x - y)$, and solve the functional equation for f.) (20.4.2)

15. Prove the uniqueness statement of Theorem 11. (20.4.2)

16. Prove Theorem 12. (20.4.2)

17. Let R be the closure of $Ra^{+} \cup \{e\} \cup \{\pi\}$ under addition and multiplication by positive rationals. Show that $\mathscr{R} = \langle R, \geqslant, + \rangle$ is a PCS and that its group of automorphisms is order dense but not homogeneous. (This establishes that in part (iii) of Theorem 14, one cannot simply drop the hypothesis that the PCS is Dedekind complete.) (20.4.3)

18. Suppose $\mathscr{A}$ is a PCS. Show that if $\mathscr{A}$ satisfies automorphism density, then the automorphism group is dense in the sense that for each $a \in A$, the set $\{\alpha(a) \mid \alpha \in \mathscr{G}\}$ is dense in $\langle A, \succsim \rangle$. (20.4.3)

19. Let $\mathscr{R} = \langle \mathrm{Re}^{+}, \geqslant, \bigcirc \rangle$, where $x \bigcirc y = (x + y)/2$. Let θ be as defined in Theorem 17. Show that $\theta(x, y, n) = [x + (2^n - 1)y]/2^n$, and prove that for any translation τ, $\theta(\tau, \iota, n)$ is also a translation. (20.4.4)

20. Prove Theorem 20. (20.4.6)

21. Prove Theorem 21. (20.4.6)

22. Suppose $\mathscr{C}$ is a conjoint structure. Show that (i) if $\mathscr{C}$ is component M-point homogeneous, then $\mathscr{C}$ is M-point homogeneous; and (ii) if $\mathscr{C}$ is N-point unique, then $\mathscr{C}$ is component N-point unique. (20.6.1)

23. Let $F(x, y) = ax + by + cxy + dx^2 + ey^2$. Show that F is additive if and only if either $c = 0$ or $d = e = 0$. (20.6.3)

Chapter 21 Axiomatization

In preceding chapters we have given numerous examples of axiom systems for various kinds of measurement. On several occasions we have distinguished between kinds of axioms, especially between necessary axioms (Section 1.4.1) and structural or nonnecessary axioms (Section 1.4.2). We have also given special notice to Archimedean axioms and the various forms they can assume. However, we have saved for the present chapter our consideration of general problems about the axiomatic approach to representation in measurement. We begin with a section devoted to general issues involved in selecting primitives, axioms, and representing structures. We then turn to more specific questions of definability and axiomatizability.

The mathematical analysis of the *form* of measurement axioms requires some specific methods. The mathematical concepts that we use in this chapter are drawn from mathematical logic and are relatively special compared with the mathematical ideas common to most of the chapters. For this reason we shall attempt to give a fairly self-contained account, but the reader who wants a deeper and more extensive orientation is urged to look at the excellent volume on the theory of models by Chang and Keisler (1977) or the appropriate chapters in the handbook on mathematical logic edited by Barwise (1977). As we shall see, what logicians call the theory of models is the part of logic most pertinent to the subject matter of the present chapter. Good textbook discussions of the theory of models are to be found in Kleene (1967), Schoenfield (1967), and Mendelson (1964). As a necessary background, in Section 2 we develop the important concept of an elementary formalization of a theory (for extensive details, see Kleene, 1952).

The main significant results that we survey for theories of measurement applying methods of logic are the following. In Section 3 we outline criteria of definition and methods of proving that concepts are not definable in a given axiom system. In Section 4 we give various positive and negative theorems on the axiomatizability of certain theories of measurement. Proofs are in Section 5. In Section 6 we consider finite axiomatizability, i.e., axiomatization with a finite set of axioms. In Section 7 we concentrate on the Archimedean axiom and prove the result, due to Abraham Robinson (1951), that this axiom in any of its familiar forms is strictly more powerful than any finite or infinite collection of *elementary* axioms (the exact meaning of "elementary" will become clear in the sequel). In Section 8 we discuss in some detail the relation between particular kinds of axioms, especially solvability and Archimedean axioms, and their testability in finite or infinite data structures, relations between constructability and testability, and the contrast between diagnostic and global tests.

21.1 AXIOM SYSTEMS AND REPRESENTATIONS

21.1.1 Why Do Scientists and Mathematicians Axiomatize?

Axiomatization is an activity that is sometimes found in mathematics and science, but not always. Its origins are ancient (Euclid's geometry), and it is widely used in contemporary mathematics and to a lesser degree in physics. But many scientific theories have not been developed axiomatically. For example, during the eighteenth and nineteenth centuries, classical mechanics, dynamics, thermodynamics, and electromagnetic theories were developed within the framework of analysis, and there was little attempt to axiomatize them. So it is natural to ask under which conditions axiomatizations appear to be useful and whether they appear to apply to the study of measurement.

We can distinguish four clear-cut roles played by axiomatizations though most specific examples exhibit some mix of them. The first is to make explicit the exact details of an argument or theory that, before the axiomatization, was either incomplete or unclear. The second is to isolate, abstract, and study more fully a class of mathematical structures that have recurred in many important contexts, often with quite different surface forms. Usually, particular methods of argument are discovered to be more general and useful than was apparent before the axiomatization was carried out.

The third is the study of what sorts of things can be axiomatized in particular ways. And the fourth is to provide the scientist with a compact way to summarize and represent an empirical situation, one that allows a systematic exploration of the implications of the postulates. We give examples in this section of the first three uses of axiomatization. Some of them concern major developments in science and mathematics, and some of the later ones are drawn from the narrower confines of measurement theory. The fourth use is illustrated by a large fraction of the results in these three volumes.

After the development of ancient Greek mathematics, perhaps the most striking examples of clarification occurred in the nineteenth century when, in general, the standards of rigor in mathematical argument were raised, and arguments that had previously been accepted had to be reexamined. We mention three.

1. A major reexploration of the axiomatic underpinnings of geometry took place, with an eye to bringing out implicit assumptions and exploring what happened as these were altered. The great flowering of alternative axiom systems (many reviewed in Chapter 13) arose in part from such efforts, especially those concerned with the development of non-Euclidean geometry.

2. An effort began, and continued into this century, to make completely explicit the basic nature of logical and mathematical reasoning, i.e., the development of mathematical logic. Here, careful clarification has led to such famous results as those of Gödel, which showed that the intuitions of the best earlier mathematicians were wrong about certain fundamental issues.

3. One of the most dramatic events in the history of science was Einstein's careful reexamination of certain basic (and usually implicit) postulates of Newtonian kinematics. The key was his recognition that no precise meaning can be given to the statement that two separated events have occurred simultaneously. The implication of these ideas leading to the special theory of relativity were explored in Section 13.9.

We mention several additional, if more modest, examples that are closer to measurement issues, all of which are drawn from the behavioral and social sciences.

4. Before the work on additive conjoint measurement (see Chapter 6) it was widely accepted that one should (or could) not rescale observations in factorial situations before checking for additivity. This belief stemmed in part from the implicit assumption that the measures used often form interval or ratio scales because they are physical quantities, such as re-

sponse times. What occurred was, first, a clarification of what it means for an ordinal structure to have an additive representation and, second, a gradual realization that indices of order for an attribute do not automatically have particular scale types just because they rest on physical measures, such as time.

5. Another case in which an axiom system was developed in response to incomplete arguments was an analysis of the explanation of why ordinal multidimensional scaling works as well as it does. Shepard (1962a, 1962b) demonstrated by computations that a configuration of points lying in a Euclidean space could be recovered with surprising accuracy given only knowledge of the ordinal properties of the interpoint distances. He remarked on the conversion from an ordinal (called "nonmetric" in this context) to a metric representation as a puzzling phenomenon, tantamount to getting something for nothing (1962b, p. 238). He specified some conditions under which the method fails, but it remained unclear, despite the utterly convincing computations, why it so often succeeded. Beals and Krantz (1967) explored the problem axiomatically for the case of metrics with additive segments. Their uniqueness theorem showed how to recover the metric from sufficiently rich ordinal data (Sections 14.2.2 and 14.2.3). Complementing the uniqueness theorem is an axiom system that, if satisfied by interpoint distances, guarantees the existence of some metric with additive segments that can be recovered from them.

6. Without going into detail, we remind the reader of other examples of clarification effected by axiomatizations. Chapter 10 and the subsequent developments reported in Section 22.7 on axiomatizing dimensional analysis were motivated entirely by the lack of clarity in the literature on the empirical base leading to the familiar physical interlock of extensive and conjoint measures as products of powers. The development of color theory in Chapter 15 was based on making explicit the implicit assumptions involved in the cancellation procedure of Jameson and Hurvich (1955), and it has led to several empirical discoveries. The work on threshold representations reported in Chapter 16 stemmed originally from two sources: from some lack of rigor in arguments made about the representation of thresholds in psychophysical research and from a concern about the unrealistic idealization that judged indifference be treated as a transitive relation.

A second major role for axiom systems—the isolation of common structures and arguments—is the preeminent one in modern mathematics, and its use has spilled over considerably into modern science. We cite three major developments that took place largely in the century beginning about 1850.

1. As Bishop Berkeley correctly observed, the arguments used by Newton in developing the calculus were flawed. This difficulty did not receive clarification until much later, when mathematicians began to note that similar arguments were based on the fact that any sequence—of numbers, points in space, functions, or operators—must contain a convergent subsequence. This property was ultimately isolated as an axiom (now called *sequential compactness*), and the argument could then be applied to the abstract class of spaces satisfying this axiom.

2. An abstraction of utmost significance was that of sets as objects. Sets had long been used informally, and certain kinds of arguments involving induction on members of sets were found in many contexts. Increasingly it came to be realized that much, perhaps all, of mathematics could be viewed from the perspective of sets and that it was important to state their properties clearly. Many of the axioms were not controversial; but some were, the most important being the axiom of choice, which led to trouble when used too expansively. Great clarification, much interlocked with developments in mathematical logic, and abstraction began just before the turn of the century and continues today. (Our entire enterprise in this work is based on informal set theory, as is most contemporary mathematics.)

3. Another major abstraction of importance to measurement theory is the algebraic concept of a group. It became evident during the nineteenth century that there are a number of mathematical systems of direct importance—addition and multiplication of numbers and of matrices being the simplest—with the common property of having an operation that is associative, an identity, and for each element, an inverse. But such operations arose in various ways as abstract systems were axiomatized and their inherent structure studied. For example, the automorphisms of geometries and later of other systems became a way of exploring certain abstract and common features, and these automorphisms formed mathematical groups. Ultimately, the subject of groups became a topic unto itself, one that has found applications throughout the sciences, especially modern physics.

Ordered groups, both as the basis of measurement and as a way of classifying measurement systems in terms of automorphisms, have figured large in this work. As a basis of constructive measurement, one abstraction that we have seen clearly and repeatedly is that of Archimedeanness, which is a property that ordered groups may exhibit. Even if a structure with an operation does not satisfy associativity, Archimedeanness continues to play a crucial role as a property of the automorphism group. It is what permits construction of a certain type of approximation to a measurement representation whenever an operation is available. The common aspect is reflected in the identical wording of the Archimedean axiom as "every bounded standard sequence is finite," once standard sequence and boundedness have been defined appropriately for the context.

Somewhat more abstract is the role of Archimedean-ordered groups of translations (automorphisms with no fixed point, together with the identity) in structures for which no empirical operation is a primitive nor is one easily defined in terms of the primitives. In several special cases—extensive, conjoint, general concatenation, and in general relational systems on the real numbers—it was found that the translations formed an Archimedean-ordered group. In the homogeneous case that fact provides a schema for constructing a representation and for characterizing the family of possible representations as generated by a subgroup of the positive affine group. A certain unity is thus achieved by abstracting an important common feature from special cases that have arisen more or less naturally.

Still another measurement structure that often recurs is the additive conjoint structure with identical or closely related component factors. Such structures were treated abstractly in Section 6.11.2; but they came up again in the treatment of conditional expected utility (Section 8.2.4), in the additive difference metric (Theorem 14.9), in invariant proximity-feature structures (Theorem 14.11), and in generalized linear operators (Krantz, 1973). Such common abstractions are, of course, what gives the area of measurement the comparative unity that it has.

The third use of axiomatization (cited earlier) is to clarify what is and is not possible in axiomatizations, thereby clarifying the nature of various classes of mathematical structures. An example especially pertinent to measurement is the negative result that Archimedean axioms cannot be expressed in first-order logic (Section 21.7). This result makes clear one sense in which Archimedean axioms are not elementary in character and why they possess substantial power. Another kind of example is to decide whether the structures satisfying certain properties can be axiomatized by a finite list of axioms. A classical result, mentioned in Section 21.6, is that of Ryll-Nardzewski showing that the Peano axiomatization of elementary number theory cannot be expressed by a finite number of first-order axioms. Rather a schema of induction is required that corresponds to an infinite list of first-order axioms. The difference between this and the Archimedean result is that in the latter case no first-order formulation whatsoever is possible whereas for number theory a first-order recursive theory is possible, but no finite theory is.

Results along similar lines will be given for particular theories of measurement in later sections of this chapter. For example, the fact that various classes of finite measurement structures cannot be finitely axiomatized goes a considerable distance in showing conclusively just how complex finite structures can be (Section 21.6). An important insight into the nature of measurement is to realize that merely because measurement structures of a certain type are all finite does not imply the existence of a simple and transparent axiomatization of them.

21.1.2 The Axiomatic-Representational Viewpoint in Measurement

The most pervasive abstraction in measurement theory consists in formalizing basic observations as a relational structure, that is, a set with some primitive relations and operations. This abstraction arises from considering the nature of empirical, qualitative observations. Typically, we study some restricted class of relational structures, characterized by axioms that are satisfied either by the primitive relations or by the automorphism group of the structure. It is usually difficult to comprehend at all fully the implications of the axioms. Nevertheless, if a representation theorem exists that establishes structural identity (isomorphism) or structural similarity (homomorphism) to an easily described, familiar numerical structure, then the implications become clearer. They often become transparent (e.g., all of the cancellation properties of an additive conjoint representation in Chapter 6) or reduce to well-understood numerical calculations (e.g., the unit representations of homogeneous concatenation structures in Section 20.4). Typically, these representing structures are expressed in terms of operations or functions of real numbers, or in a real vector space, or in some well studied geometrical system.

A second useful feature of such an axiomatic approach is that it lays down the basis for numerical calculations—for the use of classical analysis including calculus, differential equations, functional equations, and probability methods. A full understanding of the structure and its numerical representation makes clear which calculations lead to meaningful conclusions that can be interpreted substantively.

In developing such an axiomatic-representational theory, one encounters three main problems, the first of which is largely conceptual and the other two largely mathematical. The conceptual task is to select the abstraction to be studied, namely, the primitives and the axioms. Any class of structures can be characterized in a variety of ways, and the scientist has options for characterizing any particular case. Throughout this work, we have assumed one primitive to be an ordering; but we know that is not enough to warrant the use of the strong techniques of analysis. So a number of additional primitives have been explored, the most prominent being operations, factorial structure, and (in a sense) automorphism groups. Once the primitives are selected, the question of axioms arises. Some axioms, such as transitivity and associativity, are universal statements; others, such as solvability and closure of operations, are existential. The former typically are necessary properties in the sense that they must be satisfied if the structure is to have a particular representation. The latter rarely are necessary properties and so constitute a restriction on our choice of the class of structures to consider that have the intended representation; they are invoked largely because they make the two purely mathematical problems more tractable, which means that in applying the theory one must go to some pains to ensure that they are satisfied either by the choice of application or by the empirical procedures used.

A major limitation that is frequently invoked, often indirectly, is to axiomatize only structures with continuum representations, eliminating finite and countable ones. The reasons are the following. First, the behavior of structures that can be represented on the continuum is usually far more regular and easily understood than is the behavior of the myriad possibilities that arise in the other cases; indeed, axiomatizations for finite structures often require additional universal axioms while omitting solvability and Archimedean ones. Second, most techniques of analysis are based on the continuum, in particular, on the existence of limits of sequences that enable the calculation of such quantities as the derivatives and integrals of functions.

The first mathematical problem is to establish the existence of an isomorphism between the structures delineated by the primitives and axioms and some familiar, usually numerical, mathematical structure. Moreover, if possible, the mathematical construction should be adapted to actually calculating the representation for a prescribed sample of data to some specified degree of accuracy. The second, and closely related, mathematical problem is to understand the entire family of possible representations. The two are called the representation and uniqueness problems, respectively.

In Sections 21.4–21.7 we state and prove a number of results on the problem of representation. Almost none of these results bear on the uniqueness problem, but that has received extensive analysis in Chapter 20.

21.1.3 Types of Representing Structures

In much of the work reported in these three volumes, the representation was assumed to be *onto* the positive, multiplicative real numbers. It was common in classical physics to treat almost all variables as measured by real numbers, capable of assuming any value; this proved to be highly convenient even if in particular cases, such as mass in particle mechanics, only a finite number of values are possible. Alternatively, such discrete structures can sometimes be modeled by an equally spaced structure that corresponds to a single standard sequence, in which case the isomorphism is often into or onto the integers. Another important class of representing structures for bounded qualitative structures is that of intervals of numbers. The best-known examples are probability, usually mapped into $[0, 1]$, and relativistic velocity, mapped onto $[0, c]$, where c is the velocity of light.

Going in the other direction, to representations in structures that are richer than the real numbers, we encounter the work of Narens (1985) on representation in "nonstandard" real numbers. In such structures, each real number has an associated class of nonstandard numbers that differ infinitesimally from one another. The Archimedean axiom fails. Hence these structures might be useful if one ever encountered qualitative empirical structures in which the Archimedean axiom seems to be false. We are not aware of plausible applications along this line, but the study of such representations is interesting because it isolates the implications of the first-order axioms.

Still another important class of representing structures consists of the various analytic geometries described in Chapter 12 and used in the representation theorems cited in Chapters 13–15. One feature of geometrical representations differs appreciably from one-dimensional measurement, namely, the intimate and rather surprising relation between certain geometric properties and certain algebraic properties of the representing structure. The classical example, given in Chapter 13, is that a Desarguesian projective plane satisfies Pappus' proposition if and only if the multiplication operation of the representing division ring is commutative. Less familiar examples were also given, such as representing the relation between the Desarguesian property of a plane and the associativity of the representing operations of addition or multiplication. In these cases the focus is not on direct isomorphism between the primitive concepts of the geometry and the representing concepts of the "numerical" structure. The representing structure has, for example, operations of addition and multiplication as primitive, and then new operations or relations are defined in terms of them, isomorphic to the primitives of the geometric structure. We ordinarily anticipate, in thinking about analytical geometry, that such representing structures will be fields with the elementary properties of the field of real numbers. It has been important and clarifying to show that in the kinds of examples just mentioned this is not the case. As far as we know, no similarly surprising results have yet been found for one-dimensional measurement structures.

One of the most important aspects of a representation concerns the existence of limits of sequences within the representation. Limits are relevant when measurement is a precursor to the use of continuum mathematics, e.g., differential equations. In structures defined on the rational numbers such limits do not necessarily exist; that incompleteness is exactly how and why the real numbers came about. The irrational numbers can be thought of as the missing limits of some sequences of rational numbers. Within the framework of general, ordered, relational structures—not just numerical ones—the existence of limits of sequences is called Dedekind completeness in the measurement literature. Since this is a useful property, it is interesting to know when a structure can be Dedekind completed, i.e., embedded densely into a structure that is Dedekind complete. The classical example of this is the dense embedding of the ordered rational numbers in the ordered real numbers. Such an embedding may also justify the use of continuum mathematics. This topic was discussed in some detail in Chapters 19 and 20 for generalized concatenation structures.

21.2 ELEMENTARY FORMALIZATION OF THEORIES

In this section we provide some general background from logic that is available in more detail in sources cited above, but the exposition here is meant to be essentially self-contained. First, we characterize elementary or first-order languages (21.2.1); then we characterize elementary logic (21.2.2) and give some examples of elementary theories (21.2.3). In the next subsection (21.2.4), we state some of the general theorems on axiomatizability of elementary theories.

21.2.1 Elementary Languages

In this section and in following ones we refer to elementary languages, elementary logics, and elementary theories. There are at least two other terminologies in the literature: one is to speak of *first-order* languages, etc.; another is to speak of *standard* languages, etc. We believe that the term *elementary* is most suggestive, and it is the one we shall adopt here. The exact characterization of elementary languages is given below, but the intuitive idea is easy to state: essentially, the variables range over a domain of individuals but not over sets of properties of the individuals. Thus, for example, in an elementary language we cannot say that for every object x there is a property P such that $P(x)$. Matters are more subtle than they might seem because we can express in an elementary language a statement such as *For every nonempty set there is an element x that is a member of that set*. The reason for this is that we can let sets count as individuals, and thus the variables range over sets as well as over individuals. In practice, we can stay within elementary languages as long as we have a finite number of different sorts of objects in mind, for example, individuals and sets, which, we emphasize, is no restriction on having an infinite number of objects as such.

We now turn to a formal development of these ideas. First, we begin with a fourfold classification of the symbols of an elementary language L as logical symbols, relation symbols, function symbols, and individual constants. The logical symbols themselves are divided into five categories: parentheses, sentential connectives, quantifiers, variables and logical predicates.

The parentheses are just the familiar left and right parentheses.

The sentential connectives are the five most common sentential connectives (theoretically we could use just two of the five). In particular, the *negation* of a formula P is written as $\neg P$; the *conjunction* of two formulas P and Q is written as $P \mathbin{\&} Q$; the *disjunction* of P and Q as $P \vee Q$; the *implication* with P as antecedent and Q as consequent as $P \rightarrow Q$; the *equivalence* P if and only if Q as $P \leftrightarrow Q$.

The notation for the quantifiers is also standard and familiar. The universal quantifier *for every* v is written as $(\forall v)$, the existential quantifier *for some* v as $(\exists v)$; we occasionally use the symbol $(\exists! v)$ for *there is exactly one* v *such that*. The variables are lowercase Latin letters with primes and subscripts as required; it is important that we have a countable infinity of such variables available, so as not to restrict, relative to some arbitrary integer n, the expressive power of the language.

Following the usual practice, we also include as our only logical predicate, the identity symbol $=$.

From an official standpoint, which we shall informally deviate from on various occasions, the relation symbols of an elementary language L will be the capital Latin letter R with and without subscripts and possibly with primes; function symbols will be f with and without subscripts and possibly with primes. Individual constants will be denoted by the letter c with and without subscripts and possibly with primes. For the purposes of the developments in this chapter we shall assume that the relation symbols, function symbols, and individual constant symbols of an elementary language L form a finite set.[1] (It should be noted that for many other purposes of logic in the last several decades it has been important to study languages with denumerably many such symbols, and readers can find exposition of these matters in Barwise (1977); one application of such ideas is made in the analysis of Archimedean axioms in Section 21.6.)

[1] The symbols in this set we often refer to as the *nonlogical* symbols of a theory, or as the *nonlogical constants* of a theory.

Each relation symbol of L is assumed to be an n-placed relation for some natural number n; also, each function symbol of L is an n-placed function symbol. To distinguish individual constants from 0-placed relation or function symbols, the latter shall be excluded. When using such familiar symbols as, for example, an ordering relation, we shall use the standard informal infix notation rather than the formal prefix notation defined above; that is, instead of $R(x, y)$, we shall write

$$x \geqslant y.$$

The next step is to define *terms* of an elementary language L. We use the usual recursive definition. Throughout the definition it is understood that the objects discussed are objects of the language L. Thus, the first clause could be expanded to read "A variable of L is a term of L."

(i) A variable is a term.

(ii) A constant symbol is a term.

(iii) If f is an n-placed function symbol and $t_1, \ldots, t_n$ are terms, then $f(t_1, \ldots, t_n)$ is a term.

(iv) A string of symbols is a term if and only if it can be shown to be a term by a finite number of applications of (i)–(iii).

Having defined terms, we now define atomic formulas of an elementary language L. As should be obvious, what we are doing is giving the syntax of L in a certain standard recursive form. The atomic formulas of L are strings of one of the following two forms:

(i) $t_1 = t_2$ is an atomic formula, where t_1 and t_2 are terms.

(ii) If R is an n-placed relation symbol and $t_1, \ldots, t_n$ are terms, then $R(t_1, \ldots, t_n)$ is an atomic formula.

As can be seen from this characterization, there are just two ways of building up atomic formulas: by using the logical predicate of identity and by using one of the nonlogical relation symbols of L.

We next define formulas in general.

(i) Every atomic formula is a formula.

(ii) If P is a formula, then $\neg(P)$ is a formula.

(iii) If P and Q are formulas then $(P \,\&\, Q)$, $(P \vee Q)$, $(P \rightarrow Q)$, and $(P \leftrightarrow Q)$ are formulas.

(iv) If P is a formula and v is a variable, then $(\forall v)(P)$, $(\exists v)(P)$, and $(\exists ! v)(P)$ are formulas.

(v) A string of symbols is a formula if and only if it can be shown to be a formula by a finite number of applications of (i)–(iv).

A few remarks concerning quantifiers may also be helpful. The *scope* of a quantifier is the quantifier itself together with the smallest formula immediately following the quantifier. What the smallest formula is, is always indicated by parentheses. (In practice, we omit parentheses when they are unnecessary.) Thus in the formula

$$(\forall x)(\exists y)(x < y) \vee y = 0,$$

the scope of the quantifier $(\forall x)$ is the formula $(\forall x)(\exists y)(x < y)$, and the scope of the quantifier $(\exists y)$ is the formula $(\exists y)(x < y)$. In a few places we shall need the notions of *bound* and *free* variables. An occurrence of a variable in a formula is bound if and only if this occurrence is within the scope of a quantifier using this variable. An occurrence of a variable in a formula is free if not bound. Finally, a variable is a *bound variable* in a formula if and only if at least one occurrence is bound; it is a free variable in a formula if and only if at least one occurrence is free. A *sentence* of the language L is a formula of the language in which no free variables occur. This is an important distinction because among formulas only sentences can be considered as true or false simpliciters in a given interpretation or model of the language, a point amplified later.

Here are some examples of elementary languages. First, one version of the elementary language of group theory has, in addition to the logical symbols (which we shall not mention again), the infix binary operation symbol $\bigcirc$ and the identity element, that is, the individual constant e. Examples of formulas of this elementary language are the following:

$$x \bigcirc y = z$$
$$(\forall x)(\exists y)(x \bigcirc y = e)$$
$$e \bigcirc e = e$$

Note that the "official" elementary language of group theory as defined above would use, in place of the infix operation symbol, a prefix binary function symbol f, so that $f(x, y) = x \bigcirc y$. No confusion will result in moving back and forth from the official to the more familiar notation.

Another example is the elementary language of real numbers. In this case we have two infix operation symbols $(+, \cdot)$, the familiar relation symbol $<$, and the two individual constants 0 and 1.

Examples of a rather different sort can be found by looking at some of the geometrical theories in Chapter 13. For example, the elementary language of affine geometry requires only the single ternary relation symbol B

for betweenness. The elementary language of the projective plane requires two one-place relation symbols, one for the property of being a point and the other for the property of being a line, and a two-place relation symbol for the binary relation of incidence.

21.2.2 Models of Elementary Languages

Intuitively, the models of elementary languages are the kind of set-theoretical objects or relational structures that we have used in characterizing, within a set-theoretical framework, theories of measurement. To make such relational structures models of particular elementary languages, we need to associate relations, operations, and individuals with the corresponding nonlogical symbols of the language.

First, we can always give the finite set of nonlogical symbols—relations, functions, and constants—of an elementary language L as a set:

$$\{R_1, \ldots, R_m, f_1, \ldots, f_n, c_1, \ldots, c_p\}.$$

Each relation symbol R_i of L is assumed to denote an n_i-placed relation for some integer $n_i \geqslant 1$, and each function symbol f_j of L denotes an n_j-placed function symbol where $n_j \geqslant 1$. It should be apparent that a language L need not have all the three types of symbols, that is, relation symbols, function symbols, or individual constants. Of course, if none of the three types appears, then the language would be what is sometimes called a pure language of elementary logic with identity.

When the nonlogical symbols of the language are familiar, as in the examples given at the end of the last section, we shall show the finite set of symbols of the language in the usual notation. Thus the elementary language of groups has as its nonlogical symbols the finite set $\{\circ, e\}$.[2] The elementary language of real numbers has as its finite set of nonlogical symbols $\{<, +, \cdot, 0, 1\}$. When standard symbols are used, the number of places of various relation symbols and function symbols will be understood without explicit remark.

A *model* for an elementary language L consists of an ordered pair, $\mathfrak{A} = \langle A, v \rangle$. The first member of $\mathfrak{A}$ is a nonempty set A, the *universe* or *domain* of the model, and the second member is a *valuation function* v that maps the symbols of L to appropriate relations, functions, and individuals

[2] Here and elsewhere in this chapter, for simplicity of notation we do not strictly differentiate use from mention of a symbol when no confusion seems likely except in the next paragraph.

in A. The domain of v is always the set of nonlogical symbols of the language L. Thus an example of a model for the elementary language of groups is the pair $\langle\{0,1\}, v_1\rangle$, where $v_1(`\bigcirc') =$ the binary operation on the set $\{0,1\}$ given by the following matrix

$$\begin{array}{c|cc} & 0 & 1 \\ \hline 0 & 0 & 1 \\ 1 & 1 & 0 \end{array}$$

and $v_1(`e') = 0$. Note that the sentential connectives, the universal and existential quantifiers, and the logical predicate of identity have the same meaning in all models, but for a given universe A there can be many interpretations (i.e., valuations) of the symbols of L and therefore many different models. As a trivial modification of this example, we can take $\mathfrak{A}_2 = \langle A, v_2\rangle$, where $v_2(`\bigcirc') = v_1(`\bigcirc')$, and $v_2(`e') = 1$. It is also important to note that the models of the language L do not necessarily satisfy in an intuitive sense the intended theory of that language. Thus the models of the elementary language of groups need not necessarily be groups; $\mathfrak{A}_2$, for example, is not. Later on we shall talk about models of the *theory* of groups. These are models that also satisfy the axioms of the theory of groups. To refer to models as groups or other standard mathematical objects, we have in mind the purely set-theoretical structure without explicit reference to the language L. From this standpoint, a group is a structure $\langle A, \bigcirc, e\rangle$, not $\langle A,v\rangle$, but no confusion will result from conflating the two and calling both *models* in the context of this chapter.

It is useful to characterize explicitly some standard notions and operations on models. Most of these notions have already been characterized in the set-theoretical framework used in other chapters. Given two models, $\mathfrak{A} = \langle A, v\rangle$ and $\mathfrak{A}' = \langle A', v'\rangle$, we say that R' in model $\mathfrak{A}'$ is the *relation corresponding* to R in the model $\mathfrak{A}$ if they are the interpretations of the same relation symbol in L. We characterize in a similar way *corresponding functions* and *individuals* for two models of a language.

Two models $\mathfrak{A}$ and $\mathfrak{A}'$ for L are *isomorphic* if and only if there is a one-to-one function φ mapping A on A' satisfying the following:

(i) For each n-placed relation R of $\mathfrak{A}$ and the corresponding relation R' of $\mathfrak{A}'$,

$$R(x_1,\ldots,x_n) \quad \text{if and only if} \quad R'(\varphi(x_1),\ldots,\varphi(x_n))$$

for all $x_1,\ldots,x_n$ in A.

(ii) For each m-placed function f of $\mathfrak{A}$ and the corresponding function f' of $\mathfrak{A}'$,

$$\varphi(f(x_1,\ldots,x_m)) = f'(\varphi(x_1),\ldots,\varphi(x_m)),$$

for all $x_1,\ldots,x_m$ in A.

(iii) For each individual c of $\mathfrak{A}$ and the corresponding individual c' of $\mathfrak{A}'$,

$$\varphi(c) = c'.$$

The model $\mathfrak{A}$ is said to be *homomorphically embeddable* in, or homomorphic to, $\mathfrak{A}'$ if the function φ is not known to be one-to-one and the mapping is *into* not necessarily *onto* A'. [This definition is in essence just the one given in Section 1.2.2. It is a stronger definition of homomorphism than is standard in model theory, where in the case of homomorphisms condition (i) is weakened to: if $R(x_1,\ldots,x_n)$, then $R'(\varphi(x_1),\ldots,\varphi(x_n))$.][3]

A model $\mathfrak{A}'$ is called a *submodel* of $\mathfrak{A}$ if $A' \subseteq A$ and:

(i) Each n-placed relation R' of $\mathfrak{A}'$ is the restriction to A' of the corresponding relation R of $\mathfrak{A}$; that is,

$$R' = R \cap \Big(\underbrace{A' \times \cdots \times A'}_{n \text{ times}}\Big).$$

(ii) Each m-placed function f' of $\mathfrak{A}'$ is the restriction to A' of the corresponding function f of $\mathfrak{A}$; that is,

$$f' = f \Big| \Big(\underbrace{A' \times \cdots \times A'}_{m \text{ times}}\Big).$$

(iii) Each individual of $\mathfrak{A}'$ is the corresponding individual of $\mathfrak{A}$. We use $\mathfrak{A}' \subseteq \mathfrak{A}$ to denote that $\mathfrak{A}'$ is a submodel of $\mathfrak{A}$.

We shall often refer to a model of a language as a *finite* model or as an *infinite* model. This terminology simply means that the universe of the model is finite or infinite as the case may be.

The central logical notion for models of an elementary language L is the concept of a formula of L being satisfied in a model $\mathfrak{A}$ of L. The precise

[3] This stronger definition of homomorphism does introduce an asymmetry when functions are replaced by relations, for the functions satisfy the weaker model-theoretic definition of homomorphism but the equivalent relations do not. For example, if we define the relation R by $R(a,b,c)$ iff $a \bigcirc b = c$, under a homomorphism φ we do not get: $R(a,b,c)$ iff $R'(\varphi(a),\varphi(b),\varphi(c))$ because $\varphi(a \bigcirc b) = \varphi(c)$ does not imply $a \bigcirc b = c$.

definition of satisfaction originates in the classical work of Tarski (1930b) on the concept of truth. We describe informally the central ideas. Let $\mathfrak{A} = \langle A, v\rangle$ be a model of a language L. An *assignment* in $\mathfrak{A}$ for L is a function s that maps each variable of L into a member of A. Consider, as an example we shall expand upon further in a moment, the elementary language L_1 of ordering relations for which the only nonlogical symbol is $\geqslant$. Let $\mathfrak{A}_1 = \langle A_1, v_1\rangle$ be a model of this language such that

$$A_1 = \{0,1\}$$
$$v_1(`\geqslant') = \{\langle 0,0\rangle\}.$$

Let s be the assignment such that for every variable x of L_1, $s(x) = 0$. Then the assignment s *satisfies* the formula '$x \geqslant y$' in the model $\mathfrak{A}_1$, for example, because $s(x) \geqslant s(y)$, i.e., $0 \geqslant 0$ in $\mathfrak{A}_1$.[4]

As is apparent from the details given in footnote 4, if φ is a sentence, i.e., a formula with no free variables, the truth or falsity of φ in a model $\mathfrak{A}$ is independent of an assignment s. In other words, a sentence φ is true in $\mathfrak{A}$ for all assignments, or false in $\mathfrak{A}$ for all assignments.

There is one point that we want to reiterate. When we say that a sentence of a language is true in a model, we are restricting the variables of the language to take as values the individuals of the universe of the model; and the other logical constants (i.e., the sentential connectives, quantifiers, and logical predicate of identity) are interpreted in the usual way. It is also important to note that the concept of truth is relativized to a model of the language.

Thus, to illustrate the concept of a sentence of a language being true in a model of the language, we can look again at the elementary language of ordering relations for which the only nonlogical symbol is the familiar

[4] More generally, a formula φ of L_1 is satisfied by an assignment s in a model $\mathfrak{A}$ of L_1 if under the assignment s the following are the case: If t_1 and t_2 are variables of L_1,

(i) $t_1 = t_2$ is satisfied by assignment s in $\mathfrak{A}$ iff $s(t_1) = s(t_2)$ in $\mathfrak{A}$;

(ii) $t_1 \geqslant t_2$ is satisfied by s in A iff $s(t_1) \geqslant s(t_2)$ in $\mathfrak{A}$;

(iii) $\neg\varphi$ is satisfied by s in $\mathfrak{A}$ iff φ is not so satisfied;

(iv) $\varphi \,\&\, \psi$ is satisfied by s in $\mathfrak{A}$ iff φ is so satisfied and also φ is so satisfied, and similarly for the other sentential connectives;

(v) $(\exists x)\psi$ is satisfied by s in $\mathfrak{A}$ iff there is an a in A such that φ is satisfied by s' in $\mathfrak{A}$, where s' is the same function as s except $s'(x) = a$;

(vi) $(\forall x)\varphi$ is satisfied by s in $\mathfrak{A}$ iff for all a in A φ is satisfied by s_a in $\mathfrak{A}$, where s_a is the same as s except that $s_a(x) = a$.

Extending this definition of satisfaction to an arbitrary elementary language L is straightforward (see, e.g., Barwise, 1977).

binary relation symbol $\geqslant$. We gave one model, $\mathfrak{A}_1 = \langle A_1, v_1 \rangle$, of this language. We also introduce a second, $\mathfrak{A}_2 = \langle A_2, v_2 \rangle$, where A_2 is the set of real numbers, and $v_2(\text{'}\geqslant\text{'})$ is the standard relation $\geqslant$ for the real numbers. Now consider the following sentences of L.

(1) $(\exists x)(x \geqslant x)$

(2) $(\exists x)(\forall y)(x \geqslant y \;\&\; x \neq y)$

(3) $(\forall x)(\forall y)(x \geqslant y \vee y \geqslant x)$

It is obvious that sentence (1) is true in both models, sentences (2) and (3) are false in the first model, sentence (2) is false in the second model, and sentence (3) is true in the second model.

Given a set S of sentences of a language L, we say that a model $\mathfrak{A}$ of L is a model of S if and only if each sentence in S is true in $\mathfrak{A}$. It is also convenient to say that $\mathfrak{A}$ is a model of a sentence, which is just another way of saying that the sentence is true in $\mathfrak{A}$. In a similar vein, instead of saying that a sentence is true in a model, we shall sometimes say that the sentence *holds* in the model, that the model *satisfies* the sentence, and that the sentence *is satisfied* in the model. A sentence of L that holds in every model of L is called *valid*. A notion complementary to that of validity is that of satisfiability. A sentence or a set of sentences of L is *satisfiable* if and only if it has at least one model.

A sentence s is a *logical consequence* of a sentence s' if and only if every model of s' is a model of s. Generalizing this relation between sentences, we say that a sentence s is a logical consequence of a set of sentences S if and only if every model of S is a model of s.

Two models $\mathfrak{A}$ and $\mathfrak{A}'$ of L are *elementarily equivalent* if and only if any sentence of L is true in $\mathfrak{A}$ when and only when it is true in $\mathfrak{A}'$. It is straightforward to show that if two models of a language are isomorphic in the sense defined earlier, then they are elementarily equivalent. In general, the converse is true only when the models are finite. It is beyond the scope of this chapter to develop the theory of elementary equivalence in general; but for our later discussions, an important case of models that are elementarily equivalent but not isomorphic are two algebraic fields, one of which is Archimedean and the other not. As we shall see, the proof that the Archimedean axiom cannot be expressed in an elementary language essentially rests on this fact.

Proofs of independence. An important application of the concept of model is to prove that a formula is *logically independent* of a given set of formulas, i.e., is not a logical consequence of the given set. The most common application is to the problem of showing that a particular axiom of a system is independent of the remaining axioms.

The technique is straightforward: construct a model of the given set of formulas, i.e., a model in which they are all satisfied, but have the formula in question not be satisfied in the model. If the formula in question could then be derived from the given set of formulas, we would have a violation of the concept of logical consequence, as defined earlier in this section.

As a simple example, much used in this work, consider the two axioms for a weak order:

(1) $(\forall x)(\forall y)(\forall z)(x \succsim y \; \& y \succsim z \to x \succsim z)$.

(2) $(\forall x)(\forall y)(x \succsim y \vee y \succsim x)$.

We want to show they are mutually independent. First we show that (2) cannot be derived from (1):

$$A_1 = \{0\}$$
$$\succsim_1 = \varnothing .$$

Clearly, transitivity holds trivially in model $\mathfrak{A}_1 = \langle A_1, \succsim_1 \rangle$, and connectivity fails. To show that (1) cannot be derived from (2):

$$A_2 = \{0,1,2\}$$
$$\succsim_2 = \{\langle 0,1\rangle, \langle 1,2\rangle, \langle 2,0\rangle, \langle 0,0\rangle, \langle 1,1\rangle, \langle 2,2\rangle\}.$$

Transitivity fails in model $\mathfrak{A}_2 = \langle A_2, \succsim_2 \rangle$ because we have $0 \succsim_2 1$ and $1 \succsim_2 2$ but not $0 \succsim_2 2$. On the other hand, connectivity holds.

This simple example has been given for illustrative purposes. Already in Volume I this familiar technique for proving axioms independent was used to show mutual independence of the axioms of Definition 3.1.

21.2.3 General Theorems about Elementary Logic

To have a logic for elementary languages we need to introduce the syntactical notions of being a theorem, being consistent, etc., corresponding to the semantic concepts of being valid, having a model, etc. There are many equivalent ways of characterizing elementary logic. The references given at the beginning of the chapter exemplify various alternatives. One way, shown in Suppes (1957), is by giving rules of natural deduction. Thus a familiar rule is the rule of inference, called the rule of detachment or *modus ponendo ponens*: from formulas P and $P \to Q$, infer Q. Another is the rule of universal generalization: from P infer $(\forall x)P$ (but subject to certain restrictions on prior occurrences of the variable x). Unfortunately, it is a rather delicate matter to state the full set of rules and their restrictions in any of the various systems of inference available in such a form that no fallacious conclusions are derivable. It is not to the point to construct detailed formal inferences in this chapter. We shall assume that logical rules of the sort just mentioned, as well as rules for logical identities, are available, and we also shall assume that the notions of a proof and of a theorem formally defined in terms of an elementary language L and a fixed set of rules of inference are already available (see Barwise, 1977). If S is a set of sentences of L and P is a formula of L such that there is a proof of P from S using the logical rules of inference, we say that P is *derivable* from S. The formula P is a *logical theorem* if it is derivable from the empty set of sentences. A set of sentences S is *inconsistent* if and only if every formula of L can be derived from S. Otherwise, S is *consistent*. This language of consistency is extended to individual sentences simply by considering the unit set consisting of that sentence. A set of sentences S is *maximally consistent* in L if and only if S is consistent and S is not a proper subset of any consistent set of sentences of L.

We now state a number of general theorems without proof. These theorems represent classical results about elementary logic. Some of them will be essential in the more special results proved later. The first theorem states two important properties about sets of consistent sentences. The second half of the theorem about the extension of a consistent set of sentences is known in the literature as Lindenbaum's theorem. It was first stated in Tarski (1930a).

THEOREM 1.

(i) *A set S of sentences of L is consistent if and only if every finite subset of S is consistent.*

(ii) *Any consistent set of sentences of L can be extended to a maximal consistent set of sentences of L.*

The next theorem is Gödel's completeness theorem (1930). The logical rules of inference we adopt can and must satisfy this theorem.

THEOREM 2 (Gödel's completeness theorem). *Given any sentence P of L, P is a logical theorem if and only if P is valid.*

The next two theorems are also due to Gödel.

THEOREM 3 (Extended completeness theorem). *If S is any set of sentences of an elementary language L, then S is consistent if and only if S has a model.*

THEOREM 4 (Compactness theorem). *A set of sentences S of L has a model if and only if every finite subset of S has a model.*

The next theorem is about infinite models. The restricted formulation that if a set of sentences has an infinite model, then it has a countably infinite model was first proved by Löwenheim (1915) and Śkołem (1920). The general case as stated in the theorem for models of every infinite cardinality is due to Tarski and is sometimes called the upward Löwenheim-Śkołem theorem.

THEOREM 5 (Löwenheim-Śkołem-Tarski). *If a set of sentences of an elementary language has an infinite model, i.e., a model whose domain is an infinite set, then the set of sentences has a model of every infinite cardinality, i.e., given any infinite cardinal α, the set of sentences has a model whose domain is of cardinality α.*

Some of the paradoxical consequences of this theorem are worth remarking. It is well known that axiomatic set theory, for example, Zermelo-Fraenkel set theory, can be formulated in an elementary language, and within this formalization the classical theory of real numbers can be developed, including the theorem that the set of real numbers is uncountable. How then can the set of axioms of Zermelo-Fraenkel set theory have a countable model as the theorem asserts? The answer is this. The mapping that proves that the set of real numbers must also be countable is itself not part of the countable model of the theory; it lies outside the model and therefore does not contradict the theorem of the theory that there is no one-to-one mapping of the real numbers onto the natural numbers.

21.2.4 Elementary Theories

An *elementary theory* T of an elementary language L is a collection C of sentences of L closed under logical consequence, that is, if C_1 is a subset of C and P is a logical consequence of C_1, then P is in C. Because of the equivalence of logical consequence and inference in elementary logic we could also require that a theory be a collection of sentences that is closed under logical inference. This closure requirement is obvious and natural and, in fact, tacitly assumed in mathematical and scientific practice. Suppose, for example, that a certain sentence is a logical consequence of the axioms for Euclidean geometry. It would seem paradoxical to exclude this sentence from the theory. Elementary theories are also often called in the literature *first-order theories*, or *theories with standard formalization*, in accordance with the alternative descriptions of elementary languages stated in Section 21.2.1.

We can extend our terminology for a model of a language in an obvious way to models of a theory. Thus we shall treat as equivalent a model of an elementary language L being a *model of a theory* T of this language, *satisfying the theory* T, or *the theory* T *being satisfied in the model.* This extension is direct because of our earlier characterization about sets of sentences having a model, etc.

Sometimes in the literature the requirement is not made that a theory be closed; then an explicit terminology is introduced for the theory being closed. For the applications we have in mind it is reasonable always to require that theories be closed under logical consequence. On the other hand, the concept of completeness is often not satisfied by a theory. An elementary theory is called *complete* if and only if its set of consequences is maximally consistent.

We also want to extend our earlier concept of validity to the more restricted notion of *being valid in a theory* T. It is especially important that we be explicit about the notion of validity of a sentence in a given theory in order to separate out questions of axiomatizability. The most common method is to give a model, or class of models, in which sentences of the theory T must hold in order to be valid. In our case, interesting questions can be asked about the class of models of a theory of measurement that are embeddable in a given numerical model. We define the validity of sentences in the theory as sentences that hold in every model of this class. For example, we can define valid sentences of the theory of weak orders as those sentences that are true for every weak order that is homomorphically embeddable in the numerical model $\langle Re, \geqslant \rangle$.

Of course, when the basic assumptions of a theory are already well understood, and when these assumptions can be expressed within the elementary language of the theory, we simply single out these sentences, call them axioms, and say that a sentence is valid in the theory if and only if it is derivable from the axioms. If a theory can be characterized by a set of elementary sentences, we say that it is *axiomatizable* (in an elementary language). We often do place restrictions on the axioms. For instance, we usually require that a set of axioms be recursive; that is, that there is a decision procedure or algorithm for recognizing whether or not a formula of the language of the theory is an axiom. When a recursive set of axioms can be given, we say that the theory is *recursively axiomatizable*; and when a finite set of axioms can be given, we say that the theory is *finitely axiomatizable*. When the axioms can be given using only universal quantifiers standing at the front of a formula with their scopes being the remainder of the formula, we say that the theory is *universally axiomatizable*; and in case the number of universal sentences is finite, we also often say that the theory is *axiomatizable by a universal sentence*.

To illustrate these ideas and to anticipate some later theories, we now give some examples of theories of the various axiomatic types. Any measurement theory (e.g., the theory of closed extensive structures characterized in Definition 3.1) that requires an Archimedean axiom is not axiomatizable in an elementary language. Roughly speaking, this is true of any measurement theory whose models are infinite. Later we prove a theorem about measurement theories with only finite models that are axiomatizable but not recursively axiomatizable. Perhaps the best known example of a theory that is recursively axiomatizable but not finitely axiomatizable is the elementary theory of numbers. The induction "axiom" is given as a schema that can be used to generate an infinite set of elementary axioms. Measurement theories that are finitely axiomatizable in an elementary language must be ones whose models are finite. Examples from Volume I are the theory of finite, equally spaced, additive conjoint structures (Definition 1.5), the theory of finite, equally spaced difference structures (Definition 4.7), and the theory of finite probability structures with equivalent atoms (Theorem 5.6). The theory of finite weak orders (Theorem 1.1) is an example of a theory that is axiomatizable by a universal sentence, as can be seen by taking the conjunction of the two axioms of Definition 1.1. In this connection it is important to note that the condition that the models be finite is *not* formulated in the elementary axioms but is used to characterize the class of models being axiomatized.

Here are some examples of elementary theories whose elementary language was discussed earlier.

Example of group theory. As stated earlier, the elementary language has two symbols, the infix binary operation symbol $\bigcirc$ and the identity element, that is, the individual constant e. One form of the axioms of the theory is the following set; as usual, the universal quantifiers are omitted from the front:

Axiom 1. $x \bigcirc (y \bigcirc z) = (x \bigcirc y) \bigcirc z$.

Axiom 2. $x \bigcirc e = x$.

Axiom 3. $(\exists y)(x \bigcirc y = e)$.

By adding to the elementary language of group theory the one-place function symbol for the inverse operation, which we denote in the usual way by a negative superscript, we can simplify the axioms so that their conjunction is a universal sentence.

Axiom 1. $x \bigcirc (y \bigcirc z) = (x \bigcirc y) \bigcirc z$.

Axiom 2. $x \bigcirc e = x$.

Axiom 3. $x \bigcirc x^{-1} = e$.

This example illustrates a general principle. The transparency and simplicity of axioms very much depend on the language of the theory that is available for expression. Of course, in many cases, defined symbols are introduced. For example, we could have introduced by definition the inverse operation, but we have not yet discussed the theory of definability and its ramifications, which we consider in the next section. In the case of the elementary theory of groups, it is a familiar fact that we could have simply used the single binary operation symbol as the only nonlogical symbol of the elementary language. In this case the axioms are the following three:

Axiom 1. $x \bigcirc (y \bigcirc z) = (z \bigcirc y) \bigcirc z$.

Axiom 2. $(\exists z)(x = y \bigcirc z)$.

Axiom 3. $(\exists z)(x = z \bigcirc y)$.

Still other versions of the axioms for groups are possible, but the three we have given are the three main variants ordinarily considered, depending upon the concepts that are taken as primitive or, to put it linguistically, the symbols of the elementary language of groups that are taken as primitive. Strictly speaking, the three sets of symbols and their associated axioms characterize three different but closely related elementary theories. Because the differences are usually of little significance we speak of *the* theory of groups.

Example of the theory of semiorders. The theory of semiorders, introduced in Chapter 16, can be easily formulated as an elementary theory when restricted to finite models. The only symbol of the elementary language of the theory is the usual binary relation $\succ$ of ordering. The three axioms of the theory are the following:

Axiom 1. $\neg(x \succ x)$.

Axiom 2. $(x \succ y \ \& \ x' \succ y') \rightarrow (x \succ y' \vee x' \succ y)$.

Axiom 3. $(x \succ y \ \& \ y \succ z) \rightarrow (x \succ w \vee w \succ z)$.

21.3 DEFINABILITY AND INTERPRETABILITY

Other elementary theories that we considered in earlier chapters require the definition of symbols, and we now turn to that topic.

21.3.1 Definability

The concept of definability plays a role in our subsequent developments. As we have remarked, three kinds of nonlogical constants may occur in an elementary theory: relation symbols, operation symbols, and individual constants. In the earlier discussion we were tacitly assuming that the nonlogical constants are the primitive or basic constants introduced at the beginning of our formulation of a theory. In mathematical practice, we also introduce a large number of additional, nonlogical constants by definitions formulated in terms of the primitive constants or previously defined constants. In studying problems of axiomatizability it is important to have a theoretical view of such definitions. Here we shall sketch the basic ideas; a detailed introduction is to be found in Suppes (1957, Chapter 8).

The motivation for the rules of definition we give is that a defined symbol should satisfy the criteria of eliminability and noncreativity. The form of the definition should be such that the defined symbol can always be *eliminated* in favor of the primitive notation of the theory. Generally speaking, the form of the definition must be that of an equivalence or identity to make this elimination in all contexts possible. The definition must also be *noncreative* in the sense that it cannot be used to prove theorems formulated entirely in the primitive notation but unprovable from the axioms of the theory alone.

For example, suppose we have as the single axiom of a theory T_1, whose only nonlogical constant is the binary operation symbol '$\bigcirc$', the axiom:

$$(\forall x)(\forall y)(\forall z)(x \bigcirc (y \bigcirc z) = (x \bigcirc y) \bigcirc z).$$

Then the identity

$$(\forall x)(x \bigcirc 0 = x)$$

cannot serve as a definition of the individual constant '0' because we can use this improper definition to prove, in the primitive notation only, the theorem

$$(\exists y)(\forall x)(x \bigcirc y = x),$$

which certainly does not follow from the axiom alone. In short, the proposed definition is creative and therefore improper.

For each of the three types of nonlogical symbols we can state rules of definition that guarantee that any definition satisfying the appropriate rule also satisfies the criteria of eliminability and noncreativity.

The rule for defining relation symbols is this: *An equivalence D introducing a new n-placed relation symbol R is a proper definition in a theory iff D is of the form $R(v_1, \ldots, v_n) \leftrightarrow S$, and the following restrictions are satisfied*:

(i) *$v_1, \ldots, v_n$ are distinct variables*;

(ii) *S has no free variables other than $v_1, \ldots, v_n$: and*

(iii) *S is a formula in which the only nonlogical constants are primitive symbols and previously defined symbols of the theory.*

In the case of operation symbols it is also necessary to require that the existence and uniqueness of the result of performing the operation be proved. Without this requirement we can easily make the definition maximally creative by deriving from it a contradiction. For example, we might introduce in arithmetic the pseudo-operation $*$ by the following equivalence:

$$x * y = z \quad \text{iff} \quad x < z \quad \text{and} \quad y < z.$$

Now, since $1 < 3$ and $2 < 3$, we then have $1 * 2 = 3$; but also, since $1 < 4$ and $2 < 4$, we also have $1 * 2 = 4$, whence $3 = 4$, which is absurd.

The rule for defining operation symbols is this: *An equivalence D introducing a new n-placed operation symbol f is a proper definition in a theory iff D is of the form*

$$f(v_1, \ldots, v_n) = w \leftrightarrow S,$$

and the following restrictions are satisfied:

(i) $v_1, \ldots, v_n, w$ *are distinct variables;*

(ii) *S has no free variables other than* $v_1, \ldots, v_n, w$;

(iii) *S is a formula in which the only nonlogical constants are primitive symbols and previously defined symbols of the theory; and*

(iv) *the formula* $(\exists! w)S$ *is derivable from the axioms and preceding definitions of the theory.*

We can also introduce operation symbols by identities rather than equivalences. For instance, we can define the binary operation symbol of subtraction in terms of the binary operation symbol of addition and the unary negative operation symbol:

$$x - y = x + (-y).$$

We shall not give explicit rules for such defining identities or for definitions of individual constants, but leave these matters to the reader as an exercise.

From a logical standpoint, a definition is a noncreative axiom. Thus, for example, defined symbols that are used in the statement of the axioms of an elementary theory should properly first be defined explicitly and the axioms defining them labeled as definitions.

It is, of course, understood that as a new defined symbol is added to the language, the definition of terms and atomic formulas of the language is thereby extended.

Example of conjoint structures. These ideas can be nicely illustrated by considering the elementary theory of finite, equally spaced, additive conjoint structures introduced in Section 1.3.2 of Volume I. The only nonlogical primitive symbol of the theory is a quaternary relation $\succsim$. We introduce, as defined symbols, binary ordering relations on each of the conjoint components. We also introduce by definition a binary relation J_i on each component for the immediate successor. The appropriate axioms and definitions as given in Definitions 3–5 of Section 1.3.2 are then embodied in the following elementary axiomatization, with universal quantifiers implicitly understood for free variables:

Axiom 1. $(x, y) \succsim (u, v) \vee (u, v) \succsim (x, y)$.

Axiom 2. $(x, y) \succsim (u, v) \& (u, v) \succsim (w, z) \rightarrow (x, y) \succsim (w, z)$

Axiom 3. $(x, y) \succsim (x', y) \rightarrow (x, z) \succsim (x', z)$.

Axiom 4. $(x, y) \succsim (x, y') \rightarrow (z, y) \succsim (z, y')$.

Definition 1.1. $x \succsim_1 y \leftrightarrow (\forall x)(x, z) \succsim (y, z)$.

Definition 1.2. $x \succsim_2 y \leftrightarrow (\forall z)(z, x) \succsim (z, y)$.

Definition 2.1. $x J_1 y \leftrightarrow \neg y \succsim_1 x \& (\forall z)(z \succsim_1 x \vee y \succsim_1 z)$.

Definition 2.2. $x J_2 y \leftrightarrow \neg y \succsim_2 x \& (\forall z)(z \succsim_2 x \vee y \succsim_2 z)$.

Axiom 5. $(x J_1 y \ \& \ x' J_2 y') \rightarrow ((x, y') \succsim (x', y) \& (x', y) \succsim (x, y'))$.

Comparing this elementary theory of conjoint measurement with that given in Section 1.3.2 shows that one distinction made there has been neglected here. This is the possibility of having the two components drawn from different collections of objects. However, it is easy to introduce this distinction in our elementary theory. We add two one-place relation symbols to the symbols of the language, namely, P_1 and P_2, expressing the property of being from the collection of objects composing the first or second component, respectively. Each of the axioms must then be modified to introduce the appropriate restriction imposed by these predicates. Thus, for example, Axiom 1 would now be formulated as follows:

Axiom 1′. $P_1(x) \& P_1(u) \& P_2(y) \& P_2(v) \rightarrow (x, y) \geqslant (u, v) \vee (u, v) \geqslant (x, y)$.

Independence of primitive symbols. When the primitive symbols of a theory are given, it is natural to ask if it would be possible to define one of them in terms of the others. Alessandro Padoa (1902) formulated a principle, in terms of models of a theory, that can be used to show that the primitive symbols are independent, i.e., one cannot be defined in terms of the other. The principle is simple: to prove that a given primitive symbol is independent of the remaining primitives, find two models of the axioms of the theory such that the given primitive is extensionally different in the two models and the remaining primitive symbols are the same in both models. For instance, consider the theory of preference based on the primitive relation symbols '$\succ$' (for strict preference) and '$\sim$' (for indifference). The axioms of the theory are as follows:

1. *If* $x \succ y$ *and* $y \succ z$, *then* $x \succ z$.
2. *If* $x \sim y$ *and* $y \sim z$, *then* $x \sim z$.
3. *Exactly one of the following*: $x \succ y$, $y \succ x$, $x \sim y$.

We want to show that '$\succ$' is independent of '$\sim$', that is, $\succ$ cannot be defined in terms of '$\sim$'. Let the domain of both models be the set $\{1, 2\}$. Let

$$\sim_1 = \sim_2 = \{\langle 1,1\rangle, \langle 2,2\rangle\},$$
$$\succ_1 = \{\langle 1,2\rangle\},$$

and

$$\succ_2 = \{\langle 2,1\rangle\}.$$

Now if '$\succ$' were definable in terms of '$\sim$', then '$\succ$' would have to be the same in both models since '$\sim$' is. However, $\succ_1 \neq \succ_2$, and we conclude that '$\succ$' cannot be defined in terms of '$\sim$'.

To clarify the procedure for applying Padoa's principle to any theory formalized in first-order predicate logic with identity, we want to make more precise the general definition of independence of a primitive symbol and also to characterize more sharply the notion of two models of a theory being *different* for a given primitive symbol of the theory.

Let R be an n-place primitive relation symbol of a theory. Then we say that R is *dependent* on the other primitive symbols of the theory if a formula of the form

$$R(v_1, \ldots, v_n) \leftrightarrow S$$

can be derived from the axioms, where (i) $v_1, \ldots, v_n$ are distinct variables, (ii) the only free variables in S are $v_1, \ldots, v_n$, and (iii) the only nonlogical constants occurring in S are the other primitive symbols of the theory.

The close relation between this definition of dependence of a primitive relation symbol and the role for defining new relation symbols in a theory is obvious and expected. The definitions of dependence of operation symbols and individual constants are similar and will be left as exercises.

We now want to use the definition of dependence for relation symbols to sharpen the description of Padoa's principle for proving independence. To prove an n-placed primitive relation symbol 'R' independent of the other primitive symbols of a theory, we need to find two models of the theory such that the following are the case.

(i) The domain of both models is the same.

(ii) The two models are the same for all other primitive symbols of the theory.

(iii) Let R_1 be the denotation of R in the first model and R_2 the denotation in the second. Then R_1 and R_2 must be different in the following respect: there are elements $x_1, \ldots, x_n$ in the common domain of the two models such that either $R_1(x_1, \ldots, x_n)$ and not $R_2(x_1, \ldots, x_n)$, or not $R_1(x_1, \ldots, x_n)$ and $R_2(x_1, \ldots, x_n)$.

To see that two such models establish the independence of R, suppose that R is dependent on the other primitive symbols of the theory; that is, suppose that there is a formula:

$$R(x_1, \ldots, x_n) \leftrightarrow S \tag{1}$$

of the kind demanded by the definition of dependence such that Equation (1) is derivable from the axioms of the theory. In both models we must have

$$R_1(x_1, \ldots, x_n) \leftrightarrow S_1, \tag{2}$$

and

$$R_2(x_1, \ldots, x_n) \leftrightarrow S_2, \tag{3}$$

since Equation (1) is a logical consequence of the axioms of the theory. Moreover, since all primitive symbols except R are the same in both models, we also have

$$S_1 \leftrightarrow S_2. \tag{4}$$

From Equations (2), (3), and (4) we infer that

$$R_1(x_1, \ldots, x_n) \leftrightarrow R_2(x_1, \ldots, x_n), \tag{5}$$

which contradicts (iii) and proves that our supposition of dependence is absurd.

Without going into details it may be mentioned that Padoa's principle can be easily extended to theories that assume in their formalization not only first-order predicate logic with identity but also classical mathematics. Probably some of the most interesting applications of Padoa's principle to such richer theories are in the domain of empirical science, for there is a fair amount of confused discussion regarding the interdefinability of various empirical concepts.

On the other hand, there is a natural completeness of definability for elementary theories that does not hold in general. As in the case of other concepts we have studied, such as finite axiomatizability, definitive general results hold only for elementary theories. The intuitive content of the following theorem, due to Beth (1953) is this: *If a primitive symbol of an elementary theory cannot be proved independent by Padoa's principle, which is a semantic result, then it can be defined explicitly in terms of the other primitive symbols, which is a syntactic result.*

21.3.2 Interpretability

We next come to the notion of interpretability, the last general notion we introduce for elementary theories.

To begin with, a theory T_1 is a *subtheory* of a theory T_2 if and only if every sentence that is valid in T_1 is also valid in T_2; the theory T_2 is called an *extension* of T_1. It follows from this definition that any nonlogical constant of T_1 is also a nonlogical constant of T_2; but, of course, not necessarily vice versa. The theory T_2 can have nonlogical constants that are not part of T_1. Suppose now that T_1 and T_2 are any two theories that have no nonlogical constants in common. Then we say that T_2 is *interpretable* in T_1 if and only if there is a theory T satisfying the following conditions:

(i) T is a common extension of T_1 and T_2.

(ii) Every constant of T is a constant of T_1 or T_2.

(iii) The valid sentences of T include definitions of the nonlogical constants of T_2 in terms of the nonlogical constants of T_1.

(iv) Every valid sentence of T is derivable from a set of sentences that are either valid in T_1 or are definitions of the sort described in (iii).

As a trivial example of these concepts, suppose we axiomatize the elementary theory F of ordered fields in terms of the nonlogical constants $+$, $\cdot$, 0, and 1. And we axiomatize the elementary theory G of groups in terms of the binary operation $\bigcirc$, the identity symbol e, and the unary inverse operation symbol $^{-1}$, as was done in the preceding section. We now define in the theory F (omitting in the definitions universal quantifiers)

$$x \bigcirc y = x + y$$
$$e = 1$$
$$x^{-1} = 0 - x.$$

Then the axioms of G can be proved in this extension of F, and therefore G is interpretable in F. On the other hand, it is easy to see that F is not interpretable in G.

A subtler example can be exhibited in the relation between simple orders and weak orders. The theory of weak orders is interpretable in the theory of simple orders, but not vice versa because the antisymmetry property of simple orders cannot be derived from the axioms for weak orders.

21.4 SOME THEOREMS ON AXIOMATIZABILITY

In Chapter 1 we defined theories of measurement in an informal set-theoretic manner. To obtain any definite metamathematical results about the axiomatizability of such theories it is necessary to have a more formal definition.

DEFINITION 1. *Let L be an elementary language. Then a nonempty class K of models of L is a* class of measurement structures (*with respect to L*) *iff*

(i) *K is closed under isomorphism, i.e., if $\mathfrak{A}$ is in K and $\mathfrak{A}'$ is isomorphic to $\mathfrak{A}$, then $\mathfrak{A}'$ is in K; and*

(ii) *there is a numerical model $\mathfrak{R}$ of L, i.e., a model of L whose domain is the set of real numbers, such that all models in K are homomorphically embeddable in $\mathfrak{R}$.*

Note that the numerical model $\mathfrak{R}$ cannot be a member of K; for example, all models in K might be finite, but $\mathfrak{R}$ is obviously an infinite model. Of course, not all numerical models are really of interest from a measurement viewpoint, and it would be possible to put some restrictions on the numerical models considered suitable for embedding empirical models. We have not attempted to give such restrictions here.

The first results we want to state are negative ones that follow rather directly from the standard theorems in logic stated in Section 21.2.3. The central idea is this. If an elementary theory has an infinite model, then as a direct consequence of the Löwenheim-Skolem-Tarski theorem the theory also has models of every infinite cardinality. If the theory imposes much structure at all on its models (e.g., the sort of structure characteristic of extensive measurement), then they will not be homomorphically embeddable in a numerical model.

As a first simple instance of this fact, we note that the models of the elementary theory of weak orders do not form a class of measurement structures, as we established in Chapter 2 by a direct mathematical argument that used the lexicographic ordering of the plane.

We can generalize this particular result to the following theorem, which is stated for simple orders rather than weak orders.

THEOREM 6. *The class of models of any elementary theory that has infinite models and is such that the theory of simple orders is interpretable in it, is not a class of measurement structures.*

This theorem is not satisfactory because most of the theories of measurement considered in earlier chapters postulate a weak order rather than a simple order because, of course, two distinct objects can have a given attribute to the same quantitative degree. We can, however, extend the theorem to weak orders, provided the theory in question has a consistent extension that converts the weak order to a simple order.

We can illustrate these ideas in the simplest possible context by beginning with only the theory of weak order, which has already been stated as an elementary theory T. Let T_1 be the extension of T obtained by adding the axiom of antisymmetry:

$$(\forall x)(\forall y)((x \succsim y \;\&\; y \succsim x) \rightarrow x = y). \tag{1}$$

So T_1 is the theory of simple orders. Then by Theorem 6, the models of T_1 do not form a class of measurement structures because a model of cardinality 2^{c_1} of T_1 is also a model of T, and if it is homomorphically embedded in the set of real numbers, it must be isomorphically embedded, which is impossible. (Here c_1 is the cardinality of the set of real numbers or the continuum. The fact that every homomorphism of a simple order is an isomorphism is obvious.) But the situation is not always this simple. In principle, a theory of measurement might have as an axiom beyond those for weak orders the axiom

$$(\forall x)(\exists y)(x \succsim y \;\&\; y \succsim x \;\&\; x \neq y). \tag{2}$$

This sentence asserts that every object is matched in the degree of the attribute or attributes in question by at least one other object. In this case the theory could not be consistently extended by adding Equation (1) because Equations (1) and (2) together lead to a contradiction, but in practice Equation (2) is not ordinarily assumed. We are thus led to the following theorem.

THEOREM 7. *Let T be any elementary theory such that*

(i) *some of the models are infinite,*

(ii) *the theory of weak orders is interpretable in T, and*

(iii) *the theory of simple orders as an extension of the theory of weak orders (with the same nonlogical constant) is interpretable in a consistent extension of T that has an infinite model.*

Then the class of models of T is not a class of measurement structures.

We can see how Theorems 6 and 7 apply to various axiomatizations given in Volume I. Of the axioms for extensive measurement given in Definition 1 of Chapter 3, only one is not elementary. It will perhaps be useful to spell out this point in somewhat greater detail. As an elementary theory, the theory of extensive measurement expressed in this definition has one binary relation symbol $\succsim$ and one binary operation symbol $\bigcirc$. Avoiding any defined relation symbols, we can reformulate Axioms 1–3 and 5 as follows:

1a. $(\forall a)(\forall b)(\forall c)(a \succsim b \;\&\; b \succsim c \rightarrow a \succsim c)$

1b. $(\forall a)(\forall b)(a \succsim b \vee b \succsim a)$

2. $(\forall a)(\forall b)(\forall c)(a \bigcirc (b \bigcirc c) \succsim (a \bigcirc b) \bigcirc c \;\&\; (a \bigcirc b) \bigcirc c \succsim a \bigcirc (b \bigcirc c))$

3. $(\forall a)(\forall b)(\forall c)(a \succsim b \leftrightarrow a \bigcirc c \succsim b \bigcirc c \;\&\; a \succsim b \leftrightarrow c \bigcirc a \succsim c \bigcirc b)$

5. $(\forall a)(\forall b)(\neg(a \succsim a \bigcirc b))$.

It is, of course, Axiom 4, the Archimedean axiom, that cannot be given an elementary formulation because to express this axiom it is necessary to quantify over the positive integers:

4′. $(\forall a)(\forall b)(\exists n)(na \succsim b)$,

or, corresponding to the more complex form given as part of Definition 1 of Chapter 3,

4. $(\forall a)(\forall b)(\forall c)(\forall d)(\exists n)(\neg b \succsim a \rightarrow na \bigcirc c \succsim nb \bigcirc d)$.

The situation just described is essentially the same for the axioms of closed periodic extensive structures given in Definition 2 of Chapter 3; once again only the Archimedean axiom is not elementary.

On the surface, the situation seems to have changed when we turn to Definition 3 of Chapter 3, that is, the axioms for extensive structures with no essential maximum. Here a new primitive relation B is part of the structure, and the continual expression in the axioms that certain pairs are members of B suggest a need for second-order, or nonelementary, axioms to express the assumptions about the set B. It is easy to see, however, that this is not so. From the viewpoint of an elementary formalization, we simply replace the set B by a binary relation B' and, at the same time, replace the binary operation $\bigcirc$ on B by a ternary relation. This move is slightly awkward but necessary to avoid having in the elementary formalization a binary operation symbol that is not defined for elements of the domain of a model. (Some complications arise from changes in the characterization of homomorphisms in passing from functions to relations, as described in footnote 3, p. 210. These complications can of course always be avoided by formally considering only equivalence classes of objects, which map isomorphically into the representing numerical structure, rather than the objects themselves.) With these changes, we can formulate all but the Archimedean axiom as elementary axioms. Axiom 3, for instance, which is stated in Definition 3 as

$$\textit{If } (a, c) \in B \textit{ and } a \succsim b,$$
$$\textit{then } (c, b) \in B \textit{ and } a \bigcirc c \succsim c \bigcirc b,$$

would have this rather awkward elementary formulation:

$$(\forall a)(\forall b)(\forall c)(aBc \;\&\; a \succsim b \rightarrow cBb \;\&\; (\exists d)(\exists e)$$
$$(R(a, c, d) \& R(c, b, e) \;\&\; d \succsim e))$$

Of course, the point of such an elementary formulation is not to provide a new technique for proving elementary theorems but rather to provide a formulation within a framework whose general metamathematical properties are understood and thereby to make possible the establishment of certain general metamathematical results such as the theorems of this section on the limits of formalization of theories of measurement. Significant results of this limiting kind will be given in Sections 6 and 7.

On the other hand, we can obtain some positive results for *finitary* classes of measurement structures. A class of measurement structures is finitary if and only if each model in the class has a finite domain.

THEOREM 8. *Any finitary class of measurement structures with respect to an elementary language L is axiomatizable but not necessarily recursively axiomatizable in L.*

The importance of this theorem is that it shows the expressive power of elementary languages is adequate for finitary classes of measurement structures. In contrast, Theorems 6 and 7 show that the expressive power of such languages is not adequate for any of the standard classes of measurement structures, which contain infinite models.

It would be too much to expect that any finitary class of measurement structures is recursively axiomatizable, because there are too many such structures, namely, a continuum. A more special positive result is the following.

THEOREM 9. *Let K be the finitary class of measurement structures with respect to elementary language L and with respect to numerical model* $\mathfrak{R}$ *of L such that K includes all finite models of L homomorphically embeddable in* $\mathfrak{R}$. *If the domain, relations, functions, and constants of* $\mathfrak{R}$ *are definable in elementary form in terms of* $\langle Re, \leqslant, +, \cdot, 0, 1\rangle$ *then the set of sentences of L that are satisfied in every model of K is recursively axiomatizable.*

Neither in this section nor in the later ones on particular questions of axiomatizability do we give any results on uniqueness of numerical representations, a central question for measurement structures, as we emphasized repeatedly in Volume I and Chapter 20. This absence of results is not accidental. Questions about uniqueness of numerical representation involve in an essential way the automorphisms of the representing numerical structure. The set of these automorphisms, which is a group in the standard cases, cannot be analyzed in a natural way within an elementary formulation of the theory of measurement itself, and there are no standard metamathematical results about elementary theories that can be used in any direct way. For this reason, general questions about uniqueness fit more naturally into the framework of Chapters 20 and 22 and are discussed there.

21.5 PROOFS

21.5.1 Theorem 6 (p. 226)

The class of models of any elementary theory that has infinite models and is such that the theory of simple orders is interpretable in it is not a class of measurement structures.

PROOF. We proceed by contradiction. Suppose there were an elementary theory T such that the class of models of T forms a class of measurement structures. Then by hypothesis there is a definition in an extension T' of T of the form

$$x \succsim y \quad \text{iff} \quad S(x, y),$$

where $S(x, y)$ is a formula of T', and it is possible to prove in T' that

$$x \succsim y \ \& \ y \succsim x \rightarrow x = y$$
$$x \succsim y \ \& \ y \succsim z \rightarrow x \succsim z$$
$$x \neq y \rightarrow x \succsim y \vee y \succsim x$$

(with universal quantifiers understood). Now using Theorem 5, let $\mathfrak{A}$ be a model of T that has cardinality 2^{c_1}, where c_1 is the cardinality of the continuum. Let f be a homomorphic embedding of $\mathfrak{A}$ into Re. We want to show that it must be an isomorphism, but this leads to a contradiction because there cannot be a one-to-one function from a set of cardinality 2^{c_1} to a set of cardinality c_1. Suppose then that f is not one-to-one. This means that there are two elements x and y in the domain of $\mathfrak{A}$ such that $f(x) = f(y)$, but $x \neq y$. Without loss of generality we can assume that $x \succsim y$. Then $S(x, y)$, and since f maps $\mathfrak{A}$ into a numerical model of T, $S(f(x), f(y))$, whence $f(x) \geqslant f(y)$, where $\geqslant$ is a simple ordering of Re. But since $f(x) = f(y)$, we have also $S(f(y), f(x))$, and thus $S(y, x)$ and consequently $y \succsim x$, and thus $x = y$, contrary to our supposition. ◇

21.5.2 Theorem 7 (p. 227)

Let T be any elementary theory such that

(i) *some of the models are infinite,*

(ii) *the theory of weak orders is interpretable in T, and*

(iii) *the theory of simple orders as an extension of the theory of weak orders (with the same nonlogical constants) is interpretable in a consistent extension of T that has an infinite model.*

Then the class of models of T is not a class of measurement structures.

PROOF. Let T be an elementary theory satisfying the hypothesis of the theorem, and let T' be a consistent extension of T with T' satisfying hypothesis (iii) of the theorem. Let $\mathfrak{A}$ be an infinite model of T' of cardinality 2^{c_1} (such a model exists by the Löwenheim-Śkołem-Tarski theorem and the hypothesis of the theorem). Now $\mathfrak{A}$ is a model of T, yet $\mathfrak{A}$ cannot be embedded in the real numbers because the embedding must be isomorphic not merely homomorphic, and there can be no such embedding function from 2^{c_1} to c_1. ◇

21.5.3 Theorem 8 (p. 228)

Any finitary class of measurement structures with respect to an elementary language L is axiomatizable but not necessarily recursively axiomatizable in L.

PROOF. There is an elementary sentence of L describing the set of models isomorphic to each finite model of L not in the finitary class K of measurement structures. The negations of these sentences can be taken as the axioms of the elementary theory of K. On the other hand, in general the set of axioms is not recursive because there is obviously a continuum of distinct finitary classes of measurement structures. ◇

21.5.4 Theorem 9 (p. 229)

Let K be the finitary class of measurement structures with respect to elementary language L and with respect to numerical model $\mathfrak{R}$ *of L such that K includes all finite models of L homomorphically embeddable in* $\mathfrak{R}$. *If the domain, relations, functions, and constants of* $\mathfrak{R}$ *are definable in elementary form in terms of* $\langle Re, \leqslant, +, \cdot, 0, 1\rangle$ *then the set of sentences of L that are satisfied in every model of K is recursively axiomatizable.*

PROOF. For a finitary class of measurement structures satisfying the hypothesis of the theorem with respect to an elementary language L, the problem of a recursive axiomatization is equivalent to the problem of showing that the class of universal sentences true in the given numerical model is recursively enumerable, i.e., is the range of a recursive function. The equivalence follows from the hypothesis of closure under finite models of L. For theories whose representing numerical model $\mathfrak{R}$ is definable (in elementary logic) in terms of $\langle Re, \leqslant, +, \cdot, 0, 1\rangle$ the positive solution is immediate because the elementary theory of the latter structure is decidable by a well-known result of Tarski (1951). [This decidability result implies that there is a recursive procedure for testing the validity of any sentence of the elementary theory of $\langle Re, \leqslant, +, \cdot, 0, 1\rangle$. A positive decision procedure for the important case of the elementary theory of $\leqslant$ and $+$ was given earlier by Presburger (1930).] ◇

21.6 FINITE AXIOMATIZABILITY

As the theorems of Section 21.3 show, since we do not have axiomatizability in general, we can scarcely expect finite axiomatizability of most theories of measurement. The situation changes radically, however, if we restrict ourselves to finitary classes of measurement structures.

Many examples of theories with elementary formalization that have a finite set of axioms, with the restriction that the axioms apply to finite models, are given in Volume I. The most familiar examples were mentioned earlier in this chapter. Of these examples, the only one that is not obviously a theory with elementary formalization is the theory of finite probability structures with equivalent atoms (Section 5.4.3). In this case, an algebra of sets with the ordinary set operations is assumed, and the qualitative probability axioms are added as additional apparatus. In the formulation in Volume I, the axioms on set operations are naturally not included among the axioms of the theory of such structures. To formulate the theory as a theory with elementary formalization, the axioms for Boolean algebra can be made a part of the theory itself. In formulating the axioms we have, as before, omitted universal quantifiers that stand at the beginning of a formula and whose scopes include the rest of the formula. The nonlogical constants of the theory are the usual symbol for Boolean union, the symbol for Boolean intersection, the symbol for Boolean complementation, the individual constant 1 for the "universe," and the usual symbol $\varnothing$ for the empty event; and finally we add the qualitative ordering relation $\succsim$ as introduced in Definition 4 of Section 5.2. The theory is based on the following 18 axioms.

1. $A \cup B = B \cup A$
2. $A \cap B = B \cap A$
3. $A \cup (B \cup C) = (A \cup B) \cup C$
4. $A \cap (B \cap C) = (A \cap B) \cap C$
5. $(A \cap B) \cup B = B$
6. $(A \cup B) \cap B = B$
7. $A \cap (B \cup C) = (A \cap B) \cup (A \cap C)$
8. $A \cup (B \cap C) = (A \cup B) \cap (A \cup C)$
9. $A \cup \varnothing = A$
10. $1 \cap A = A$
11. $A \cap - A = \varnothing$
12. $A \cup - A = 1$
13. $A \succsim B \;\&\; B \succsim C \rightarrow A \succsim C$
14. $A \succsim B \vee B \succsim A$
15. $X \succsim \varnothing \;\&\; \neg \varnothing \succsim X$
16. $A \succsim \varnothing$
17. $A \cap B = A \cap C \;\&\; A \cap B = \varnothing \rightarrow (B \succsim C \leftrightarrow A \cup B \succsim A \cup C)$
18. $A \succsim B \rightarrow (\exists C)(A \succsim B \cup C \;\&\; B \cup C \succsim A)$

It should be apparent that these axioms are not independent; there are several redundancies among the first 12 axioms characterizing Boolean algebra. Above all, axioms 11 and 12 are just definitions of the empty event and of the certain event. The first 8 axioms are stated in dual form, and some reduction is possible. The 12 axioms as they stand, however, form an extremely transparent and easily comprehended characterization of Boolean algebras. For the finite models of this theory we can prove Theorem 6 of Section 5.5. The proof, of course, is unchanged and provides a standard representation theorem for finite models. Thus restricted to finite models, the theory is a genuine theory of measurement.

Other interesting examples of finite axiomatizations of theories with standard formalization have been given in various chapters of Volume II, and some of these will be referred to later.

Given a theory with elementary formalization that is recursively axiomatizable, it might be thought straightforward to settle whether a finite axiomatization of the theory is possible. Unfortunately, this in general turns out to be a difficult problem, and there are few tools available for a standard attack. One of the most interesting examples, mentioned earlier, of a theory that has been shown to be not finitely axiomatizable is Peano's (really Dedekind's) formulation of elementary number theory with recursive definitions of addition and multiplication taken as axioms. Since these axioms are formulated to make them a theory with elementary formalization, the usual axiom of induction must be formulated as an axiom schema in the following manner, where $F(n)$ is the schema that is replaced by a particular formula of the elementary language to obtain an instance of the induction schema:

$$F(0) \mathbin{\&} (\forall n)(F(n) \to F(n+1)) \to (\forall n)F(n)$$

It has been shown by Ryll-Nardzewski (1952) that this axiom schema cannot be replaced by a finite number of axioms, and therefore Peano's theory is not finitely axiomatizable as an elementary theory. On the other hand, it would be a mistake to think that the difficulty here is simply the power of Peano's axioms leading to a theory that is too rich for finite axiomatization in elementary terms because a still richer theory—namely, von Neumann-Bernays-Gödel (NBG) set theory—based just on the nonlogical relation symbol of membership is finitely axiomatizable in an elementary manner; and from the axioms, standard set theory including both number theory and classical analysis can be derived. The point is that the latter system is rich enough to let many familiar objects be denoted by terms in the theory, especially in the use of proper classes that are not sets. For example, in Zermelo-Fraenkel set theory the operation symbol for union of sets does not denote a set, but it does denote a proper class in NBG set theory.

Turning back to theories of measurement, we shall find that some basic intuitions are not borne out by results that are stated in the remainder of this section. We have seen that certain theories of measurement can have satisfactory finite axiomatizations when the models of the theory are restricted to finite models. Using the notion of a finitary class of measurement structures as defined in Section 21.4, we examine a number of familiar classes and ask if we can finitely axiomatize the elementary theory of any of these finitary classes. For example, in the case of conjoint structures, let K be the finitary class of measurement structures consisting of all finite models homomorphically embeddable in the set of real numbers under the intended numerical interpretation of the quaternary relation. Most of the models in K do not satisfy axiom 5, the structural axiom given in Section 21.3.1 in the example of conjoint structures. Of course, given the intended numerical interpretation of the quaternary relation, it follows at once from Theorem 9 of Section 21.3 that the theory K is recursively axiomatizable, but it is a problem of another sort to settle whether it is finitely axiomatizable. This same situation is true of many of the finitary classes of interest.

21.6.1 Axiomatizable by a Universal Sentence

What can be settled in many cases is that the theories are not axiomatizable (in the given elementary language) by a universal sentence or, what comes to the same thing, by a finite set of universal sentences. For classes of the type just defined in terms of finite models of conjoint measurement, it would seem that the passage from a result about not being axiomatizable by a universal sentence to a result about not being finitely axiomatizable would be fairly direct. The intuitive reason for thinking this is that the class of models is closed under submodels; that is, if a model is in K, then any submodel of the given model is also an element of K. With such closure under submodels, one does not expect to be able to express much by existential quantifiers; thus it would seem intuitively that axiomatization by a universal sentence should be essentially equivalent to finite axiomatization. Moreover, in the general case, when we do not restrict ourselves to finite models, there is a well-known theorem of Tarski's that says if a theory is finitely axiomatizable and closed under submodels, then it is axiomatizable by a universal sentence (Tarski, 1954, 1955). As an extension of Tarski's theorem it might seem reasonable to conjecture the following. If S is a sentence such that if it is satisfied by a finite model it is satisfied by every submodel of the finite model, then there is a universal sentence finitely equivalent to S. (We define two sentences of an elementary language L as finitely equivalent if and only if they are satisfied by the same finite models of L.) Exactly this was conjectured by Scott and Suppes (1958), but the conjecture was proved false by Tait (1959). Subsequently we shall examine this example because it affords insight into the difficulties of moving from axiomatizable by a universal sentence to finitely axiomatizable.

As opposed to the situation for finite axiomatizability of finitary theories of measurement, we do have a powerful and simple criterion stated in explicit form for finitary theories of measurement by Scott and Suppes (1958) which is directly derivable from an earlier theorem of Vaught (1954). The criterion is stated in the following theorem.

THEOREM 10 (VAUGHT'S CRITERION). *Let L be an elementary language without function symbols. A finitary class K of measurement structures (with respect to L) is axiomatizable by a universal sentence iff K is closed under submodels and there is an integer n such that if any finite model $\mathfrak{A}$ of L has the property that every submodel of $\mathfrak{A}$ with no more than n elements is in K, then $\mathfrak{A}$ is in K.*

For the reasons stated at the beginning of this chapter, we shall omit the proof of this theorem because it would require development of additional concepts from mathematical logic.

The intuitive basis of Vaught's criterion for finitary classes of measurement structures is easy to explain. Consider, for example, the axioms for a weak order. Because the axioms involve three distinct variables, it is sufficient to check triples of objects in the domain of a model of the language to determine if it is a model of the theory. Generally speaking, the number of distinct variables determines the size of the submodels that must be checked to see if universal axioms are satisfied. To have a universal axiom for a theory, or what is equivalent, a finite set of universal axioms, it is necessary that the number of distinct variables be some definite number, say n. A check on submodels involving no more than n objects is then sufficient to determine satisfaction of the axiom or axioms, and this is the intuitive source of Vaught's criterion.

It is also useful to state a corollary that simply states the negation of Vaught's criterion for determining when a finitary class of measurement structures is not axiomatizable in an elementary language by a universal sentence.

COROLLARY. *Let L be an elementary language without function symbols, and let K be a finitary class of measurement structures (with respect to L) closed under submodels. Then K is not axiomatizable by a universal sentence of L iff for every integer n there is a finite model $\mathfrak{A}$ of L that has the property that every submodel with no more than n elements is in K, but $\mathfrak{A}$ is not in K.*

It is the corollary that we shall use, of course, in proving negative results about the axiomatizability by a universal sentence of various finitary classes of measurement structures. We interpret these negative results as showing that the complexity of relationships in various finitary classes is unbounded, at least unbounded in the sense that it cannot be caught by a fixed finite number of variables used in open formulas of elementary logic.

As the first application we show that finite indifference graphs, as defined in Section 16.2.7, are not axiomatizable by a universal sentence. To bring the discussion within the framework of this chapter, let L be the elementary language whose only symbol is the binary relation symbol '$\sim$' of indifference or indistinguishability. We can define the finitary class J of measurement structures that consists of all finite models $(A, \sim)$ of L such that there is a real-valued function f defined on A such that for all x and y in A,

$$|f(x) - f(y)| \leqslant 1 \qquad \text{iff} \qquad x \sim y.$$

We then have the following theorem, originally due to Roberts (1968, 1969).

THEOREM 11. *Let L be the elementary language whose only nonlogical symbol is the binary relational symbol* $\sim$. *Then the finitary class J of measurement structures for indifference graphs is not axiomatizable in L by a universal sentence.*

PROOF. The proof will be given here because it is a straightforward application of results from Chapter 16, using the basic corollary of Theorem 10. As stated in Chapter 16, the cycles exhibited as type I of Figure 1 must be prohibited for each n. However, for cycles of length n, we have at once a proof by use of the corollary that no axiomatization in terms of a universal sentence of the language L is possible because if we remove any one link from one of these subgraphs with a cycle, then the remaining subgraph is in the class J, but the entire subgraph is not. ◇

It is somewhat surprising that the theory of indifference or indistinguishability is not axiomatizable by a universal sentence because the theory of semiorders is, as are also the simpler structures of weak ordering and simple ordering. On the other hand, Fishburn (1981b) has shown that the class of all finite interval orders that can be interval-represented using no more than n interval lengths or threshold levels is not axiomatizable by a universal sentence for $n \geqslant 2$. For $n = 1$ the answer is positive because this is just the class of all finite semiorders.

We turn now to the case of difference measurement, for which the proof of nonaxiomatizability by a universal sentence is considerably more complex. The proof is given in a subsequent section and follows that given by Scott and Suppes (1958).

Let L be the elementary language whose only symbol is the quaternary relation symbol 'D'. (We write $xy\,D\,uv$, etc.) Let Δ be the quaternary numerical relation such that for real numbers x, y, u, and v,

$$xy\,\Delta\, uv \qquad \text{iff} \qquad x - y \geqslant u - v.$$

We then define the finitary class $\mathscr{D}$ of measurement structures for (algebraic) difference as consisting of all models $\mathfrak{A} = \langle A, D\rangle$ of L such that

(i) A is a nonempty finite set,

(ii) D is a quaternary relation on A,

(iii) $\mathfrak{A}$ is homomorphically embeddable in $\langle \mathrm{Re}, \Delta\rangle$.

THEOREM 12. *Let L be the elementary language whose only nonlogical symbol is the quaternary relation symbol D. The finitary class $\mathscr{D}$ of measurement structures for (algebraic) difference is not axiomatizable in L by a universal sentence.*

The intuitive idea of the proof can be seen from an examination of a ten-element structure, all of whose substructures have a numerical representation, but which does not itself have such a representation.

Let the ten elements $a_1, \ldots, a_{10}$ be ordered as shown in Figure 1 with atomic intervals given the designations indicated. Let α be the interval (a_1, a_5), let β be the interval (a_6, a_{10}), and let the interval γ be larger than either α or β. We suppose further that $\alpha_1, \alpha_2, \alpha_3, \alpha_4$ is equal in size to $\beta_2, \beta_4, \beta_1, \beta_3$, respectively, but α is less than β which cannot be true either geometrically or numerically. The relationships among the remaining intervals can be so chosen that any submodel of nine elements is embeddable in (Re, Δ) whereas the full system of ten elements is clearly not. The proof given in Section 21.6.2 generalizes this example to an arbitrary finite number of elements.

Furthermore, although the theorem and its proof are given for algebraic-difference structures, it is clear that the proof can be easily modified to cover positive-difference and absolute-difference structures.

A less obvious but almost immediate application to additive conjoint measurement is due to Titiev (1972). The finitary class of conjoint measure-

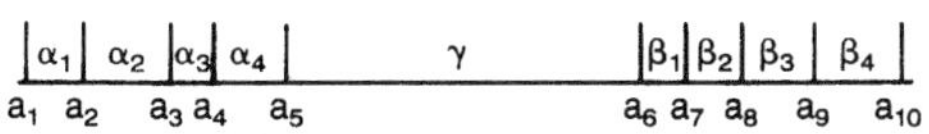

α_1 α_2 α_3 α_4 γ β_1 β_2 β_3 β_4

a_1 a_2 a_3 a_4 a_5 a_6 a_7 a_8 a_9 a_{10}

FIGURE 1. Counterexample for algebraic difference structures.

ment structures with n components is defined in the obvious way. We then have the following.

THEOREM 13. *The finitary class of conjoint measurement structures with $n \geqslant 2$ components is not axiomatizable by a universal sentence.*

PROOF. The proof is a direct application of Theorem 12. We simply take an n-dimensional case for which the preference relation $\geqslant$ is defined by the algebraic-difference relation in the first two dimensions:

$$a_1 a_2 \cdots a_n \geqslant b_1 b_2 \cdots b \quad \text{iff} \quad a_1 b_2 \, D \, b_1 a_2 .$$

Effectively this assigns a constant value to all the components of each dimension greater than 2. The reduction to Theorem 12 is then immediate. ◇

A less obvious application of Theorem 12 is to multidimensional scaling with a Euclidean metric, a special case of the proximity spaces developed in Chapter 14. To emphasize that we are now concerned with multidimensional results, we shall refer to finitary classes of multidimensional scaling structures. Let Re^n be n-dimensional space whose points are n-tuples of real numbers, and $\delta(x, y)$ for x and y in Re^n be the ordinary Euclidean metric

$$\delta(x, y) = \left[\sum_{i=1}^{n} (x_i - y_i)^2 \right]^{1/2} .$$

Let Δ_n be the four-place relation on Re^n such that

$$xy \Delta_n uv \quad \text{iff} \quad \delta(x, y) \geqslant \delta(u, v),$$

and let L be the elementary language whose only symbol is the quaternary relation D. Then we define the *finitary class K_n of n-dimensional scaling structures with Euclidean metric (with respect to L)* as the class of models $\mathfrak{A} = \langle A, D \rangle$ of L such that

(i) A is a finite nonempty set,
(ii) D is a quaternary relation on A,
(iii) $\mathfrak{A}$ is homomorphically embeddable in $\langle \mathrm{Re}^n, \Delta_n \rangle$.

THEOREM 14. *Let L be the elementary language whose only nonlogical symbol is the quaternary relation symbol D. Then the finitary class K_n of* *n-dimensional scaling structures with Euclidean metric for $n \geqslant 1$ is not axiomatizable in L by a universal sentence.*

An outline of the technical details of the proof of Theorem 14 can be found in Titiev (1972) and will not be given here, but the geometric intuition behind the method of reducing the proof to that of Theorem 12 is natural and easy to characterize. In the case of $n = 2$, i.e., the case of the plane, we construct a model so that m points are collinear and then apply Theorem 12 to this collinear substructure. We can express the collinearity of the points in terms of D by requiring the m points $a_1, \ldots, a_m$ to be equidistant from two distinct points b_1 and b_2, which of course cannot be collinear with $a_1, \ldots, a_m$. In the case of $n = 3$, the points $a_1, \ldots, a_m$ are on a line perpendicular to the plane of an equilateral triangle with vertices b_1, b_2, and b_3; of course, the line passes through the centroid of the triangle, and thus each a_i is equidistant from b_1, b_2, and b_3. Generalization of this construction to n dimensions is geometrically obvious.

Titiev (1972) also gives a similar proof for the dominance or supremum metric, which is sometimes used in multidimensional scaling investigations. This metric is defined by

$$\delta(x, y) = \max_i |x_i - y_i|.$$

On the other hand, no proof for more general metric spaces is known to us, even though we would conjecture that in all the interesting cases (e.g., with additive segments) the finitary class of structures is not axiomatizable by a universal sentence. Titiev (1980) has given a negative proof for the city-block metric in $n \leqslant 3$ dimensions. (The city-block metric in R_e^n is defined by

$$\delta(x, y) = \sum_{i=1}^{n} |x_i - y_i|,$$

where $x = (x_1, \ldots, x_n)$ and $y = (y_1, \ldots, y_n)$. The proof for $n = 3$ required computer assistance to examine 21,780 cases, each of which involved 10 equations in 12 unknowns and a related set of inequalities.

We conclude this section by examining Tait's example (1959), which shows that a sentence whose satisfaction is closed under submodels is not necessarily finitely equivalent to a universal sentence. The example shows that relatively simple finitary classes of models closed under submodels can be finitely axiomatizable but not axiomatizable by a universal sentence.

The nonlogical symbols of Tait's elementary language are two binary relation symbols $\leqslant$ and R. The axioms of his theory are the following:

1. $x \leqslant y \,\&\, y \leqslant x \rightarrow x = y$
2. $x \leqslant y \,\&\, y \leqslant z \rightarrow x \leqslant z$
3. $x \leqslant y \vee y \leqslant x$
4. $x R y \rightarrow z \leqslant x \vee y \leqslant z$
5. $x R y \,\&\, y \leqslant x \rightarrow (\exists z)(y \leqslant z \,\&\, \neg x \leqslant z \,\&\, (\forall u)(z \leqslant u \rightarrow \neg z R u))$.

For each $n > 0$, let $\mathfrak{A}_n = (\{0, 1, \ldots, n\}, \leqslant, R)$, where $\leqslant$ is the standard numerical ordering and

$$x R y \quad \text{iff} \quad \begin{cases} y = x + 1, & \text{if } x \neq n, \\ y = 0, & \text{if } x = n. \end{cases}$$

Thus the intuitive meaning of $x R y$ is that y is the immediate successor of x except when x is the last element in the ordering. The models $\mathfrak{A}_n$ exhibit the cycles ruled out in the finitary class of measurement structures for indifferent graphs. Such cycles are also ruled out by axiom 5 of Tait's counterexample. Thus no $\mathfrak{A}_n$ is a model of his axioms, but every submodel of each $\mathfrak{A}_n$ for $n > 0$ is a model of his axioms. By the corollary to Vaught's criterion (Theorem 10), the finitary class K defined by the models $\mathfrak{A}_n$ is consequently not axiomatizable by a universal sentence. On the other hand, the five axioms of Tait's counterexample do have the property that they are satisfied by the submodels of each $\mathfrak{A}_n$ but not by $\mathfrak{A}_n$ itself. Consequently the conjunction of his given axioms is not finitely equivalent to any universal sentence.

Additional insight into the subtle problem of understanding what can and cannot be done with existential statements closed under submodels can be obtained by making the obvious modification in Tait's axiom 5 to obtain a universal statement:

5′. $x R y \,\&\, y \leqslant x \rightarrow \neg x \leqslant y \,\&\, (y \leqslant u \rightarrow \neg y R u)$.

In this modification the given element y plays the role of the existentially postulated z in his axiom 5. It is obvious that axiom 5′ implies axiom 5, but of course the converse does not hold. This can be seen by considering the model $\mathfrak{A}_3$. Consider the submodel obtained by deleting 2 from the set $\{0, 1, 2, 3\}$. Then

$$R = \{(0, 1), (3, 0)\}.$$

Let $x = 3$, $y = 0$, and $u = 1$. Then axiom 5′ is not satisfied by this submodel, but axiom 5 is by taking $z = 1$.

Some general results for the theory of all substructures of a relational structure are to be found in Manders (1979).

21.6.2 Proof of Theorem 12 (p. 237)

The finitary class $\mathscr{D}$ of measurement structures for algebraic difference is not axiomatizable by a universal sentence.

PROOF. We need to show that for every n there is a finite model $\mathfrak{A}$ such that every submodel of $\mathfrak{A}$ with n elements in its domain is in $\mathscr{D}$ but $\mathfrak{A}$ is not. For every integer $n = 2m \geqslant 10$ with m odd, we construct a finite model $\mathfrak{A}$ such that every submodel of $2m - 1$ elements is in $\mathscr{D}$. (Clearly every submodel of $2m - k$ elements for $k < 2m$ is in $\mathscr{D}$.) To make the construction definite, we take numbers as elements of the domain and disrupt exactly one numerical relationship. Let n now be an even integer equal to or greater than 10. The selection of numbers $a_1, \ldots, a_{2m}$ can be described by specifying the numerical size of the atomic intervals. We define $\alpha_1 = a_{i+1} - a_i$ for $i = 1, \ldots, m - 1$ and $\beta_i = a_{m+i+1} - a_{m+i}$ for $i = 1, \ldots, m - 1$. We then set $a_i = 1, \alpha_i = 2^i$ for $i = 1, \ldots, m - 1$, and $a_{m+1} = 2^{2m}$. In fixing the size of β_i, we use the fact that m is odd and that therefore $m - 1$ is even. We set $\beta_i + \alpha_{i/2}$ for $i = 2, 4, \ldots, m - 1$ and $\beta_i = \alpha_{(m+i)/2}$ for $i = 1, 3, \ldots, m - 2$. Thus, if $n = 2m = 10$, we have $\alpha_1 = \beta_2$, $\alpha_2 = \beta_4$, $\alpha_3 = \beta_1$, $\alpha_4 = \beta_3$.

With the set $A = \{a_1, \ldots, a_{2m}\}$ defined, we now define the quaternary relation D as the expected numerical relation, except for permutations of a_1, a_m, a_{m+1}, and a_{2m}. If $x, y, z, w \in A$ and (x, y, z, w) is not some permutation of $(a_1, a_m, a_{m+1}, a_{2m})$, then $\langle x, y, z, w\rangle \in D$ if and only if

$$x - y \leqslant z - w. \tag{1}$$

Moreover, let $a = a_1$, $b = a_m$, $c = a_{m+1}$, $d = a_{2m}$. Then we put the following permutations of $\langle a, b, c, d\rangle$ in D:

$$\{\langle q, r, s, t\rangle : \langle q, r, s, t\rangle \text{ is a permutation of } \langle a, b, c, d\rangle \text{ and } q - r < s - t\} \cup \{\langle b, a, d, c\rangle, \langle b, d, a, c\rangle, \langle c, d, a, b\rangle, \langle c, a, d, b\rangle\}. \tag{2}$$

From the choice of the numbers in A and the definition of D it is obvious that $\langle A, D\rangle$ is not embeddable in $\langle \text{Re}, \Delta\rangle$; that is, $\langle A, D\rangle$ is not in $\mathscr{D}$ because the atomic intervals between a_1 and a_m must add up to a length equal to the sum of the atomic intervals between a_{m+1} and a_{2m}. But by hypothesis the interval (a_1, a_m) is less than interval (a_{m+1}, a_{2m}). It remains to show that every submodel of $2m - 1$ elements is in $\mathscr{D}$. Two cases naturally arise.

Case 1. The element omitted in the submodel is a_1, a_m, a_{m+1}, or a_{2m}. Then the permutations of Equation (2) are not in D restricted to the submodel, and the submodel is not merely embeddable in $\langle \mathrm{Re}, \Delta \rangle$ but by virtue of Equation (1) is a submodel of it.

Case 2. The element omitted is neither a_1, a_m, a_{m+1}, nor a_{2m}. Let a_i be the element not in the submodel. There are two cases to consider.

Case 2*a*. $a_i < a_m$. For this situation we can use for our numerical assignment the function f defined by $f(a_{i-j}) = a_{i-j} + 1$ for $j = 1, \ldots, i-1$, $f(a_{i+j}) = a_{i+j}$ for $j = 1, \ldots, n-i$. It is straightforward to verify that f is a numerical assignment, that is, it preserves the relation D as defined by Equations (1) and (2). Regarding atomic intervals (in the full model), we note that if

$$a_{i-j+1} - a_{i-j} = a_{k+1} - a_k \qquad \text{for } k > i,$$

then

$$\begin{aligned} &f(a_{i-j+1}) - f(a_{i-j}) \\ &\quad = (a_{i-j+1} + 1) - (a_{i-j} + 1) = a_{k+1} - a_k = f(a_{k+1}) - f(a_k). \end{aligned}$$

Case 2*b*. $a_i > a_{m+1}$. Here we can use a numerical assignment f defined, as would be expected from the previous case, by $f(a_{i-j}) = a_{i-j}$ for $j = 1, \ldots, i-1$, $f(a_{i+j}) = a_{i+j} + 1$ for $j = 1, \ldots, n-i$. This completes the proof of the theorem. ◇

21.6.3 Finite Axiomatizability of Finitary Classes

To a surprising extent the negative results of the preceding subsection can be extended to questions of finite axiomatizability. The main theorems of this subsection are due to Per Lindstrom (unpublished). As would be expected, the proofs involve methods that are more complicated than those of Theorems 11–13.

Because the theorems themselves are direct extensions of the ones just referred to, additional discussion is scarcely required, and we include the proofs with the statement of the theorems. (In fact, we really prove only one theorem, as in the previous case.)

We first need several definitions. Let L be an elementary language whose only nonlogical symbols are relation symbols. Let $\mathfrak{A}$ and $\mathscr{B}$ be two models of L. By a *Fraissé-Ehrenfeucht sequence of length* $m+1$ for $(\mathfrak{A}, \mathscr{B})$ we understand a sequence (I_k), $k \leqslant m$, of relations such that

(i) $I_k \subseteq A^k \times B^k$, for $k \leqslant m$;

(ii) $\varnothing I_0 \varnothing$;

(iii) if $k < m$ and $s I_k t$, then for every $a \in A$ $(b \in B)$ there is a $b \in B$ $(a \in A)$ such that $sa I_{k+1} tb$;

(iv) if $a_0 \cdots a_{m-1} I_m b_0 \cdots b_{m-1}$, then $\{(a_i, b_i) : i < m\}$ is an isomorphism of the submodel $A|\{a_0, \ldots, a_{m-1}\}$ onto the submodel $B|\{b_0, \ldots, b_{m-1}\}$. [See Fraissé (1955) and Ehrenfeucht (1961).]

The notation $\mathfrak{A} \equiv_m \mathscr{B}$ means there is a Fraissé-Ehrenfeucht sequence of length m for $(\mathfrak{A}, \mathscr{B})$. The following two theorems follow from results in the literature of mathematical logic. For a detailed development of the ideas, see Monk (1975, Chap. 26); but the point of such sequences is to provide a technique for proving elementary equivalence of two models of a language whose only nonlogical symbols are relation symbols.

THEOREM 15. *Let L be an elementary language whose only nonlogical symbols are relation symbols. For every m, the relation $\equiv_{m+1}$ between models of L is an equivalence relation with finitely many equivalence classes.*

THEOREM 16. *Let L be an elementary language whose only nonlogical symbols are relation symbols. Let K be any class of finite models of L. If K can be characterized by a finite set of elementary sentences of L, then there is an m such that for any finite models $\mathfrak{A}$ and $\mathscr{B}$ of L, if $\mathfrak{A} \in K$ and $\mathfrak{A} \equiv_m \mathscr{B}$, then $\mathscr{B} \in K$.*

The theorem that Lindstrom proves is slightly stronger than the simple extension of Theorem 11 to finite axiomatizability. As in the case of Theorem 11, let $\mathscr{D}$ be the finitary class of measurement structures for algebraic difference, and let $\mathscr{L}\mathscr{D}$ be the subset of such structures that have any two numerical representations related by a linear transformation.

THEOREM 17 (Lindstrom). *Let K be any class of finite measurement structures for algebraic difference such that $\mathscr{L}\mathscr{D} \subseteq K \subseteq \mathscr{D}$. Then K cannot be characterized by a finite set of elementary axioms.*

PROOF. Let n be arbitrary but fixed throughout the proof. Our approach is to define two models $\mathfrak{A}_n = (A, D)$ and $\mathfrak{A}'_n = (A', D')$ of L such that $\mathfrak{A}_n \notin \mathscr{D}$, $\mathfrak{A}'_n \in \mathscr{L}\mathscr{D}$, and $\mathfrak{A}_n \equiv_{n+1} \mathfrak{A}'_n$. By exhibition of this sequence of models it follows from Theorem 16 that a class K of models such that $\mathscr{L}\mathscr{D} \subseteq K \subseteq \mathscr{D}$ cannot be finitely axiomatized in an elementary manner.

Let $A_s = \{0, \ldots, s\}$ and $\mathfrak{A}_s = \langle A_s, \delta \cap A_s^4 \rangle$ where δ is the quaternary numerical relation defined earlier: $x - y \geqslant u - v$. It follows easily from Theorem 15 that for the language L there is an equivalence class of $\equiv_{n+1}$ with infinitely many members. So there are positive integers $k_0 > k_1 \geqslant 2$ such that

$$\mathfrak{A}_{k_0} \equiv_{n+1} \mathfrak{A}_{k_1}. \tag{3}$$

We now define numbers m_r as follows[5]:

$$\begin{aligned}
m_0 &= k_0 + k_1(4k_1 - 1), & \\
m_1 &= m_0 + k_1^2, & \\
m_{r+1} &= m_r + k_1, & \text{for } 1 \leqslant r < k_1, \\
m_{k_1} + 1 &= m_{k_1} + k_1^2, & \\
m_{k_1+2r} &= m_{k_1+2r-1} + 1, & \text{for } 1 \leqslant r \leqslant k_1, \\
m_{k_1+2r+1} &= m_{k_1+2r} + k_1, & \text{for } 1 \leqslant r < k_1, \\
m_{3k_1+1} &= m_{3k_1} + k_1(4k_1 - 1). &
\end{aligned}$$

Next let

$$\begin{aligned}
C &= \{m_r : r \leqslant 3k_1 + 1\}, \\
A &= A_{k_0} \cup C, \\
A' &= A_{k_1} \cup \{n : n + k_0 - k_1 \in C\}, \\
D' &= \delta \cap (A')^4.
\end{aligned}$$

From the definition of D' it follows at once that $\mathfrak{A}'$ is in $\mathscr{D}$. It is less obvious but straightforward to check that $\mathfrak{A}'$ is in $\mathscr{L}\mathscr{D}$. The reason is that every atomic interval (x, y) with $x > y + 1$ is D'-equivalent to an interval consisting of positive atomic intervals D'-less than (x, y). (An interval is atomic iff it properly contains no other interval of the model.)

Now for the definition of D. The intuitive idea is that $xy\,D\,uv$ is to be the same as $xy\,D'\,uv$ whenever (x, y) and (u, v) are atomic intervals and that $(m_{k_1}, 0)$ is to be D-equivalent to (m_{3k_1+1}, m_{k_1}). Given this characterization we must have $\mathfrak{A} \notin \mathscr{D}$. As for the formal definition, first we define, for x, y

[5] To get a sense of this construction, consider the simplest case: $k_0 = 3$, $k_1 = 2$. Then $A_{k_0} = A_3 = \{0, 1, 2, 3\}$, $C = \{m_r : r \leqslant 3k_1 + 1\} = \{m_0, \ldots, m_7\}$, $m_0 = 17$, $m_1 = 21$, $m_2 = 23$, $m_3 = 27$, $m_4 = 28$, $m_5 = 30$, $m_6 = 31$, $m_7 = 45$, and $A = A_{k_0} \cup C = \{0, 1, 2, 3, 17, 21, 23, 27, 28, 30, 31, 45\}$.

in A and $x > y$,

$$d(x, y) = \begin{cases} k_1 & \text{if } x = k_0 \text{ and } y = 0, \\ 2 & \text{if } 1 < x - y < k_0 \text{ and } x \leqslant k_0, \\ 1 & \text{if } x - y = 1 \text{ and } x \leqslant k_0, \\ (x - k_0) + d(k_0, y) & \text{if } y < k_0 < x, \\ x - y \quad \text{if } k_0 \leqslant y. \end{cases}$$

Let

$$X = A^4 - \left(A^4_{k_0} \cup (C \cup \{k_0\})^4\right),$$
$$D_0 = \delta \cap (A^4 - X),$$
$$D_1 = \{(x, y, z, u) \in X : x > y, z > u, (x \leqslant u \text{ or } z \leqslant y), \text{ and } d(x, y) \leqslant d(z, u)\}.$$

Then, let D be the least subset of A^4 containing D_0 and D_1 and such that D satisfies axioms 1–3 of Definition 3, Chapter 4, for algebraic-difference structures. It is tedious but straightforward to show that $\mathfrak{A}$ is not in $\mathscr{D}$.

What remains is the proof that as defined $\mathfrak{A} \equiv_{n+1} \mathfrak{A}'$. By Equation (3) there is a Fraissé-Ehrenfeucht sequence $I' = (I'_k)_{k \leqslant n}$ for $(\mathfrak{A}_{k_0}, \mathfrak{A}_{k_1})$. Now let $I = (I_k)_{k \leqslant n}$ be defined as follows:

$$(a_0, \ldots, a_{k-1})\, I_k\, (a'_0, \ldots, a'_{k-1}) \quad \text{iff}$$
$$\{i_0, \ldots, i_{s-1}\} = \{i < k : a_i \leqslant k_0\},$$

where $i_0 < \cdots < i_{s-1}$, then

$$\{i_0, \ldots, i_{s-1}\} = \{i < k : a'_i \leqslant k_1\},$$
$$(a_{i_0}, \ldots, a_{i_{s-1}})\, I'_s\, (a'_{i_0}, \ldots, a'_{i_{s-1}}),$$

and

$$\text{for} \quad i \leqslant k, \quad \text{if} \quad k_0 < a_i, \quad \text{then} \quad a_i - k_0 = a'_i - k_1.$$

As defined, it may be seen that I is a Fraissé-Ehrenfeucht sequence for $(\mathfrak{A}, \mathfrak{A}')$. Referring to the definition of such sequences, we see that it is obvious that I satisfies (i) and (ii). That I satisfies (iii) follows from the fact that I' satisfies (iii) with $\mathfrak{A}$ and $\mathfrak{A}'$, replaced of course by $\mathfrak{A}_{k_0}$ and $\mathfrak{A}_{k_1}$.

So finally, we need only to show that I satisfies (iv) of the definition. Assume that $(a_0, \ldots, a_{n-1})\, I_n\, (a'_0, \ldots, a'_{n-1})$ and $i_0, i_1, i_2, i_3 < n$. For simplification of notation let $c_k = a_{i_k}$ and $d_k = a'_{i_k}$. What must be shown is that

$$c_0c_1\, D\, c_2c_3 \quad \text{iff} \quad d_0d_1\, D'\, d_2d_3, \tag{4}$$
$$c_0 = c_1 \quad \text{iff} \quad d_0 = d_1. \tag{5}$$

Equivalence (5) follows at once from the definition of I and the fact that I' satisfies (iv). On the same grounds, (4) is obvious when $(c_0, c_1, c_2, c_3) \in A^4 - X$. This leaves the case that $(c_0, c_1, c_2, c_3) \in X$. From the definition of D this case reduces to the case that (c_0, c_1) and (c_2, c_3) are disjoint positive intervals. But in this situation, Equation (4) follows from the following facts, which can be directly checked:

$$\begin{aligned} c_i \leqslant c_j \quad & \text{iff } d_i \leqslant d_j; \\ d(c_i, c_{i+1}) = 1 \quad & \text{iff } d_i - d_{i+1} = 1; \\ d(c_i, c_{i+1}) > 1 \quad & \text{iff } d_i - d_{i+1} > 1; \\ d(c_i, c_{i+1}) = k_1 \quad & \text{iff } d_i - d_{i+1} = k_1; \end{aligned}$$

if $x, y, u \in A$, $k_0 < x \leqslant u$, and $x - k_0 \leqslant z - u$, then $(z - u) - (x - k_0) = 0$ or $= 1$ or $\geqslant k_1$;

if $x, z, u \in A'$, $k_1 < x \leqslant u$, and $x - k_1 \leqslant z - u$, then $(z - u) - (x - k_1) = 0$ or $= 1$ or $\geqslant k_1$.

Thus I satisfies (iv) and the proof of the theorem is complete. ◇

By similar methods we can prove analogues of Theorems 12 and 13, which we only state here.

THEOREM 18. *The finitary class of conjoint measurement structures with $n \geqslant 2$ components is not axiomatizable by a finite set of elementary sentences. The same negative result also holds for the finitary class of n-dimensional scaling structures with Euclidean metric for $n \geqslant 1$.*

21.7 THE ARCHIMEDEAN AXIOM

For essentially every variety of measurement structure considered in Volume I, we stressed the fundamental role of an Archimedean axiom. An axiom of this sort is required whenever the structures are infinite, that is, when they have an infinite set as domain and when a representation in terms of the real numbers is wanted. This statement must be qualified in the following sense. There is a tradition that goes back to the nineteenth century in geometry of proving representation theorems for geometric structures in terms of non-Archimedean fields. With the development of nonstandard analysis by A. Robinson (1951, 1961, 1963) and others, there has been an explicit emphasis on non-Archimedean structures in the theory of measurement (see especially, Narens, 1985; Skala, 1975). The characteristic feature of nonstandard analysis is the introduction and use of infinitesimals, which do not satisfy an Archimedean axiom. The primary reason for using infinitesimals is to simplify proofs and computations, especially of the sort that occur in the theory of stochastic processes and mathematical physics (for a recent detailed exposition of these matters, see Albeverio, Fenstad, Hoegh-Krohn, and Lindstrom, 1986). These developments do not, however, decrease the significance of an Archimedean axiom but provide a broader perspective.

On the other hand, to characterize axiomatically the system of real numbers, a completeness property stronger than the Archimedean condition is required. There are three closely related classical approaches. Historically the first is Dedekind's (1872/1901/1963) axiom of "continuity":

> If the system of real numbers breaks up into two nonempty classes A and B such that every number a of class A is less than every number b of class B, then there exists one and only one number c by which this separation is produced.

Cantor, in contrast, used an approach formulated in terms of Cauchy sequences. A sequence $\langle x_1, \ldots, x_n, \ldots, \rangle$ of real numbers is a Cauchy sequence if and only if for every positive rational number ϵ there is a natural number N such that for all $m, n > N$, $|x_n - x_m| < \epsilon$. Completeness is then expressed by the axiom that a sequence of real numbers has a limit if and only if it is a Cauchy sequence.

Finally, a common approach in this century is to use the least upper bound axiom, which is very close in spirit to Dedekind's. The axiom is that a nonempty set of real numbers that has an upper bound has a least upper bound.

Hölder (1901), in his classic article on measurement, uses a completeness axiom of the kind just described; consequently, it can be proved that any two models of his axioms for extensive measurement are isomorphic.

The theory of measurement and much of modern algebra avoid categorical systems of axioms (a system of axioms is *categorical* if and only if any two models of the axioms are isomorphic). The reason for this attitude in the theory of measurement is apparent: diverse empirical applications are intended, and it would be surprising if different applications led to isomorphic models of the theory. The same attitude is strikingly evident in physics. The theory of mechanics, or any other branch of physics, is applicable to physical situations that have quite different initial and boundary conditions.

Unless the object of physics is to create the one true model of the entire universe, categoricity of a physical theory is a defect, not a virtue.

So the situation is this. To characterize the real numbers we need an axiom of completeness, which implies *the* Archimedean axiom for real numbers but is not implied by it (see Theorem 19.1, p. 37). But such a strong completeness axiom is not appropriate for measurement structures because we do not want the theory to be categorical. In general, diverse models are required for diverse applications. On the other hand, some axiom, which we are unable to formulate in elementary logic, is required for proof that infinite structures of a given class can be embedded in the system of real numbers. *The* Archimedean axiom plays this role very well. There is, of course, no unique Archimedean axiom, as the various examples in Volume I demonstrate, but the thrust of all the variations is the same; consequently, we shall continue to speak of *the* Archimedean axiom.

The statement that the Archimedean axiom cannot be formulated in elementary logic must be carefully qualified. It is true only when the variables range over the objects in the domain of the relation structures that are the intended models of the elementary theory. These matters have been discussed in Section 21.2, but we treat them still more explicitly in this section.

If we take axiomatic set theory as our elementary theory [e.g., Zermelo-Fraenkel set theory as formulated in Suppes (1960/1972)], then all of the definitions of measurement structures in earlier chapters of this volume and of Volume I can easily be cast as elementary definitions satisfying the canons of definition laid down in Section 21.3.1, and the part of each definition that formulates the Archimedean axiom stays within this elementary framework. Explicit discussion of this set-theoretical approach to axiomatization is to be found in Suppes (1957, Chap. 12).

The taking of set theory as our standard theory is really the foundation of the mathematical developments in our earlier chapters. Nonetheless, the axiomatic structure of various theories of measurement is illuminated if we view each theory as an independent elementary theory and not as characterized by a definition within set theory. Valuable insights into the structure of theories of measurement can be gained by looking at them as elementary theories, but that point of view is not necessary for general purposes.

For those theories of measurement that have infinite models and can be formulated as elementary theories except for the Archimedean axiom, the following theorem, an immediate consequence of Theorem 7, shows that the Archimedean axiom is not equivalent to any set of elementary formulas.

THEOREM 19. *Let the class K of models of any elementary theory T formulated in a language L satisfy hypotheses* (i) *and* (ii) *of Theorem* 7. *Then* *there is no consistent extension T^* of T such that*

(i) *T^* has an infinite model,*

(ii) *the theory of simple orders as an extension of the theory of weak orders (with the same nonlogical constant) is interpretable in T^*, and*

(iii) *T^* is a theory of measurement.*

Consequently, for such a theory T there is no set of elementary formulas of L equivalent to an Archimedean axiom for T.

PROOF. If there were such an extension T^*, then the new formulas made valid by the extension would imply the necessary Archimedean axiom, but this is impossible by the argument given for the proof of Theorem 7 because there are infinite models of T^* of arbitrarily large cardinality. $\diamond$

As discussed in connection with Theorems 6 and 7, the formulation of Theorem 19 is made awkward by the necessity of dealing with homomorphisms rather than isomorphisms. A classic treatment of the Archimedean axiom that has been used in the developments given here, but not exactly followed for the reason just stated, is to be found in A. Robinson (1951).

Because of the importance of the Archimedean axiom in theories of measurement, we give a second proof using the compactness theorem (Theorem 4) that the axiom is not expressible in an elementary language. For this purpose we use (as we did in Section 21.4) the axioms for extensive measurement given in Definition 3.1, and restated after Theorem 7 of this chapter.

THEOREM 20. *Let T be the elementary theory whose two nonlogical symbols are the binary relation predicate $\succsim$ and the binary operation symbol $\bigcirc$ and whose axioms are Axioms 1–3 and 5 of extensive measurement as formulated in Section 21.3. Then there is no elementary extension T' of T in which the Archimedean axiom 4 of Section 21.4 is valid.*

PROOF. Let E be the set of all sentences of the elementary language of T holding in the ordered, additive semigroup of positive real numbers. Let α, β, γ, and δ be four constant symbols, and let

$$F = E \cup \{\alpha \succ \beta \;\&\; n\beta \bigcirc \delta \succ n\alpha \bigcirc \gamma : n \in I^+\}.$$

(The notation here should be clear: for each n, say, $n = 3$, we have in F

the elementary sentence

$$\alpha \succ \beta \;\&\; \beta \bigcirc \beta \bigcirc \beta \bigcirc \delta \succ \alpha \bigcirc \alpha \bigcirc \alpha \bigcirc \gamma, \tag{6}$$

with parentheses omitted since the operation is associative.) For every finite subset F' of F there is an expansion of the positive reals to a model of F', i.e., we can add the symbols α, β, γ, and δ to the language of T and interpret these symbols in the positive reals for the finite subset F' of sentences. (For example, let F' contain only (6) as a sentence using the symbols α, β, γ, and δ. Then we can interpret the symbols as follows in order to have a model of F' in the positive reals: $\alpha = 2$, $\beta = 1$, $\gamma = 5$, $\delta = 10$.)

Then, by compactness there is a model of F in the positive reals in which α, β, γ, and δ are interpreted as positive real numbers. But by construction of F for all finite multiples of α and β, the Archimedean axiom fails. ◇

This theorem and its proof make possible a concrete clarification of these negative results about the Archimedean axiom. Suppose someone proposed to add, as an axiom of extensive measurement, the countably infinite disjunction

$$a \succ b \rightarrow ((a \bigcirc c \succsim b \bigcirc d) \vee (2a \bigcirc c \succsim 2b \bigcirc d) \vee \cdots (na \bigcirc c \succsim nb \bigcirc d) \vee \cdots). \tag{7}$$

This infinite disjunction can indeed replace the Archimedean axiom 4 of Definition 1, Chapter 3; but (7) is not a formula of an elementary language. It belongs to the simplest infinitary logic, one that permits formulas that have a countable infinity of symbols. In principle we could have used such a language in earlier chapters to eliminate Archimedean axioms of the standard form. We would then have had to replace them with infinite disjunctions of the form of (7). But at the same time most of the powerful theorems about elementary languages no longer hold in such an infinitary logic; consequently, the advantage of such a formulation would be slight—for example, the compactness theorem is no longer valid. On the other hand, the disadvantage is obvious. Infinitary languages are familiar only to logicians, not to most scientists who have developed interests in the foundations of measurement (two introductory accounts are Keisler, 1977; Makkai, 1977).

Narens (1974) provided a general account of Archimedean axioms of various forms. By generalizing the proof of Theorem 20 along the lines developed by Narens, we can use the compactness theorem to give a different proof of Theorem 7, which has as a consequence that the theory of extensive measurement of Definition 3.1 is not axiomatizable in an elementary language.

21.8 TESTABILITY OF AXIOMS

The idea of testing axioms in many ways seems more closely associated with problems of statistical inference than directly with problems of axiomatizability. However, issues about testability raise important conceptual issues about the nature of axioms. It is therefore appropriate to consider some of the issues about testability in this chapter; though it is not the point of this section to enter into details of how particular measurement structures are to be given appropriate experimental tests or how those tests are to be set up and analyzed from a statistical standpoint.

In a general way, axioms of measurement structures of the kind we have considered in the various chapters of this treatise fall into one of four classes.

Design axioms. Certain axioms have traditionally been regarded as design in the sense that they are obviously true, given the standard intended interpretations of the primitive concepts of the measurement theory being considered. In many psychological experiments, for example, the requirement that the comparison of stimuli be connected, that is, that any two stimuli in the basic set be compared, would be tested not as an empirical law but would be satisfied by the very character of the design of the experiment. Axioms of this nature are needed for completeness of the intended mathematical results. They are part of the formal structure, but in no sense are they subject to direct empirical test.

Technical axioms. Axioms that guarantee nice mathematical properties, which are in themselves not in any obvious way directly testable, are often called technical axioms. We shall discuss later whether Archimedean axioms fall under this classification or are to be regarded as empirical laws. Our answer is neither simple nor uncontroversial.

Examples of technical axioms other than Archimedean axioms but of a related kind are to be found in Chapter 13, the various kinds of axioms of completeness or of continuity used in geometry. In fact, the discussion of axioms of completeness in the preceding section also provides axioms that are often regarded as technical axioms, scarcely subject to empirical test.

Still other examples can be found in Chapter 10. The various axioms for physical systems that require differentiability are often regarded as technical. It is commonly said that it is scarcely possible to think of any experiments that would differentiate between continuous and differentiable paths for particles or rigid bodies in classical mechanics, but from a formal standpoint the difference between continuity and differentiability is enormous. Of course, it is not simply differentiability but still stronger smoothness conditions that are needed for detailed mathematical analysis of physical structures and a variety of geometric structures, e.g., the existence of second derivatives. From one standpoint, at least, such axioms can be regarded as purely technical axioms. As already indicated, we shall have more to say about this later.

Empirical axioms. The axioms that are ordinarily tested under some standard interpretation of the primitive concepts of the measurement theory form the most important class, the one that is the real focus of analysis. It is presumably the objective of particular experiments to test particular axioms, if not the entire collection of empirical axioms of a given theory. Axioms of independence or cancellation in the theory of conjoint measurement are good examples of what are ordinarily thought of as empirical axioms, as is the axiom of monotonicity in the theory of extensive measurement.

It would be desirable to have all the axioms for a given measurement structure fall into one of the three classes stated above, but there is another distinction that is perhaps useful and that introduces a fourth class. This is the following.

Idealized axioms. It is often the case in scientific theories that axioms are introduced that are not meant to be taken literally but that, in principle, could be replaced by more complicated axioms expressing the more complicated and awkward facts as they really are. Again, good examples can be found in the theories of geometry set forth in Chapter 13. It is often maintained that the idealized concept of a point is not, from a scientific standpoint, the right concept. One should work with extended bodies as primitive objects rather than points, but the axioms stated purely in terms of such extended bodies are, it is anticipated from work that has been done thus far at least, always more complicated and a good deal more awkward to work with. The idealized concept of a point is difficult to give up in geometry, as can be seen from its tenaciously held prominence in geometric theories. What has been said about points applies in almost the same vein to the concept of particles in classical mechanics. The natural physical concept is that of an extended body, either rigid or not; but again the axioms and the important consequences of the axioms are much easier to develop in terms of particles. Moreover, many of the most important problems of mechanics are already intractable for particles. For example, the problem of determining the motions of n particles acting on each other purely in terms of gravitational attraction presents such formidable difficulties that there is no point in studying it with more realistic assumptions about the objects to which the theory is meant to apply.

It also seems to be the case that idealized axioms can be, in principle, surrogates for more complicated empirical or technical axioms. Axioms of differentiability and, in many cases, axioms of continuity provide examples of idealized axioms; and mathematically there have often been strong efforts in particular domains to weaken idealized axioms, for example, to move from differentiability to continuity assumptions only. On the other hand, movements in the reverse direction are also notable, as in the modern mathematical concern with differentiable mappings as opposed to continuous mappings.

The important point is that the status of an axiom in terms of the four classes identified cannot be inferred from its mathematical form alone. Its status depends as well on the intended interpretations of the primitive concepts and, more loosely, on prior knowledge about the empirical domain and the kinds of experiments that are conducted in this domain in which the primitive concepts are ordinarily interpreted.

Data structures. Just as there exist several natural classifications of axioms with respect to their empirical content, it is also possible to look at data structures in several different ways. Naturally, the classification of data structures affects final judgments about the testability of axioms, quite apart from the considerations already discussed.

There has been, for obvious reasons, a tendency to talk as if real experiments always need to be reduced to finite data structures. It is certainly important to analyze finite data structures, and we do so in the next section; but we also consider it important to state the case for infinite data structures.

Probably the best case for infinite data structures is to be made within a statistical framework. Why is it that in probability and statistics there is a continual use of infinite sample spaces or of random variables with continuous distributions, even though the sample spaces or random variables of most importance usually have a rather direct empirical interpretation. We can be interested, for example, in the distribution of height in a population, the distribution of a disease in another population, the distribution of radioactive decay in a substance of a certain kind, etc. It is not to the point here to go into these matters in detail, but there is one aspect that we want to stress. In talking about the relation between axioms and data structures or between theoretical and data structures, it is natural to talk in terms of infinite probability structures and to introduce finite statistical data structures only at a second step. The finite statistical structures are meant to be statistical samples of the infinite data structure.

When this approach is adopted, it is natural to study the testability of axioms in terms of the infinite data structures represented by probability spaces or continuous random variables and to state, in a variety of ways, empirical hypotheses or procedures for estimating parameters that can be referred to in the statement of axioms, etc.

21.8.1 Finite Data Structures

There are at least three senses in which finite data structures can be directly related to the testability of axioms. The first and strong sense is that the finite data structure is itself a model of the axioms, the second is that the finite data structure can be homomorphically embedded in models of the axioms, and the third is that the finite data structure bears a statistical or probabilistic relation to the models of the axioms. In this section we examine the first two senses and defer the third, the statistical sense, to the following section on infinite data structures (Suppes, 1962).

We shall say that an axiom is *finitely testable in the strong sense* if and only if the axiom is satisfied in a finite model, that is, in a finite data structure. This definition is directly extended to a set of axioms, for example, a set of axioms for a given class of measurement structures. (We could always use the term *finite model*, already introduced to refer to a model whose domain is finite, but we think that for the purposes of the intended analysis it is more suggestive to use the term *finite data structure* for such finite models.)

Axioms for a number of measurement structures defined in Volume I are testable in this strong sense. They are the finite, equally spaced structures mentioned earlier. The description of these structures uses the word *finite*, which is itself not an elementary term, but the axioms themselves are either elementary in character or can easily be expressed as elementary axioms.

The question of testing individual axioms in finite data structures is rather subtle because axioms that do not, when taken together with other axioms, have a finite model can permit individual testing in finite data structures but not in a way that is intended or interesting. As a simple example, consider the following positivity axiom, which is often assumed in extensive structures:

$$a \bigcirc b \succ a.$$

A trivial finite data structure is obtained by taking a domain of one element, letting the operation be closed in that domain, and letting the relation be that of identity. This trivial example shows that we are really not interested in the general question of finite models but only in finite data structures that have some prior intuitive interpretation. Ordinarily there is an experimental tradition or set of traditions that we intend to appeal to, and the procedures associated with these traditions should generate reasonable and intuitively acceptable finite data structures. It is only this class of data structures that we are actually interested in considering in our analysis of testability.

Adams, Fagot, and Robinson (1970) provided a detailed analysis of the relation between finite data structures and theories of measurement. We formulate their results in a somewhat different form from theirs, and we do not introduce all the distinctions that they do. We shall also proceed set-theoretically without explicit reference to an elementary language. We can illustrate the ideas by restricting ourselves to additive conjoint structures $\langle A_1, A_2, \succsim \rangle$, as defined in Definition 7 of Chapter 6. The axioms are (i) weak ordering, (ii) independence, (iii) Thomsen condition, (iv) restricted solvability, (v) Archimedean condition, and (vi) essentiality.

We now consider *generalized* additive conjoint structures $\langle A_1, A_2, \succsim \rangle$ as structures that satisfy axioms (i)–(iv) and (vi) but not necessarily the Archimedean axiom.

Let B_1 and B_2 be finite sets. A finite data structure $\langle B_1, B_2, \succsim \rangle$ is said to be *potentially additive conjoint* iff it is a submodel of an additive conjoint structure. We define *potentially generalized additive conjoint* in a similar fashion.

THEOREM 21. *A finite data structure is potentially additive conjoint iff it is potentially generalized additive conjoint. In other words, satisfaction of the Archimedean axioms for additive conjoint structures cannot be tested in a fixed finite data structure.*

Thus, if testability is restricted to finite data structures in the manner described, the Archimedean axiom for conjoint structures is a technical rather than empirical axiom. The reader can easily apply these ideas to other structures, for example, closed extensive structures, and to other axioms.

Yet, results such as Theorem 21 do not seem to tell the whole story. Here is one way to get different results. Consider the Archimedean axiom for closed extensive structures (Definition 1 of Chapter 3). This axiom, already discussed extensively in this chapter, asserts that if $a \succ b$, then for any c, d in A, there exists a positive integer n such that $na \bigcirc c \succsim nb \bigcirc d$. For a finite data structure $(B, \succsim, \bigcirc)$ that is a submodel of a closed extensive structure $\langle A, \succsim, \bigcirc \rangle$, we can think of the "operation" $\bigcirc$ only as a ternary relation to avoid closure problems for the operation. In a particular case of such a finite data structure, we are likely to find that the Archimedean axiom is not satisfied because in the given finite structure there are not enough concatenations. But for all standard cases of this kind, we intuitively have in mind a finite constructive procedure for producing more elements to satisfy the axiom. Thus, even if d is much heavier than c, we still have in mind a procedure P for producing copies of a and b such that $na \bigcirc c \succsim nb \bigcirc d$. The properties of this procedure are not known a priori, but we find in a reasonably large number of cases that it works; so we come to hold the view that tests of the Archimedean axiom relative to empirical procedure P are positive, and we accept the axiom as having a high empirical probability of being true.

This kind of procedure can be formalized, but we shall not pursue the matter here. The important conceptual point is that, in general, data structures are not simply given to us; we produce them. Moreover, the methods of production provide a natural way of extending a data structure and testing axioms by means of the extension.

21.8.2 Convergence of Finite to Infinite Data Structures

We begin by remarking that infinite data structures are a normal and natural part of many empirical sciences, and we are not in any sense restricted to finite data structures. One familiar example is the continuous recordings of physical variables such as temperature and barometric pressure in meteorology. It is possible to take the theoretical view that these data structures are discrete and finite, but this is not the ordinary view; it is just as natural to treat them as continuous and to manipulate them as continuous variables for data computations.

When such continuous structures are accepted as the underlying theory, the problem becomes how best to use finite data samples to estimate the infinite structure that obtains. In some sense, we want to be assured that the information inferred from a sample is of a certain degree of accuracy; that if we take sufficiently large samples, convergence to a unique continuous model is assured; and that the particular sampling procedure used does not affect the conclusion. A number of results of this character exist; we mention three that seem particularly relevant.

First, the constructions used to arrive at representations for both associative and nonassociative positive concatenation structures (Sections 2.2 and 19.5.4) show how to use observable equalities or inequalities to arrive at a countable sequence of observations that converges to a representation of the continuous structure. The construction in the proof of Theorem 2.4 involves solving equalities to form standard sequences and then approximating the additive representation in terms of these sequences. In particular, either by selecting ever smaller objects to generate the standard sequences or by generating a standard sequence based on the object to be measured and holding the reference standard sequence fixed, any prescribed degree of accuracy can be achieved in a known number of steps. That is, the rate of convergence is understood in the additive case on the assumption of the existence of precise standard sequences. This is the basis for much length and mass measurement, and it works despite the fact that exact solutions to equations are impossible.

The proof for a representation of a nonassociative PCS (Theorem 19.1), though similar in spirit to the proof of Theorem 2.4, differs in that only approximate standard sequences are used. This makes it much more difficult to understand the rate of convergence involved. Moreover, even if the proof were modified to use exact standard sequences, the fact that we do not know the nature of the target representation means that, in general, we do not have any estimate of the rate of convergence. This can be seen vividly in the homogeneous case, when the target representation is of the form

$$x \otimes y = yf(x/y),$$

and so any standard sequence is of the form

$$x(n) = xf^n(1),$$

where $f^1(1) = f(1)$ and $f^n(1) = f[f^{n-1}(1)]$. It is obvious that the number of steps required to approximate another element, and therefore the rate of convergence of the approximation, depends upon knowing the exact form of f—which, of course, is known in the additive case. Without f being specified, one cannot know a priori what is needed, and the obtained data have to dictate when the approximation appears to be adequate.

Second, consider the additive conjoint case, in which the situation is better in the sense that an arbitrary set of stimuli can be selected and, using the methods of Chapter 9 (also see the theorem of the alternative, Section 2.3), an additive representation can be constructed, if one exists, that preserves the empirical order of the stimuli. Note that standard sequences are not involved and that the uniqueness of such an additive representation is very difficult to characterize. But in this situation it makes sense to ask what happens as we increase the sample of stimuli so that the original data structure is a substructure of the new one. This can only restrict the class of additive representations. If this is continued indefinitely, adding more and more data, one hopes the process will converge to as unique a representation as possible, namely, an interval-scale representation. Such a theorem is provided by Narens (1985, p. 270). To state it, we first need the concept of a bounded conjoint structure. Let $\langle A \times P, \succsim \rangle$ be a conjoint structure. It is said to be bounded by ap and bq, where $a, b \in A$ and $p, q \in P$, provided that for all $c \in A$ and $r \in P$, $ap \succsim cr \succsim bq$.

THEOREM 22. *Suppose* $\langle A \times P, \succsim \rangle$ *is a conjoint structure that is bounded by ap and bq and that has an interval-scale additive representation. Let* φ_A, φ_P *be the representation with* $\varphi_A(a) = \varphi_P(p) = 1$ *and* $\varphi_A(b) = \varphi_P(q) = 0$. *Let* A_i, P_i *be a sequence of nonempty subsets of* A, P *respectively, such that*

$$A_i \times P_i \subseteq A_{i+1} \times P_{i+1} \qquad \text{and} \qquad A \times P = \bigcup_{i \in I^+} A_i \times P_i.$$

Let $\succsim_i$ *be the restriction of* $\succsim$ *to* $A_i \times P_i$, *and suppose* $\alpha_{A_i}, \varphi_{P_i}$ *is an additive representation of* $\langle A_i \times P_i, \succsim_i \rangle$ *with the same normalization as* φ_A, φ_P. *Then* $\varphi_{A_i}, \varphi_{P_i}$ *converge, respectively, to* φ_A, φ_P *in the sense that for each* $\epsilon > 0$, *there exists an integer* N *such that for all* $n > N$, *all* $c \in A$, *and all* $r \in P$, $|\varphi_A(c) - \varphi_{A_i}(c)| < \epsilon$, *and* $|\varphi_P(r) - \varphi_{P_i}(r)| < \epsilon$.

This result provides no estimate of or bound on the rate of convergence. For that, one would have to select the A_i, P_i to include at least one increasingly larger standard sequence. This has not yet been formulated, though it does not appear to be particularly difficult to do so.

Our next and last example is probably the most familiar issue of convergence, namely, that of sequences of random variable discussed in statistics. This can arise in the measurement context, for example, if we want to test transitivity of the stochastic relation defined by

$$a \succsim b \qquad \text{iff} \qquad P(a, b) \geqslant \frac{1}{2}.$$

Our infinite data structure is a sequence of outcomes (x, y)—meaning that on a particular occasion x was preferred to y—or (y, x). (For simplification, we suppose a forced-choice experiment.) In an actual experiment, a particular finite subsequence of choices is observed; but, nevertheless, the statistical theory of the infinite sequence provides the right framework to understand the observations. The ubiquitous presence of the multivariate Gaussian (normal) probability distribution in so much statistical theory and practice also suggests that any simple adoption of finite data structures as the only framework for the testability of axioms would be mistaken.

A final point concerning the role of infinite structures is the great simplicity they often exhibit, which is totally obscured in the finite case. To take just one example, there is nothing in the finite case comparably simple to the unit structure representation, mentioned above, of homogeneous concatenation structures. Not only is it easy to characterize this entire class of structures, but it provides a very clear target for the experimental scientist, whose task when confronted with a nonassociative operation is to devise sampling methods that enable an accurate estimate of f. The difficulty, as we pointed out earlier, is that the rate of convergence is unknown. In practice in other areas, such as differential equations, such lack of knowledge of rates has not proved insurmountable.

It is no accident that the theory of ordinary and partial differential equations has played a central role in the application of mathematics in the physical sciences. By assuming that such essential quantities as position and velocity are continuous, and thus in principle should be represented by infinite data structures, we can reduce problems of a great variety to the problem of solving a differential equation. Crude finite approximations usually do not work as well, in the sense of fitting observed data with solutions of finite difference equations. When the finite approximation is made very accurately, the computation involved in finding solutions for a wide class of problems is much more difficult to carry out when finite difference equations rather than differential equations are used.

There is a still deeper point to make about the importance of continuous quantities. Most of the serious experience of building mathematically formulated theories of empirical phenomena has been in terms of deriving the governing differential equations. It seems difficult and awkward to think in terms of discrete quantities when reflecting on most, but certainly not all, phenomena. Perhaps in the distant future we shall learn to think mainly in digital and discrete concepts, but it seems likely that in the foreseeable future there will be a continuing central place for infinite but continuous data structures and their associated theories.

21.8.3 Testability and Constructability

There is a tradition in mathematics of identifying proofs of theorems that are constructive. Numerous proposals have been made for giving an exact meaning to the general idea of constructivity. The reader will find an excellent survey in Troelstra (1977). One appealing simple idea is that a proof of a sentence of the form $(\exists x)S$ is constructive if the proof shows us how to produce or find an x satisfying S. Most mathematics used directly in the sciences is constructive in character. The following elementary nonconstructive proof has a flavor that is distinctive of pure mathematics and is seldom seen in applied mathematics.

Assertion. There exist two irrational numbers x and y such that x^y is rational.

PROOF. $(\sqrt{2})^{\sqrt{2}}$ is either rational or irrational. If it is rational, take $x = y = \sqrt{2}$. If it is irrational, take $x = (\sqrt{2})^{\sqrt{2}}$, and $y = \sqrt{2}$ because in this case $x^y = 2$. ◇

In the theory of measurement there is a natural desire to have constructive proofs of the numerical representations whenever possible because a constructive proof should provide a framework for actually constructing the specific assignment of numbers to empirical objects or phenomena. One approach that satisfies this demand for constructive proofs and that fits in well with much of measurement theory is that axioms be quantifier-free so that no "pure" existential axioms are admitted. Proofs within such a framework consist of giving a specific construction. Obvious examples of such quantifier-free axioms are those for closed extensive structures (Definition 3.1), with the exception of the Archimedean axiom, which is existential in character.

There is a clear intuitive sense that such a specific constructive restriction makes axioms and theorems more directly testable: just construct desired objects by performing specific constructions. Appealing as such a program may be philosophically, there are good reasons why we have not adopted it as the framework for the foundations of measurement. The restriction sharply limits what can be proved.

Such quantifier-free systems have been studied in logic for a number of years. Early in the century, Śkołem (1920) observed that such systems form a natural foundational version of constructive, finitist number theory. Kreisel (1954) pointed out that this is perhaps the only way of making a clear distinction between the Selberg-Erdos elementary proof of the prime number theorem and the classical proof. Kreisel used quantifier-free systems to give an explication of the notion of direct proof. Surprising results can be obtained by the imposition of such a strict notion of constructability on an axiom system in number theory. For example, Shepherdson (1965) showed that in a system having as axioms only defining equations for successor, predecessor, addition, multiplication, and subtraction, together wtih a rule of induction, the question raised by Śkołem (1955) about the irrationality of $\sqrt{2}$ can be answered negatively; namely, it cannot be proved in such a free-variable system. Because the irrationality of $\sqrt{2}$ is one of the most ancient and simplest results of elementary number theory, the strength of the restriction to free-variable methods, that is, quantifier-free methods, is clearly a powerful one.

An example closer to the theory of measurement (in particular, one from constructive plane geometry) can be found in Moler and Suppes (1968). They give quantifier-free axioms for the constructive geometry consisting of the two operations of finding the intersection of two lines and of laying off one segment on another. This geometry is well known, having been analyzed informally in an appendix of Hilbert's well-known *Foundations of Geometry* (1899/1956). They obtain a representation in a Pythagorean field, that is, an ordered field such that if a and b are in the field, then so is $(a^2 + b^2)^{1/2}$. Note that this representation need not be Archimedean. A second geometrical example was given as a footnote to Definition 13.3. A quantifier-free version of the axioms for a projective plane was stated, but a constructive version of the theory was not developed.

A more liberal version of constructability seems practical for the theory of measurement, given the inevitable Archimedean axioms for infinite structures and the frequent use of existential solvability axioms. This more liberal version, which aims at satisfying the construction of the existential object postulated by such axioms, is discussed in Section 22.9.3. A recent analysis emphasizing the constructability of scales is found in Niederée (1987).

21.8.4 Diagnostic versus Global Tests

The axioms in these volumes are used to characterize in qualitative terms a variety of theories that give rise to numerical representations. In addition to this role, some of the axioms serve another function by providing direct targets for empirical testing of the relevant theories. This function is served by the necessary first-order axioms such as transitivity, cancellation, or monotonicity; Archimedean and solvability axioms are usually accepted or rejected on the basis of general considerations rather than specific experimental tests. In this section we discuss the use of critical axioms for testing alternative representations and compare this approach with the more traditional method of evaluating models based on some global measure of goodness-of-fit.

It may appear that testing the validity of a numerical representation of some data structure is preferable to the testing of specific axioms because the former approach provides a global test of all the necessary axioms. Indeed, overall tests for goodness-of-fit are commonly used to evaluate the adequacy of numerical representations in both the natural and the social sciences, notably in econometrics and psychometrics.

Although global testing is, in principle, far more comprehensive than tests of individual axioms, it is often difficult to carry out successfully for two interrelated reasons. First, data are always plagued by some error, and in behavioral and social examples it often seems to be substantial. Attempts to reduce it have not been successful. Second, the representations often include one or more free functions that must be estimated from the data. It is all too easy to exploit that freedom in the representation to "account" for much of the error as well as the structure in the data, which can lead one to accept a model when in fact it is wrong. This was illustrated by Shepard (1974), who generated artificial data from a known model, distorted it with some level of noise, and then fit the result with models of a character different from the one generating the data. Often the model fit to the data was not rejected.

The problem just cited is mitigated to the degree that either the error in the data is reduced or the representation is more constrained. The former approach is the one usually followed in the physical sciences, but it has so far proved infeasible in the behavioral and social sciences. The latter approach arises in at least two ways. For example, there may be arguments outside of the measurement model that lead one to restrict a free function to some more limited class of functions, such as power functions, in which case the estimation is reduced from a free function to a few (in this case two) parameters. Second, in a few cases the measurement theory itself has led to a representation with only a few free parameters. We saw an example of that in Section 3.14.1, due to Pollatsek and Tversky (1970), in which an attempt to axiomatize relative risk judgments led to a model with a single free parameter. That model was readily rejected by a global test. Recently, Luce and Weber (1986) provided a different risk axiomatization that resulted in seven free parameters, and they carried out both global and diagnostic tests. Again, the global tests were feasible largely because the number of parameters was limited.

In the more usual situations, especially when the data are noisy, a focus on key axioms appears to provide a clearer and more powerful test of a theory than does a global test of goodness-of-fit. A few examples follow.

Before addressing specific cases, however, it is instructive to distinguish between quantitative and qualitative applications of measurement theory. Classical physical measurement (e.g., of mass, length, or time) is the prototypical example of quantitative measurement leading to the construction of numerical scales that are later used to formulate physical laws. Because the basic axioms needed to guarantee the construction of these representations (e.g., transitivity and monotonicity) are generally accepted as valid physical principles or at least as acceptable idealizations; they do not serve as targets for empirical investigations. The situation is different in the social and the behavioral sciences, where the possibility of fundamental measurement cannot be taken for granted, and the measurement model serves as a simple substantive theory. In these areas, therefore, there is often more interest in testing the empirical adequacy of the model than in the actual construction of a numerical representation. Indeed, there are very few instances in the social and the behavioral sciences in which representations derived via a fundamental measurement procedure have been used in the formulation of other laws. Economic theory, for example, incorporates many quantitative principles, but these are usually formulated in terms of attributes, such as demand and supply, that are not derived from a prior axiomatic analysis.

In the context of qualitative applications of measurement theory, the testing of critical axioms has three advantages over global tests of goodness-of-fit. First, individual axioms can be tested without actually constructing a numerical representation, which sometimes requires a major effort. Such tests are, in effect, measurement free in the sense that the theory is tested without ever assigning numbers to objects. Second, in the absence of a satisfactory statistical treatment of ordinal data structures (see Section 16.8), it is often difficult to interpret and assess the significance of global measures of goodness-of-fit. Moreover, such indices do not always distinguish between random errors due to uncontrolled factors and systematic errors that violate the underlying model. Third, the focus on key qualitative properties often gives rise to a much more powerful test of the theory than could be obtained by a global measure of correspondence, which is often insensitive to relatively small but significant deviations.

We shall briefly discuss four examples, described in the preceding chapters, that serve to illustrate these points. The first and best known example concerns the study of the expected utility model for choice under risk and uncertainty (see Chapter 8 of Volume I). Much of the research on this problem has focused on the testing of critical axioms, particularly the assumptions of independence and the extended sure-thing principle. Observed failure of these axioms, first demonstrated by the counterexamples devised by Allais (1953) for independence and Ellsberg (1961) for the extended sure-thing principle, has led to the development of many alternative theories that replace, in particular, the independence axiom by other assumptions. For reviews of recent developments see Machina (1987), Tversky and Kahneman (1986), Weber and Camerer (1987), and Section 20.4.5. The construction of clear counterexamples against the independence axiom, therefore, has provided convincing evidence against the expectation principle and has suggested new directions for development. Given the difficulties of simultaneous measurement of utility and probability and the statistical problems of evaluating goodness-of-fit, it is extremely unlikely that a global test of expected utility theory could have led to very clear conclusions.

The power of diagnostic testing is further illustrated in the analysis of polynomial conjoint measurement (discussed in Chapter 7 of Volume I). Assuming an ordering that satisfies solvability and the Archimedean axiom, we have described in Section 7.4 an algorithm for diagnosing all simple polynomials in three variables (additive, multiplicative, distributive, and dual-distributive) by testing the appropriate cancellation and sign-dependence axioms. This approach provides a method for selecting among alternative representations by focusing on their critical distinctive features. The multiplicative and the distributive rules, for example, are so close to each other (in the positive case) that with fallible data it is extremely difficult to distinguish between them on the basis of any global assessment of goodness-of-fit.

The many diagnostic and global tests of the geometry of visual space, which are discussed in some detail in Section 13.10, constitute a third example. Tests of the Luneburg theory of binocular vision, especially the experiments of Indow and his associates, are excellent examples of global tests. In many of the experiments only two numerical parameters have been estimated from large data samples. Equally informative and significant in a different way are the experiments of Foley, which tested individual geometrical axioms. The negative outcome of his 1972 homogeneity experiment, illustrated in Figure 13.8, raises critical questions about the nature of visual space.

Our last example concerns the evaluation of proximity or distance models that are widely used in the social and the behavioral sciences. Because much of the work in this field has been concerned with the construction of multidimensional representations, these models have usually been evaluated using global indices, such as stress, on the proportion of explained variance (see, e.g., Davies and Coxon, 1982). The application of measurement-theoretical analysis to proximity data, summarized in Sections 14.6 and 14.7, however, sheds new light on these representations by providing ordinal methods for testing the basic dimensional and metric assumptions underlying multidimensional scaling. In particular, we constructed an ordinal test of the triangle inequality and demonstrated some conditions under which it holds or fails.

In summary, then, global tests are more likely to be used in quantitative applications in which numerical scales are actually constructed. They are more effective when the theory is sufficiently specific to lead to the estimation of free parameters rather than free functions. Diagnostic tests of specific axioms are more likely to be usefully employed in qualitative applications that focus on tests of the underlying model.

EXERCISES

1. Which of the following structures are preserved under submodels: partial orders, weak orders, dense simple orders, Boolean algebras, closed extensive structures? (21.2.2)

2. A set is *recursively enumerable* iff it is the range of a recursive function. Show that if the set of valid sentences of an elementary theory is recursively enumerable, then the theory is recursively axiomatizable. (21.2.3)

3. Consider axioms 1–5 for extensive measurement stated just after Theorem 7. Use Padoa's principle to prove that the binary operation cannot be defined in terms of the ordering relation and that the ordering relation cannot be defined in terms of the binary operation.
(21.3.1)

4. Prove that the elementary theory of simple orders is not interpretable in the elementary theory of partial orders. (21.3.2)

5. Give an example of a finitary class of measurement structures that is closed under submodels and is recursively but not finitely axiomatizable. (21.4)

6. Consider the elementary theory whose axioms are the following (with the quaternary relation symbol D the only nonlogical constant):

Axiom 1. $ab\,D\,cd \leftrightarrow (\neg cd\,D\,cc \vee ab\,D\,aa)$.

Axiom 2. $(ab\,D\,aa \;\&\; bc\,D\,bb) \rightarrow ac\,D\,aa$.

Axiom 3. $ab\,D\,aa \rightarrow ba\,D\,bb$.

Show that any finite model of the theory is homomorphically embeddable in (Re^n, δ), where δ is the trivial metric

$$\delta(x,y) = \begin{cases} 1 & \text{if } x \neq y \\ 0 & \text{if } x = y, \end{cases}$$

[and $ab\,D\,cd$ iff $\delta(f(a), f(b)) \leqslant \delta(f(c), f(d))$, with f an embedding function]. (21.6.1)

7. Consider the ten-element model following Theorem 12 and shown in Figure 1. Delete a_4, and then show that the remaining nine-element model is embeddable in (Re, Δ). (21.6.1)

8. Outline the proof of Theorem 18. (21.6.3)

9. Use the results on Archimedean axioms to show that an elementary definition of finiteness of a model of a theory is not possible. (21.7)

10. Use the compactness theorem to prove that if an elementary theory has arbitrarily large finite models, then it has an infinite model. (21.8)

Chapter 22 Invariance and Meaningfulness

22.1 INTRODUCTION

It is widely understood that certain kinds of numerical statements involving measurement representations are meaningless in the sense that their truth value is dependent on which particular representation or representations are being employed. For example, it will not do to assert "The height of the center of the basketball team is 211"; one must supply the unit of measurement, e.g., 211 cm. The humor of "four-year olds tend to be square —height equal to weight" is lost unless it is understood that the units involved are inches and pounds. Other assertions do not require specification of a particular representation. For example, one can assert "The ratio of the height of the center to that of the guard is 1.14" or "The ratio of the height of the center to the length of a standard meter is 2.11" without concern about the unit of measurement. But not all ratio statements are meaningful in this way; for example, "The ratio of today's maximum and minimum temperatures is 1.14" is meaningless unless a particular representation, e.g., °C, is specified.

Despite the seeming obviousness of these examples, the various attempts to elucidate them have been rather technical. The best-known approach, attributed to Stevens (1946), uses the idea of permissible transformations of numerical representations and associates meaningfulness with invariance under permissible transformations, the scale type discussed in Chapter 20.

This was later formalized by R. E. Robinson (1963), Suppes (1959), and Suppes and Zinnes (1963). Pfanzagl (1968, p. 34) defined meaningfulness in terms of the invariance of the empirical relation induced by equivalent representations. He also gave a formal treatment of Stevens' invariance criterion (pp. 36–37). Yet another variant is that of Luce (1978), which requires that a numerical relation be invariant under automorphism of the representing numerical relational structure. This we shall generalize to invariance under endomorphisms. We shall cover Pfanzagl's definition in Section 22.2.2, and in Section 22.3 we review the various characterizations involving invariance, due to Stevens, Pfanzagl, and Luce.

Throughout these three volumes we have emphasized the concept of *empirical relational structure*. Objects under study are abstracted as a set A, and various conclusions concerning them are abstracted as relations $S_1, \ldots, S_n$ on A. Measurement scales are merely conventional classes of representations of such an empirical relational structure, using an isomorphic numerical structure in which the objects usually are vectors in Re^m. In such a formulation one surely must treat the defining relations $S_1, \ldots, S_n$ as being themselves meaningful, and it seems natural to hold that any other relation that can be expressed in terms of the structure $\langle A, S_1, \ldots, S_n \rangle$ is also meaningful. An assertion is meaningless if the attempt to express it in terms of the empirical relational structure shows it to be ambiguous.

Such a notion of meaningfulness is commonsensical and appealingly nontechnical. It was previously suggested by Weitzenhoffer (1951). Unfortunately, there is some vagueness in it. Just what is meant by *expressing a relation in terms of the structure* $\langle A, S_1, \ldots, S_n \rangle$? What syntactic and semantic rules are to be followed? If we attempt to remove this vagueness, we encounter another sort of technicality: the theory of definability in mathematical logic, which was outlined in Section 21.2.4. Recently Narens (in preparation) has undertaken a major attack on this problem which involves a moderately elaborate logical structure. Since this work is not yet published and is very technical, we do not attempt to repeat it here.

We shall use the commonsense notion of meaningfulness without attempting to clarify it fully. In Section 22.2, we state some sufficient conditions for a numerical relation to be meaningful in terms of an underlying empirical relational structure. We shall make no claim that these conditions are necessary, but they are broad and they seem to cover all the cases we know of in the actual practice of measurement. Then, in Section 22.3, we explicate the invariance criteria of Stevens, Pfanzagl, and Luce, cited above. Section 22.4 contains proofs of theorems, and in Section 22.5 we explore briefly the difficulties in attaining a precise characterization of meaningfulness using the theory of definability. This is the issue Narens is attempting to study axiomatically.

The subject of meaningfulness has two major controversies. We have just mentioned one, how to capture the idea of definability of relations. The other centers on the concept of "appropriate statistics." In Stevens' original development of meaningfulness and invariance (1946, 1951), he tied his ideas closely to the question of appropriate and invariant statistics. This tie-in seemed fairly straightforward and quickly became accepted, even entrenched, in textbooks on methods and statistics directed toward psychologists. It seems clear, for example, that experimenters have no business comparing the arithmetic averages of ordinal-scale measurements for two groups because such comparisons are noninvariant, and hence meaningless, when arbitrary monotone transformations of the scores are permissible. The most careful development of this idea is the paper by Adams, Fagot, and Robinson (1965).

This "obvious" tie between meaningfulness and statistics is controversial, however, because it fails to come to grips with the logic of statistical hypothesis testing. This logic depends on distributional assumptions, and if those assumptions are true, then the probability calculations follow and provide firm grounds for inferences. If scores are averaged in the course of the probability calculations, this averaging is justified by the validity of the distributional assumptions; the fact that those same scores can also be used to represent an empirical relational structure, and for the latter purpose can be replaced by any monotonically related scores, is irrelevant. This sort of argument was first offered by Lord (1953) and has often been reiterated, though it does not seem to have been widely accepted.

In Section 22.6 we examine this controversy in detail, concluding that meaningfulness issues do indeed have an impact on statistical inference simply because they affect which sorts of statistical hypotheses can be meaningfully proposed. Not much of value arises from a "valid" probability calculation based on an hypothesis that is either meaningless or almost surely false.

The last substantive section, 22.7, returns to the main topic, showing that the laws of classical physics must be dimensionally invariant to be meaningful, i.e., expressible in terms of the known empirical relations that interconnect physical attributes. This, in essence, resolves a question left hanging in Chapter 10.

22.2 METHODS OF DEFINING MEANINGFUL RELATIONS

Throughout this section we shall consider a particular relational structure $\langle A, S_1, \ldots, S_n \rangle$ and sometimes for brevity shall denote it simply as $\mathscr{A}$. We interpret $\mathscr{A}$ as an *empirical* relational structure, that is, we suppose that there is a well-established empirical interpretation for the elements of A and for each relation S_i on A. Thus we shall assume that each relation S_i, $i = 1, \ldots, n$, is a meaningful empirical relation. Our concern is this: what other meaningful empirical relations can be defined? We want to say something to the effect that a relation S on A is empirically meaningful relative to $\mathscr{A}$ if and only if S can be defined in terms of $S_1, \ldots, S_n$.

This definition is not rigorous because we do not specify the rules governing definability of S "in terms of" $S_1, \ldots, S_n$. We shall not attempt here to make it rigorous; that was done in the framework of first-order logic in Section 21.2.4. (The difficulties that would be encountered in trying to do so in general are discussed briefly in Section 22.5). Rather, we proceed informally by specifying some *sufficient* conditions for definability of S in terms of $S_1, \ldots, S_n$ and also specifying one *necessary* condition. The sufficient conditions enable us to assert that certain relations S are in fact meaningful relative to $\mathscr{A}$. Together, they cover all the obvious cases, so that, for example, we can adjudge as meaningful, relative to an extensive measurement structure $\mathscr{A} = \langle A, \succsim, \bigcirc \rangle$, both of the statements "The center's height is 211 cm" and "The ratio of the center's height to the length of a standard of meter is 2.11." The necessary condition to be formulated enables us to assert that certain statements are meaningless relative to $\mathscr{A}$, for example, the statement that the center's height is 211.

In the following subsections we shall consider four methods of defining a relation S in terms of $S_1, \ldots, S_n$. They are summarized briefly here.

The first method (Section 22.2.1) uses only the existing relations $S_1, \ldots, S_n$ and employs the rules of definition in first-order logic outlined in Section 21.2.4. For example, in an extensive measurement structure $\langle A, \succsim, \bigcirc \rangle$, we can define the following relation S: $(a, b) \in S$ iff there exists $c \in A$ such that $a \sim b \bigcirc c$.

The second method (Section 22.2.2) considers the set of all measurement representations, using a specific numerical relational structure $\mathscr{R}$. For any numerical relation T we define a corresponding empirical relation S by stating that $(a_1, \ldots, a_k) \in S$ if and only if the corresponding numerical k-tuple $[\varphi(a_1), \ldots, \varphi(a_k)]$ stands in relation T for *every* measurement representation φ. This is the method used for ratio relations in extensive measurement. For example, let $(a, b) \in S$ stand for the relation that a is 2.11 times longer than b. We use the numerical relation T: $(x, y) \in T$ iff $x = 2.11y$. Then $(a, b) \in S$ means that for *every* representation φ, $\varphi(a) = 2.11 \cdot \varphi(b)$. As will be discussed in detail for the concept of independence in probability, care must be taken using this definition since it is quite easy to define a T that does not have an invariant empirical referent.

The third method (Section 22.2.4) is like the second one except that we do not require the relation to T to hold for every representation but only for certain ones specified by means of their values for certain elements of the set A. For example, let $a \in S$ stand for the relation that the length of a is 211 cm. We use the numerical relation T: $x \in T$ iff $x = 211$. Then $a \in S$ means that $\varphi(a) = 211$ for every measurement representation φ such that $\varphi(a_1) = 1$, where a_1 is a specific one-centimeter standard.

Finally, the fourth method (Section 22.2.5) associates various numerical relations with various representations. For example, in length measurement, let relation T_φ, associated with representation φ, be defined by: $x \in T_\varphi$ iff $x = 2.11 \cdot \varphi(a_1)$, where a_1 is a specific one-meter standard. Then define $a \in S$ if and only if for every representation φ, $\varphi(a) \in T_\varphi$. Once more, this defines the empirically meaningful relation of being 2.11 meters long, but now the definition accomplishes this by specifying how the numerical measure $\varphi(a)$ must change its relation as the representation φ changes. If $\varphi(a_1) = 100$ (centimeter representation), then $\varphi(a) = 2.11 \cdot 100 = 211$, etc.

We shall now review these four methods in more detail, using additional examples that illustrate the usefulness of each method.

22.2.1 Definitions in First-Order Theories

Since we do not need to repeat the material of Section 21.2.4, we limit this discussion to a few examples. In these examples, the full theories cannot be formulated solely in first-order logic for the reasons made clear in Chapter 21, especially in Section 21.6 on the Archimedean axiom. But the definitions considered below use only their elementary axioms and thus stay within first-order logic.

In extensive measurement, the principal theorems concern empirical relational structures of form $\mathscr{A} = \langle A, \succsim, B, \bigcirc \rangle$, where $\succsim$ and B are binary relations (ordering and concatenatability) and $\bigcirc$ is a ternary relation (in fact, it is the concatenation function with domain B and range A). Various of the axioms of Definition 3 in Chapter 3 (p. 84) and of Definition 19.2 can be restated using defined relations. For example, define a ternary relation IT (intransitive triple) by:

$(a, b, c) \in IT$ *iff* $a \succsim b$, $b \succsim c$, *and* $c \succsim a$.

One way to state part of the weak-order axiom would be to use IT: *if* $(a, b, c) \in \mathrm{IT}$, *then* $(c, b, a) \in \mathrm{IT}$. The definition of IT uses only the logical operator of conjunction. For an example using universal quantification, consider the unary relation P (positivity) defined by:

$b \in P$ *iff for all* $a \in A$ *and all* $c \in A$, *if* $(a, b) \in B$ *and* $a \bigcirc b = c$, *then not* $a \succsim c$.

Our positivity axiom (axiom 5 in Definition 3.3 or in Definition 19.2) can be stated simply as $P = A$, i.e., the unary relation P is the full set A. An example using existential quantification is the relation D defined by:

$(a, b) \in D$ iff there exist c and d in A such that $(b, c) \in B$, $b \bigcirc c = d$, and $a \succsim d$.

Our axiom 4 (existence of fine-grain) in the same definition states that whenever *not* $b \succsim a$, then $(a, b) \in D$.

We note next that certain numerical relations, which might be taken as referring to representations, could also be defined directly by elementary means. For example, the relation that a is 3 times as long as b could be defined as follows:

There exist b_1, b_2, c, and $d \in A$ such that $b \succsim b_1$, $b_1 \succsim b$, $b \succsim b_2$, $b_2 \succsim b$, $(b, b_1) \in B$, $b \bigcirc b_1 = c$, $(c, b_1) \in B$, $c \bigcirc b_2 = d$, $a \succsim d$, and $d \succsim a$.

This is, less formally, $a \sim b \bigcirc b_1 \bigcirc b_2$, where b, b_1, and b_2 are all equivalent in length. Similarly, but very laboriously, we could define the relation that a is 2.11 times as long as b using a concatenation of 211 equivalent units to match a when 100 of them suffice to match b.

These definitions of length relations are not only laborious, they are also not general enough for our purposes. Suppose in fact that no b_1 and b_2 exist in A such that $a \sim b \bigcirc b_1 \bigcirc b_2$ with $b \sim b_1 \sim b_2$. It could still be the case that a can be approximated arbitrarily closely by concatenations of the form $c(n)$, that b can be approximated arbitrarily closely by concatenations of the form $c(m)$, and that the ratio n/m converges to 3 as c becomes arbitrarily small. This set of facts cannot be expressed in the first-order theory of $\succsim$, B, and $\bigcirc$. Nonetheless, it can be expressed in ordinary mathematical language and results in the fact that $\varphi(a)/(\varphi(b) = 3$ for every representation φ mapping $\langle A, \succsim, B, \bigcirc \rangle$ homomorphically into $\langle \mathrm{Re}^+, \geqslant, \mathrm{Re}^+ \times \mathrm{Re}^+, + \rangle$. This example motivates the use of measurement representations in defining empirically meaningful relations. The next three subsections are devoted to that sort of definition.

Before turning to the use of representations in capturing the concept of meaningfulness, we give one more example which leads outside of the extensive measurement structures and therefore may help the reader to see the scope of what we intend in discussing an abstract relational structure $\langle A, S_1, \ldots, S_n \rangle$. In additive conjoint measurement with two components we consider structures of the form $\langle A_1 \times A_2, \succsim \rangle$ (see Volume I, Section 6.2). The important double-cancellation axiom can be formulated as follows:

If $(a_1, b_2) \succsim (b_1, c_2)$ and $(b_1, a_2) \succsim (c_1, b_2)$, then $(a_1, a_2) \succsim (c_1, c_2)$.

This axiom involves a total of six elements of $A_1 \times A_2$, and the antecedent conditions involve two $\succsim$ statements plus no fewer than six instances of component-wise equality, e.g., (a_1, b_2) and (a_1, a_2) match on the first component, etc. To define a 6-ary relation that states that the antecedent conditions are satisfied, it is best to make the component-wise qualities explicit, defining $=_1$ to hold between two pairs that match on the first component and $=_2$ analogously. Thus we can define the 6-ary relation C by:

for $a, b, c, d, e, f \in A_1 \times A_2$, $(a, b, c, d, e, f) \in C$ *iff* $a \succsim b$, $c \succsim d$, $b =_1 c$, $e =_1 a$, $f =_1 d$, $a =_2 d$, $e =_2 c$, *and* $f =_2 b$.

The axiom asserts that if $(a, b, c, d, e, f) \in C$, then $e \succsim f$.

22.2.2 Reference and Structure Invariance

We consider the relational structure $\mathscr{A} = \langle A, S_1, \ldots, S_n \rangle$, together with a relational structure *of the same type* $\mathscr{R} = \langle R, T_1, \ldots, T_n \rangle$. Usually the set R is a subset of Re^m, $m \geqslant 1$, and each relation T_i is of the same order as the corresponding S_i, that is, T_i is a $k(i)$-ary relation on R; however, the definitions do not require $\mathscr{R}$ to be numerical. If we propose to use $\mathscr{R}$ for measurement representations of $\mathscr{A}$, we define the set of all homomorphisms $\Phi(\mathscr{A}, \mathscr{R})$. An element φ of $\Phi(\mathscr{A}, \mathscr{R})$ is a function with domain A, range contained in R, and satisfying, for $i = 1, \ldots, n$,

$$(a_1, \ldots, a_{k(i)}) \in S_i \quad \text{iff} \quad (\varphi(a_1), \ldots, \varphi(a_{k(i)})) \in T_i.$$

In general, of course, $\Phi(\mathscr{A}, \mathscr{R})$ can be empty, or alternatively, it can contain a variety of representations bearing no simple relation to one another. The nonemptiness of $\Phi(\mathscr{A}, \mathscr{R})$ is proved in a representation theorem; limitations on its contents are proved in a uniqueness theorem. Such theorems are only provable for empirical relational structures $\mathscr{A}$ that satisfy certain properties (axioms), and naturally, the numerical structure $\mathscr{R}$ must also be suitably chosen.

Let us now consider a fixed (perhaps, numerical) structure $\mathscr{R}$. Each possible structure $\mathscr{A}$ determines a set $\Phi(\mathscr{A}, \mathscr{R})$ of homomorphisms into $\mathscr{R}$: *so, with* $\mathscr{R}$ *fixed, we can say that* $\Phi(\mathscr{A}, \mathscr{R})$ *is defined by* $\mathscr{A}$.

Now let T be any other numerical relation on R of order k. For each representation φ in $\Phi(\mathscr{A}, \mathscr{R})$ we can define an induced k-ary relation, $S(\varphi, T)$ on A:

$$(a_1, \ldots, a_k) \in S(\varphi, T) \quad \text{iff} \quad (\varphi(a_1), \ldots, \varphi(a_k)) \in T. \tag{1}$$

If the relation $S(\varphi, T)$ actually depends on the choice of φ, then T by itself does not define any relation on A; we say that *the numerical relation T is not reference invariant relative to $\mathscr{R}$*. But if $S(\varphi, T)$ does not depend on the choice of φ, we can write $S = S(T)$; we say that T defines an empirical relation $S(T)$ that is *reference invariant*.

Among all the possible homomorphic images of $\mathscr{A}$, one is very special, namely, $\mathscr{A}$ itself. As noted earlier, such homomorphisms of $\mathscr{A}$ into $\mathscr{A}$ are called *endomorphisms*. Because this special case of reference invariance is so important and provides a concept of meaningfulness that is intrinsic to $\mathscr{A}$ itself, we use the term *structure invariance* for it.

An automorphism of a relational structure is an endomorphism that is one-to-one and *onto* (see Section 20.2.3); thus $x \to rx$ is an automorphism of $\langle \mathrm{Re}^+, \geqslant, + \rangle$, where $r > 0$, but is *not* an automorphism of $\langle \mathrm{I}^+, \geqslant, + \rangle$, since even if r is a positive integer, not every integer is of form rn, where n is an integer; i.e., the mapping is not *onto*. In fact, $\langle \mathrm{I}^+, \geqslant, + \rangle$ has no automorphisms except the identity transformation. The corresponding fact is that not every homomorphism into $\langle \mathrm{I}^+, \geqslant, + \rangle$ can be onto I^+; if φ is such a homomorphism, then 2φ maps only onto the even integers, etc. Thus the hypothesis that every φ maps A onto R has considerable strength.

As we recall from Chapter 20, the automorphisms of a relational structure form a group with the usual function composition as the operation, the identity of automorphisms as the group identity and the inverse of φ as the group inverse.

We summarize this discussion as a formal definition for any two relational structures.

DEFINITION 1. *Suppose $\mathscr{A}$ and $\mathscr{R}$ are two relational structures of the same type with a nonempty set of homomorphisms, $\Phi(\mathscr{A}, \mathscr{R})$. A relation T on $\mathscr{R}$ is $\mathscr{A}$-reference invariant relative to $\mathscr{R}$ iff $S(\varphi, T)$ defined in Equation* (1) *is independent of φ for all φ in $\Phi(\mathscr{A}, \mathscr{R})$. A relation S on A that is $\mathscr{A}$-reference invariant relative to $\mathscr{A}$ itself is said to be* structure invariant.

When there is no loss of clarity, we simply speak of a relation T as being reference invariant, a term first used by Adams *et al.* (1965). The formal definition is due to Pfanzagl (1968). Narens (1985, Definitions 14.2, 14.5) calls reference invariance, "quantitatively $\mathscr{R}$-meaningful" and structure invariance "endomorphism meaningful."

Another way to think about a reference invariant relation T is that the induced relation, $S(T)$, can be added to the empirical relational structure, and represented by T in the numerical relational structure without reducing the homomorphism set. In other words, if we form the relational structures

$$\mathscr{A}' = \langle A, S_1, \ldots, S_n, S(T) \rangle \qquad \text{and} \qquad \mathscr{R}' = \langle R, T_1, \ldots, T_n, T \rangle,$$

then $\Phi(\mathscr{A}', \mathscr{R}') = \Phi(\mathscr{A}, \mathscr{R})$. No new constraints on homomorphisms are introduced by S, T.

THEOREM 1. *Suppose $\mathscr{A}$ is a relational structure and S is a relation on A. S is structure invariant iff it is invariant under the endomorphisms of $\mathscr{A}$, i.e., for all $\alpha \in \Phi(\mathscr{A}, \mathscr{A})$, and all $a_1, \ldots, a_k \in A$,*

$$(a_1, \ldots, a_k) \in S \quad \textit{iff} \quad (\alpha(a_1), \ldots, \alpha(a_k)) \in S.$$

The elementary proof of this result is left as Exercise 2.

As an example of Definition 1, let $\mathscr{A} = \langle A, \succsim, \bigcirc \rangle$ be a closed extensive structure (Definition 3.1) and let $\mathscr{R} = \langle \mathrm{Re}^+, \geqslant, + \rangle$ be the ordered additive group of real numbers. By Theorem 3.1, $\Phi(\mathscr{A}, \mathscr{R})$ is nonempty, and any two elements φ, φ' are related by $\varphi' = r\varphi$, where r is a positive constant. Therefore, if we let the numerical relation T be a specific value of the ratio—e.g., $(x, y) \in T$ iff $x/y = 211$—the induced relation $S(\varphi, T)$ is just $\varphi(a)/\varphi(b) = 211$, and it is independent of the choices of φ. In fact, the relation $T\colon (x/y) - 211 = 0$ is meaningful in $\mathscr{R}$, and the induced empirical relation is defined in terms of $\mathscr{A}$ and the relation T. The same holds for a relation T_f defined by any function of f of k variables that is homogeneous of order 0. Let $(x_1, \ldots, x_k) \in T_f$ hold iff $f(x_1, \ldots, x_k) = 0$; then $S_f = S(T_f)$ is independent of φ, since by homogeneity, $f(rx_1, \ldots, rx_k) = f(x_1, \ldots, x_k)$.

Pfanzagl (1968, p. 35) pointed out that if we accept the procedure of measurement itself as "real," then the statements that $(\varphi(a_1), \ldots, \varphi(a_k)) \in T$ for every homomorphism φ is a statement about reality. We would rephrase this: if we accept that the set of possible representations, $\Phi(\mathscr{A}, \mathscr{R})$, is defined in terms of $\mathscr{A}$ and is hence meaningful, then the relation

$$S(T) = \{(a_1, \ldots, a_k) | \forall \varphi \in \Phi(\mathscr{A}, \mathscr{R}), (\varphi(a_1), \ldots, \varphi(a_k)) \in T\}$$

is also defined in terms of $\mathscr{A}$ and, hence, is a meaningful empirical relation. The definition of structure invariance in terms of reference invariance relative to the same structure is an intrinsic concept, and it is characterized in Theorem 1 as invariance under endomorphisms. This generalizes the definition given by Luce (1978).

In scientific contexts broader than measurement, a still wider concept of invariance, and therefore of meaningfulness, is often used. Familiar examples are to be found in geometry. For example, Euclidean or affine geometry is invariant under change of direction, as represented in Cartesian coordinates. But because the qualitative primitives are invariant under such transformations, such examples are not really the most interesting ones.

In classical or relativistic mechanics, as axiomatized in Section 10.4 of Volume I, a quite different situation is found in which transformations, in addition to endomorphisms, are used in studying invariance. For example, Galilean or Lorentz transformations of the position function of a particle are not endomorphisms of that function. They change the position function, especially its derivative with respect to time. A striking fact is that the primitive relations and functions of classical or relativistic mechanics are not themselves invariant under Galilean or Lorentz transformations, which immediately places us in a different perspective when thinking about invariance. Definition 1 of structure invariance, for example, must be generalized to deal properly with mechanics.

Finally, we note that this different, wider concept of invariance, formalized in Definitions 10.10 and 10.11, was used much earlier by McKinsey and Suppes (1955) as the appropriate characterization of meaningfulness in classical mechanics.

22.2.3 An Example: Independence in Probability Theory

An interesting example, which we can carry through the next two subsections, is offered by the case of probability measurement. The question is, to what extent is independence of events definable in terms of probability, using the usual formula $P(A \cap B) = P(A)P(B)$?

The qualitative discussion of probability (see Chapter 5) usually begins with an algebra $\mathscr{E}$ of subsets of a given universal set X, on which, of course, $\cup$ and intersection $\cap$ are defined. In addition, a weak ordering $\succsim$ on $\mathscr{E}$ is assumed. One is initially tempted to treat $\langle \mathscr{E}, \succsim, \cup, \cap, X \rangle$ as the empirical relational structure, but that is not really correct. The typical representation theorem actually defines a mapping from $\mathscr{E}$ into Re^+ that preserves $\succsim$ but does not state a representation for either $\cup$ or $\cap$. Rather, if we define the subset $\mathscr{B}$ of $\mathscr{E} \times \mathscr{E}$ to consist of all pairs $(A, B) \in \mathscr{E} \times \mathscr{E}$ for which $A \cap B = \varnothing$, and if we let $\cup_{\mathscr{B}}$ consist of the union of these pairs, then the empirical relational structure is

$$\mathscr{A} = \langle \mathscr{E}, \succsim, \mathscr{B}, \cup_{\mathscr{B}} \rangle.$$

The method of proof of the representation is basically to show that the equivalence classes of this structure fulfill the conditions of Theorem 3.3 and to show there are homomorphisms into $\langle \mathrm{Re}^+, \geqslant, \mathrm{Re}^+ \times \mathrm{Re}^+, + \rangle$. For the moment, we do not impose the usual condition that X is mapped into 1; we treat that assumption in the next subsection. The proof given of

Theorem 5.2 (Volume I, p. 213–214) shows that any two such structures are related by a positive constant.

Given that, let us attempt to define pair-wise "independence" on $\mathscr{E}$ by using the ternary relation T of the multiplication on Re: $(x, y, z) \in T$ iff $z = xy$. By the method Section 22.2.2, T induces a ternary relation $S(\varphi, T)$ on $\mathscr{A}$, namely,

$$(A, B, C) \in S(\varphi, T) \quad \text{iff} \quad \varphi(C) = \varphi(A)\varphi(B).$$

It is easy to see that this is not independent of the choice of φ since $S(\varphi, T) = S(\alpha\varphi, T)$ iff $\alpha = 1$. Thus, within this framework, the attempt is to define $\perp$ by

$$A \perp B \quad \text{iff} \quad (A, B, A \cap B) \in S(\varphi, T),$$

which, however, is not meaningful in the sense of being reference invariant.

Note the structure of the discussion. Starting with the empirical structure $\mathscr{A}$, which shows the lattice-relations and ordering of events, we use elementary means to define an associated extensive measurement structure that has a measurement representation unique up to multiplication by a positive constant. The existence and uniqueness results depend on the assumptions of Theorem 5.2. Because of the ratio-scale uniqueness, we are able to show that multiplication does not induce an invariant trinary relation S on $\mathscr{A}$; and so, when we use simple elementary or first-order methods again to define $\perp$ in terms of S, we do not have a meaningful empirical relation in the original structure $\mathscr{A}$.

22.2.4 Definitions with Particular Representations

Perhaps the most common numerically defined relations are those that specify a numerical relation and a particular representation, e.g., "Jill is 3 cm taller than Beth," or "The center is 211 cm tall," or "The probability of a die coming up 2 is 1/6." In terms of the preceding subsection, we can identify such relations as having the form $S(\varphi, T)$, where T is a numerical relation on R and φ is a particular element of $\Phi(\mathscr{A}, \mathscr{R})$. In the examples given, it suffices to identify φ by means of its value at a single element of A, e.g., in the first two $\varphi(a_1) = 1$, where a_1 is a standard 1-cm length, or $\varphi(a_1) = 0.01$, where a_1 is a standard meter, and in the third, $\varphi(X) = 1$ where X is the universal event. The general form of such a definition is

$$S = S(T; (a_1, x_1), \ldots, (a_r, x_r)),$$

where

$$(b_1, \ldots, b_k) \in S \textit{ iff, for every } \varphi \in \Phi(\mathscr{A}, \mathscr{R}) \textit{ such that } \varphi(a_j) = x_j,$$
$$j = 1, \ldots, r, \textit{ then } (\varphi(b_1), \ldots, \varphi(b_k)) \in T.$$

There are no doubt other ways of identifying a particular homomorphism or subset of homomorphisms, but giving the numerical values x_j for each of a finite set of special standards a_j is simple and broad in scope. For example, in Chapter 15 (Volume II), a particular color-metric representation can be identified by specifying the vectors x_1, x_2, x_3 assigned to three specific primary lights, a_1, a_2, a_3. (Note that each x_i is an element of Re^3 in this example.) It is often useful to define a relation among colors, such as a proposed color-distance formula, in terms of a numerical relation using particular coordinates.

Returning to the probability example in the preceding subsection, we again let T represent multiplication. We now specify $S[T; (X, 1)]$ as the relation on $\mathscr{A}$. Since there is only one homomorphism with $\varphi(X) = 1$, this S is invariant (indeed, any numerical relation whatsoever induces an invariant qualitative relation), and so the independence relation $\perp$ is reference invariant.

It should be pointed out that though everyone describes physical measurement by referring to specific physical representations, no physicist infers meaningfulness in this way. It is only in elementary axiomatizations of probability that the convention of an absolute scale has arisen, which in our view is a mistake steming from the convenience of setting $P(X) = 1$. It strikes us as misguided to make particular numerical assignments to particular objects as a means to increase the uniqueness of the set of homomorphisms, despite its being highly useful in reporting scientific data. On the other hand, a more complex axiomatization that includes random variables, their sums, and their products, leads to a natural unit.

22.2.5 Parametrized Numerical Relations

The final procedure that we discuss abandons the use of a fixed numerical relation T and employs a parametrized family of relations T_φ, where the index or parameter φ is any homomorphism in $\Phi(\mathscr{A}, \mathscr{R})$. This method, another idea of Pfanzagl (1968, pp. 50–56), is closely related to the concept of dimensionally invariant law which we discussed in Chapter 10 (Volume I, p. 509; Luce, 1971).

As an example, we return once more to independence in probability. Instead of using the ternary relation of multiplication $z = xy$, we use the family T_φ:

$$(x, y, z) \in T_\varphi \qquad \text{iff} \qquad z = xy/\varphi(X).$$

If we now use ratio-scale probability, so that every homomorphism φ has the form $r\varphi_1$, where $\varphi_1(X) = 1$, then it is easy to see that the triple $(\varphi(a), \varphi(b), \varphi(c))$ is in the relation T_φ if and only if $\varphi_1(c) = \varphi_1(a)\varphi_1(b)$. Thus, requiring that $(\varphi(a), \varphi(b), \varphi(c))$ be in T_φ defines the same relation S on $\mathscr{A}$ that we previously defined using absolute-scale probability.

We generalize the probability example. Given any empirical relational structure $\mathscr{A}$, a numerical relational structure $\mathscr{R}$ of the same type, and the homomorphism set $\Phi(\mathscr{A}, \mathscr{R})$, a *family* (of numerical) *relations* $\{T_\varphi\}$, $\varphi \in \Phi(\mathscr{A}, \mathscr{R})$ is *$\mathscr{A}$-reference invariant relative* to $\mathscr{R}$ if the induced relation $S = S(\varphi, T_\varphi)$ is independent of φ. If this is the case we call S the *empirical relation induced by T_φ*.

Observe that this approach can be applied universally to ratio-scale structures. Let T be any numerical relation on n variables, and for φ in $\Phi(\mathscr{A}, \mathscr{R})$ define T_φ as

$$(x_1, \ldots, x_n) \in T_\varphi \qquad \text{iff} \qquad (x_1/\varphi(X), \ldots, x_n/\varphi(X)) \in T.$$

Then, as above, this means that the relations T_φ define exactly the same relation S, independent of φ, on A. Moreover, given any relation S on A, one can construct T_φ of this form for which S is the empirical relation induced by T. This is equivalent to saying, for example, that because probability (as ordinarily construed) is an absolute scale (no automorphisms), all relations of the structure are invariant and, thus, are meaningful.

In our opinion, this device should be invoked with caution. Superficially, it is similar to the property exhibited by a dimensionally invariant law holding among several variables, namely, that it can be cast as a function of dimensionless combinations of the several variables (see Section 10.3). As we saw in Section 10.10.2 and as we shall take up again in the next subsection, such laws can sometimes be formulated in a dimensionally invariant way only by attaching dimensional constants to the entities being discussed. Such a device seems to be useful and important in a way that making ratio scales dimensionless seems not to be. It has to do with the fact that the automorphism groups of structures on the components of a conjoint structure can be closely related to the automorphism group of the conjoint structure (see Section 22.7.2); and that fact, which is very powerful, is lost (ignored) when all of the variables are expressed in dimensionless form.

Another example of this procedure is the definition of the empirically meaningful predicate "k meters long." We let $T_\varphi = \{x | x = k\varphi(a_1)\}$, where a_1 is a standard meter.

A more complicated example is the definition of an orthogonality relation in a color space (Chapter 15) or other vector space. If φ is a homomorphism into Re^m, then we can choose a matrix $G(\varphi)$ defining a relation

$$(x, y) \in T_\varphi \quad \text{iff} \quad \sum_{i=1}^{m} \sum_{j=1}^{m} G_{ij}(\varphi) x_i y_j = 0;$$

and if we choose the matrices $G(\varphi)$ properly (all cogredient to one another), then $S(\varphi, T_\varphi)$ is independent of φ.

A broad class of examples consists of empirical relations defined by numerical laws with several interrelated representations. If $\mathbf{x} = (x_1, \ldots, x_k)$ is a vector of scale values and $\boldsymbol{\varphi}$ a corresponding vector of homomorphisms, then

$$\mathbf{x} \in T_{\boldsymbol{\varphi}} \quad \text{iff} \quad f[\mathbf{x}, \boldsymbol{\theta}(\boldsymbol{\varphi})] = 0,$$

where $\boldsymbol{\theta}$ is a mapping from the family Φ of homomorphism into Re^m for some integer m. We speak of this as a parametrized family of numerical relations. More explicitly we can write

$$(x_1, \ldots, x_k) \in T_{\boldsymbol{\varphi}} \quad \text{where} \quad \boldsymbol{\varphi} = (\varphi_1, \ldots, \varphi_k) \quad \text{iff}$$
$$f[x_1, \ldots, x_k; \;\; \theta_1(\varphi_1, \ldots, \varphi_k), \ldots, \theta_m(\varphi_1, \ldots, \varphi_k)] = 0.$$

This latter form makes explicit that the parameter vector $\boldsymbol{\theta}$ is m-dimensional and that each component $\theta_1, \ldots, \theta_m$ may depend on all k homomorphisms $\varphi_1, \ldots, \varphi_k$.

To define a meaningful empirical relation S we must have $S(\varphi, T_\varphi)$ independent of φ; that is, for any a in A, if $f[\varphi(a), \theta(\varphi)] = 0$ holds for one φ in $\Phi(\mathscr{A}, \mathscr{R})$, then it holds for every φ. A function f with this property could be termed "dimensionally invariant" with respect to $\Phi(\mathscr{A}, \mathscr{R})$; this is a generalization of the concept of dimensionally invariant function in Chapter 10 (I, pp. 466 and 509).

22.2.6 An Example: Hooke's Law

Perhaps the simplest example of a dimensionally invariant function involving parameters is Hooke's law, which states that the force on a spring is proportional to the length of the spring, the constant of proportionality being called the spring constant for that particular spring. If x_1 is length and x_2 is force, T_{φ}, where $\varphi = (\varphi_1, \varphi_2)$ is defined by

$$x_2 - \theta(\varphi_1, \varphi_2)kx_1 = 0.$$

Here the one-dimensional parameter θ is defined by

$$\theta(\varphi_1, \varphi_2) = \varphi_2(a^*)/\varphi_1(a^*),$$

where a^* is some standard spring extended to a standard length by a standard force, e.g., a spring with length 1 cm and force 1 dyne. If $\varphi_1(a^*) = 1$ and $\varphi_2(a^*) = 1$ (length measured in centimeters and force in dynes), then $\theta = 1$ dyne/cm and the spring constant is k dyne/cm. If the units are changed so that $\varphi_1(a^*) = 10^{-2}$ and $\varphi_2(a^*) = 10^{-5}$ (length measured in meters and force in newtons), then $\theta = 10^{-3}$ newton/meter, and the spring constant is $10^{-3}k$ newton/meter. The function is dimensionally invariant because the only homomorphisms in $\Phi(\mathscr{A}, \mathscr{R})$ have the form $\varphi_i' = \gamma_i\varphi_i$, $i = 1, 2$; and if

$$\varphi_2(a) - [\varphi_2(a^*)/\varphi_1(a^*)]k\varphi_1(a) = 0,$$

then the same equation holds substituting $\gamma_i\varphi_i$ for φ_i, $i = 1, 2$.

The reader may wonder exactly how to describe the empirical relational structure $\mathscr{A}$ and the numerical relational structure $\mathscr{R}$ in this case, and what assumptions are needed for the uniqueness assertion for $\Phi(\mathscr{A}, \mathscr{R})$. Roughly speaking, the elements are springs extended to various lengths: we denote by (a_1, a_2) the element in which a_1 is the (qualitative) length of spring a_2. There is an ordering relation $\succsim_1$ and a concatenation operation $\bigcirc_1$ defined for lengths, ignoring which spring is involved; thus we can form $(a_1 \bigcirc_1 b_1, a_2)$ in which lengths a_1 and b_1 are concatenated and a_2 is stretched to that length. There is also an ordering relation (force ordering) on $A_1 \times A_2$, i.e., we assert $(a_1, a_2) \succsim (b_1, b_2)$ if the force needed to maintain a_2 at length a_1 exceeds the force needed to maintain b_2 at length b_1. Finally, there is a concatenation operation $\bigcirc$ defined on forces applied to the same spring, i.e., we write $(a_1, a_2) \bigcirc (b_1, a_2)$ for the result of applying simultaneously to a_2 both the force needed to maintain it at a_1 and the forced needed to maintain it at length b_1. Thus we can write

$$\mathscr{A} = \langle A_1 \times A_2, \succsim, \succsim_1, \bigcirc, \bigcirc_1 \rangle.$$

A homomorphism (φ_1, φ_2) into $\mathrm{Re}^+ \times \mathrm{Re}^+$ has the property that φ_1 yields an additive representation for $\langle A_1, \succsim_1, \bigcirc_1 \rangle$, φ_2 yields an additive representation for $\langle A_1 \times A_2, \succsim, \bigcirc \rangle$, and $\varphi_2 = \varphi_1\gamma$, where (φ_1, γ) is a multiplicative conjoint-measurement representation for $\langle A_1 \times A_2, \succsim \rangle$. The relation S on $A_1 \times A_2$, defined using the Hooke's law function, is a unary relation that selects all pairs (a_1, a_2) in which spring a_2 has a spring constant that is k times the spring constant of the standard a_2^*. In detail,

$$(a_1, a_2) \in S \quad \text{iff} \quad \varphi_2(a_1, a_2) - \frac{\varphi_2(a_1^*, a_2^*)}{\varphi_1(a_1^*)} k\varphi_1(a_1) = 0$$

$$\text{iff} \quad \varphi_1(a_1)\gamma(a_2) - \frac{\varphi_1(a_1^*)\gamma(a_2^*)}{\varphi_1(a_1^*)} k\varphi_1(a_1) = 0$$

$$\text{iff} \quad \gamma(a_2) = k\gamma(a_2^*).$$

Obviously a different such relation exists for each value of k.

A very similar example is given by the law that the mass of a cube made of a homogeneous material is proportional to the length of the side cubed. That is, T_φ is defined by

$$x_2 - \theta(\varphi_1, \varphi_2) d x_1^3 = 0,$$

where x_2 is mass, x_1 is length, and d is the density of the homogeneous material of which the cube consists. Again, a different relation holds for each value of d. The parameter θ has the form $\varphi_2(a^*)/\varphi_1(a^*)^3$, where a^* is a standard cube made of material with density 1 (e.g., an ice cube). The unary relation S consists of all cubes with density d, independent of size.

Section 22.7 contains further material about the class of examples just illustrated; it relates the definability of empirical relations to the methods of classical dimensional analysis.

22.2.7 A Necessary Condition for Meaningfulness

We have seen that the meaningfulness of a numerical relation T depends on some convention regarding the use of T to define an empirical relation S. We have suggested three such conventions: define S to hold for $(a_1, \ldots, a_k)$ if T holds for $(\varphi(a_1), \ldots, \varphi(a_k))$ for every homomorphism φ in $\Phi(\mathscr{A}, \mathscr{R})$; or specify a specific homomorphism or class of homomorphisms; or specify how T itself changes with the homomorphism φ. If the convention used is not specified, then the relation S is ill-defined or ambiguous.

If only a single relation T is specified, rather than a family T_φ, and if no representation is singled out, then the usual assumption is that the first convention, reference invariance, detailed in Section 22.2.2, is intended.

However, T must be a structure invariant, numerical relation in the sense of Definition 1, otherwise S is vacuous, i.e., there is no $(a_1, \ldots, a_k)$ such that the measures satisfy T for every φ.

A useful criterion that helps identify meaningless or ill-defined numerical relations is based on the endomorphisms of the structure $\mathscr{R}$, $\Phi(\mathscr{R}, \mathscr{R})$. It is very easy to see that the endomorphisms are *permissible* transformations in the following sense: if φ is in $\Phi(\mathscr{A}, \mathscr{R})$ and γ is in $\Phi(\mathscr{R}, \mathscr{R})$, then $\gamma\varphi$ is also in $\Phi(\mathscr{A}, \mathscr{R})$. Now, if T is $\mathscr{A}$-reference invariant relative to $\mathscr{R}$, the k-tuple $(\varphi(a_1), \ldots, \varphi(a_k))$ is in T, then since $\gamma\varphi$ is another homomorphism, we know $(\gamma\varphi(a_1), \ldots, \gamma\varphi(a_k))$ is also in T. That is, every k-tuple $(x_1, \ldots, x_k)$ in T that is the representation of $(a_1, \ldots, a_k)$ for some homomorphism φ is mapped into T by every homomorphism. This result has considerable practical usefulness: it is easy to recognize T as meaningless, relative to $\mathscr{A}$, if the restriction of T to $\varphi(A)$ is mapped outside of T by an endomorphism of $\mathscr{R}$. Put positively, a necessary condition for T to be $\mathscr{A}$-reference invariant relative to $\mathscr{R}$ is that, for all φ in $\Phi(\mathscr{A}, \mathscr{R})$, the restriction of T to $\varphi(A)$ must be mapped into T by any endomorphism of $\mathscr{R}$. This observation is generalized in Theorem 3 below.

For example, let $\mathscr{R} = \langle \mathrm{Re}^+, \geqslant, + \rangle$, as in extensive measurement; then for every $r > 0$, the multiplication $x \to rx$ is an automorphism, and these are the only endomorphisms of $\mathscr{R}$. The relation T: $x_2 - x_1^3 = 0$ is mapped outside itself by every multiplication except the identity ($r = 1$). Thus this relation is not meaningful for any structure $\mathscr{A}$ that is mapped homomorphically into $\mathscr{R}$ (except for very trivial cases in which the restriction of T to $\varphi(A)$ is empty for every φ). The same is true of the ternary relation defined by $x_1 - x_2 x_3 = 0$, which of course is the reason that the usual definition of independence has to be modified when probability is treated as extensive measurement.

A similar criterion can be formulated for a parametric family of numerical relations T_φ: if $(\varphi(a_1), \ldots, \varphi(a_k)) \in T_\varphi$, and γ is an endomorphism of $\mathscr{R}$, then $(\gamma\varphi(a_1), \ldots, \gamma\varphi(a_k)) \in T_{\gamma\varphi}$. Another way to write this is $\gamma[T_\varphi \cap \varphi(A)^k] \subseteq T_{\gamma\varphi}$.

It is not at all clear when this necessary condition is too broad and includes relations that for other reasons are thought not to be meaningful. The fact that every relation satisfies the necessary condition when there are no nontrivial endomorphisms makes it likely that in such cases this invariance concept is not sufficiently narrow to capture meaningfulness. Is it sufficiently broad always to capture meaningfulness? Although it seems to be, there is a disturbing result discussed at the end of Section 22.3.3, namely, that an automorphism of a structure, which of course is a relation of the structure, need not be meaningful according to this invariance condition. Intuitively, one might have expected something so intimately involved with a structure as one of its automorphisms to be meaningful within the structure, but that is not necessarily the case.

In the next section (22.3) we give a more thorough treatment of these invariance criteria.

22.2.8 Irreducible Structures: Reference Invariance of Numerical Equality

A simple but very important question is the following: Is the binary relation $=$ on R $\mathscr{A}$-reference invariant relative to $\mathscr{R}$? Suppose not. Then for some a, a' in A, there is a homomorphism φ into R such that $\varphi(a) = \varphi(a')$ and another, equally good, homomorphism ψ such that $\psi(a) \neq \psi(a')$. The first homomorphism guarantees that there is complete equivalence between a and a' with respect to the empirical relations $S_1, \ldots, S_n$—any place in any k_i-tuple in S_i that can be filled by a can also be filled by a' and vice versa. The second homomorphism, however, takes advantage of the fact that a and a' are empirically distinct and assigns them distinct numerical values, which, however, must be fully equivalent with respect to the numerical relations $T_1, \ldots, T_n$. Since a, a' are just two different objects with exactly the same properties so far as the relevant relations of $S_1, \ldots, S_n$ are concerned, it makes little sense to assign them distinct scale values. It therefore would seem reasonable to consider only homomorphisms that do not make such irrelevant distinctions, i.e., ones that assign unequal scale values only when a, a' differ with respect to some relation S_i. In such a case, $\varphi(a) = \varphi(a')$ would be meaningful; it would be interpreted as equivalence of a, a' with respect to all the empirical relations of S_i.

The simplest device that prevents homomorphisms from making irrelevant distinctions is to define an equivalence relation, denoted $\approx$ and called *congruence*, in the structure of $\mathscr{A}$, as follows: $a \approx a'$ iff a and a' are substitutable for one another in the defining relations of S_i of $\mathscr{A}$. Corresponding to $\approx$, we add $=$ to the numerical relational structure $\mathscr{R}$ and impose on any homomorphism the condition that $=$ be a representation for $\approx$, i.e.,

$$a \approx a' \quad \text{iff} \quad \varphi(a) = \varphi(a').$$

An alternative mathematically equivalent device is to replace A by the set of equivalence classes, $A/\approx$ and to consider the induced relations, denoted $S_i/\approx$, on $A/\approx$. The resulting structure $\mathscr{A}/\approx$ has the property that every homomorphism on $\mathscr{A}/\approx$ is one-to-one. Such a structure is called *irreducible*. When all homomorphisms are one-to-one then $=$ is $\mathscr{A}/\approx$-reference invariant relative to $\mathscr{R}$, since $S(\varphi, =)$ is just the equality relation on the irreducible structure. Instead of considering all homomorphisms of the original structure, we consider only those that correspond to homomorphisms of the reduced structure $\mathscr{A}/\approx$. When the structure is not irreducible, we refer to it as *reducible* and to its homomorphisms as reducible ones. Exercise 3 formulates the theorem that shows $=$ is $\mathscr{A}/\approx$-reference invariant relative to $\mathscr{R}$.

The concept of irreducible relational structure was introduced by Pfanzagl (1968), and its importance in getting definitions just right has been emphasized by Roberts and Franke (1976) and Roberts (1980). In later sections of this chapter we shall often assume that equality is reference invariant; in effect this assumes that we are dealing with irreducible structures or with only those homomorphisms induced by the homomorphisms of the reduced structures.

22.3 CHARACTERIZATIONS OF REFERENCE INVARIANCE

In Section 22.2.7 we argued that a necessary condition for a numerical relation T to represent an empirically meaningful relation is the *invariance* of T under permissible transformations of the numerical structure $\mathscr{R}$. The reason is that if φ is a homomorphism and γ a permissible transformation, then $\gamma\varphi$ is also a homomorphism. For φ and $\gamma\varphi$ to induce the same relation on A, i.e., $S(\varphi, T) = S(\gamma\varphi, T)$, we must have $\gamma[\varphi(A)^k \cap T] \subseteq T$.

The present section has two main thrusts. First, we must identify the concept of permissible transformation. Second, we are interested in the converse question: under what circumstances is invariance of T under permissible transformations sufficient to guarantee that T is reference invariant?

22.3.1 Permissible Transformations

The most natural concept of permissible transformations is a mapping of the numerical set R into itself, which takes a representation into an equally good one. More precisely, γ is *permissible* for $\Phi(\mathscr{A}, \mathscr{R})$ if γ maps R into itself, and for every φ in $\Phi(\mathscr{A}, \mathscr{R})$, $\gamma\varphi$ is also in $\Phi(\mathscr{A}, \mathscr{R})$. It is easily shown that any endomorphism of $\mathscr{R}$ satisfies this definition; conversely, any γ satisfying the definition is an endomorphism at least on the subset of possible values in R, i.e., if we restrict $\mathscr{R}$ to $\bigcup\varphi(A)$, where the union is over all φ in $\Phi(\mathscr{A}, \mathscr{R})$. Usually this is all of R, and so any permissible transformation in this sense would be an endomorphism. It is this concept that Stevens probably had in mind, and that we think of, when we say that the permissible transformations are similarity transformations, $x \to rx$ or monotone transformations $x \to \gamma(x)$.

A more general concept, introduced by Pfanzagl (1968), associates a family of permissible transformations not with the whole of $\Phi(\mathscr{A}, \mathscr{R})$ but rather with each homomorphism φ. Thus γ is permissible for φ if γ maps $\varphi(A)$ into R and $\gamma\varphi$ is another homomorphism. In this case, we must restrict $\mathscr{R}$ to just the range of φ in order that γ be an endomorphism. Following Pfanzagl, we call γ a *partial endomorphism* of $\mathscr{R}$ into itself with domain equal to $\varphi(A)$. This concept gains additional generality if there are cases where φ and $\gamma\varphi$ are both homomorphisms but where γ cannot be extended to an endomorphism of the full structure $\mathscr{R}$.

We do not know of any very interesting cases in which this more general concept is needed, but there are examples. For instance, the main theorem on extensive measurement in Chapter 3 (Volume I, p. 85) establishes ratio-scale uniqueness only for nonmaximal elements. Thus we can have two homomorphisms φ and φ' that are identical except for maximal element of A; and consequently, $\varphi' = \gamma\varphi$, where γ is the identity mapping on $\varphi(A)$ except at $\max[\varphi(A)]$. Such a γ cannot be extended to an endomorphism of $\mathscr{R} = \langle \mathrm{Re}^+, \geqslant, + \rangle$ since the only endomorphisms of the latter are multiplications. Another example comes from ordinal measurement, in which one could have homomorphisms φ, φ' with ranges of $(0,1) \cup [2,3)$ and $(0,2)$ respectively. We can have $\varphi' = \gamma\varphi$, where

$$\gamma(x) = \begin{cases} x, & 0 < x < 1 \\ x - 1, & 2 \leqslant x < 3. \end{cases}$$

Obviously γ cannot be extended to a strictly increasing function on all of Re.

In view of the general level at which we are operating in this chapter, Pfanzagl's concept of partial endomorphisms seems needed, and we shall use it, though in practice, invariance can usually be considered only for full endomorphisms in making determinations of meaningfulness, i.e., the numerical relation must be structure invariant.

The following definition and theorem summarize our ideas about partial endomorphisms.

DEFINITION 2. *Let $\mathscr{R} = \langle R, T_1, \ldots, T_n \rangle$ be a relational structure in which T_i is of order $k(i)$, and let P be a subset of R. The restriction of T_i to P, denoted $T_i|_P$, is $T_i \cap P^{k(i)}$. The* restriction *of $\mathscr{R}$ to P, denoted $\mathscr{R}|_P$, is $\langle P, T_1|_P, \ldots, T_n|_P \rangle$. Any homomorphism $\gamma \in \Phi(\mathscr{R}|_P, \mathscr{R})$ is called* a partial endomorphism of $\mathscr{R}$ with domain P.

THEOREM 2. *Let $\mathscr{A}$ and $\mathscr{R}$ be relational structures such that equality in $\mathscr{R}$ is $\mathscr{A}$-reference invariant. Let φ be a homomorphism of $\mathscr{A}$ into $\mathscr{R}$. A function $\phi': A \to R$ is a homomorphism of $\mathscr{A}$ into $\mathscr{R}$ iff there is a partial endomorphism γ of $\mathscr{R}$, with domain $\varphi(A)$, such that $\varphi' = \gamma\varphi$.*

The proof is left as Exercise 8.

The hypothesis that equality is reference invariant is essential to this theorem, because it allows γ to be well-defined by $\gamma[\varphi(a)] = \varphi'(a)$.

22.3.2 The Criterion of Invariance under Permissible Transformations

The following theorem is a slight generalization of Pfanzagl's basic result (1968, p. 36).

THEOREM 3. *Let $\mathscr{A} = \langle A, S_1, \ldots, S_n \rangle$ and $\mathscr{R} = \langle R, T_1, \ldots, T_n \rangle$ be relational structures, let φ be a homomorphism of $\mathscr{A}$ into $\mathscr{R}$, and suppose that equality on R is $\mathscr{A}$-reference invariant. A k-ary relation T on R is $\mathscr{A}$-reference invariant relative to $\mathscr{R}$ iff for every partial endomorphism γ of $\mathscr{R}|_{\varphi(A)}$ into $\mathscr{R}$, and every $(x_1, \ldots, x_k) \in \varphi(A)^k$.*

$$(x_1, \ldots, x_k) \in T \quad \textit{iff} \quad (\gamma(x_1), \ldots, \gamma(x_k)) \in T.$$

A proof is given in Section 22.4.1.

Once again, the hypothesis that equality is reference invariant can be replaced by stronger ones such as Pfanzagl's condition, i.e., that $\mathscr{A}$ be irreducible. The hypothesis is used only to make Theorem 2 apply; actually this hypothesis is needed only in the proof of sufficiency of the invariance criterion.

22.3.3 The Condition of Structure Invariance

Theorem 3 characterizes reference invariance in terms of invariance under permissible transformations for any one representation—partial endomorphisms. It does not, however, give a characterization that is intrinsic to either $\mathscr{A}$ or $\mathscr{R}$ alone. The following partially rectifies that.

THEOREM 4. *Suppose $\mathscr{A}$ and $\mathscr{R}$ are two relational structures for which $\Phi(\mathscr{A}, \mathscr{R})$ is nonempty and $\mathscr{A}$ is irreducible. If T is a relation on R that is $\mathscr{A}$-reference invariant relative to $\mathscr{R}$, then $S(T)$ is structure invariant in $\mathscr{A}$. If, in addition, $T \subseteq \bigcup_{\varphi \in \Phi} \varphi(A)^k$, then T is structure invariant in $\mathscr{R}$.*

This is proved in Section 22.4.2.

Exercise 9 asks for an example to show that, in general, structure invariance in $\mathscr{A}$ is not the same as $\mathscr{A}$-reference invariance; however, Theorem 5 below formulates sufficient conditions for them to be the same.

More satisfactory results—showing that structure and reference invariance are exactly the same—can be obtained if we require a condition of compatability. By Theorem 1 we know that if α is an endomorphism of $\mathscr{A}$ and $\varphi \in \Phi(\mathscr{A}, \mathscr{R})$, then $\varphi\alpha \in \Phi(\mathscr{A}, \mathscr{R})$, and by Theorem 2 there exists a partial endomorphism γ of $\mathscr{R}$ such that $\gamma\varphi = \varphi\alpha$. We define this formally.

DEFINITION 3. *Suppose $\mathscr{A}$ and $\mathscr{R}$ are relational structures. $\mathscr{R}$ is said to be* compatible with *$\mathscr{A}$ iff for every $\varphi \in \Phi(\mathscr{A}, \mathscr{R})$ and every partial endomorphism γ on $\mathscr{R}|_{\varphi(A)}$, there exists an endomorphism α of $\mathscr{A}$ such that $\varphi\alpha = \gamma\varphi$.*

THEOREM 5. *Suppose $\mathscr{A}$ and $\mathscr{R}$ are relational structures for which $\Phi(\mathscr{A}, \mathscr{R})$ is nonempty, $\mathscr{A}$ is irreducible, and $\mathscr{R}$ is compatible with $\mathscr{A}$. Then,*

(i) *Suppose that for some $\varphi \in \Phi(\mathscr{A}, \mathscr{R})$, φ is onto R. A relation T on R is $\mathscr{A}$-reference invariant relative to R iff T is structure invariant in $\mathscr{R}$.*

(ii) *A relation S on A is structure invariant iff there exists a relation T on R such that T is $\mathscr{A}$-reference invariant relative to $\mathscr{R}$ and $S(T) = S$.*[1]

An important special case of this result is when every homomorphism is onto $\mathscr{R}$.

COROLLARY 1. *Suppose $\mathscr{A}$ and $\mathscr{R}$ are relational structures such that $\mathscr{A}$ is irreducible, $\Phi(\mathscr{A}, \mathscr{R})$ is nonempty, and $\varphi(A) = R$ for every $\varphi \in \Phi(\mathscr{A}, \mathscr{R})$, then*

(i) *Every endomorphism of $\mathscr{A}(\mathscr{R})$ is an automorphism of $\mathscr{A}(\mathscr{R})$.*

(ii) *The two groups of automorphism are isomorphic.*

(iii) *$\mathscr{R}$ is compatible with $\mathscr{A}$.*

COROLLARY 2. *Suppose $\mathscr{A}$ and $\mathscr{R}$ are relational structures meeting the following conditions: $\mathscr{A}$ is irreducible; $\Phi(\mathscr{A}, \mathscr{R})$ is nonempty and 1-point unique in the sense that if for some $\varphi, \psi \in \Phi(\mathscr{A}, \mathscr{R})$ and some $a \in A$, $\phi(a) = \psi(a)$, then $\varphi = \psi$; for each $\varphi, \psi \in \Phi(\mathscr{A}, \mathscr{R})$, $\varphi(A) \cap \psi(A) \neq \varnothing$; and $\mathscr{A}$ is 1-point homogeneous. Then for each $\varphi, \psi \in \Phi(\mathscr{A}, \mathscr{R})$, $\phi(A) = \psi(A)$.*

We observe that if the conditions of Corollary 2 are satisfied and if there is a homomorphism from $\mathscr{A}$ onto $\mathscr{R}$, then those of Corollary 1 are also satisfied; and so Theorem 5 applies. This basically includes the contents of Theorem 14.7 of Narens (1985; he did not explicitly state the assumption $\varphi(A) \cap \psi(A) \neq \varnothing$, but he implicitly used it).

[1] The latter property Narens (1985) called "qualitatively $\mathscr{R}$-meaningful."

This result, which is proved in Section 22.4.3, is an improved version of Theorems 1 and 2 of Luce (1978) who confined his attention to the case of the corollary, defined meaningfulness only to be invariance under automorphisms, and did not define irreducibility correctly.

The major thrust of the result is that structure invariance in $\mathscr{A}$, structure invariance in $\mathscr{R}$, and reference invariance are all the same when $\mathscr{R}$ is compatible with $\mathscr{A}$, and one case of this is when all the homomorphisms are onto $\mathscr{R}$.

Given the results of Chapter 20 on automorphism groups of a broad class of measurement structure, it is natural to ask for a complete characterization of the relations that are invariant under various scale types. This appears not to have been worked out fully except in two cases. The earliest was for the class of dimensional structures that arise in physics (we take that up in Section 22.7); the other is the ordinal case, which Roberts (1984a, 1984b) has explored fully. We do not describe it here.

An interesting and, as was mentioned earlier, surprising example of a relation failing to be meaningful in the sense of being structure invariant is any individual automorphism of a structure. This fact was pointed out by Narens (1981a). Consider a structure for which structure invariance is equivalent to invariance under the automorphisms of the structure. Let α be a specific automorphism; it is a relation on A. Then α is structure invariant iff for each a, b in A and each automorphism β,

$$\alpha(a) = b \quad \text{iff} \quad \alpha[\beta(a)] = \beta(b),$$

which in turn holds iff

$$\beta[\alpha(a)] = \alpha[\beta(a)]$$

In other words, α is structure invariant iff α commutes with every automorphism. This property is clearly satisfied when the automorphism group is isomorphic to a subgroup of a ratio scale, but it does not hold for discrete interval and interval scale structures.

This state of affairs seems, on the face of it, extremely odd. An automorphism of a structure appears to be something defined for and intimately involved with the structure, and intuitively one would have expected it to be meaningful. What this simple observation says, in effect, is that except for subgroups of ratio scales it is impossible to give a structural definition of an individual automorphism. That fact makes clear why it has not been possible to arrive at any explicit automorphism for general concatenation structures, though we do have some for PCSs, which involve commutative automorphisms, namely, the n-copy operators.

The existence of this oddity has led Narens (in preparation) to pursue ideas involving meaningfulness of more complex concepts than relations. For example, the group of all automorphisms, which is not a relation although each of its members is, exhibits invariance in the sense that for each a, b in A and α in $\mathcal{G}$, if $(a, b) \in \alpha$, then for $\beta \in \mathcal{G}$, $(\beta(a), \beta(b)) \in \gamma = \beta\alpha\beta^{-1}$, which is an automorphism.

22.4 PROOFS

22.4.1 Theorem 3 (p. 287)

Suppose $\varphi \in \Phi(\mathcal{A}, \mathcal{R})$, $=$ on R is $\mathcal{A}$-reference invariant, and T is a relation on R. Then T is $\mathcal{A}$-reference invariant relative to $\mathcal{R}$ iff, for every partial endomorphism γ of $\mathcal{R}|_{\varphi(A)}$, $\gamma(T|_{\varphi(A)}) \subseteq T$ and $\gamma(-T|_{\varphi(A)}) \subseteq -T$.

PROOF. First, suppose T is $\mathcal{A}$-reference invariant relative to $\mathcal{R}$. Let γ be a partial endomorphism of $\mathcal{R}|_{\varphi(A)}$ into $\mathcal{R}$; then $\gamma\varphi \in \Phi(\mathcal{A}, \mathcal{R})$, and so $S(\varphi, T) = S(\gamma\varphi, T)$. Suppose $x_i = \varphi(a_i)$, $i = 1, \ldots, k$. If $(x_1, \ldots, x_k) \in T$, then $(a_1, \ldots, a_k) \in S(\varphi, T)$; hence $(a_1, \ldots, a_k) \in S(\gamma\varphi, T)$; so $(\gamma(x_1), \ldots, \gamma(x_k)) \in T$. And if $(x_1, \ldots, x_k) \notin T$, then $(a_1, \ldots, a_k) \notin S(\varphi, T)$; hence $(a_1, \ldots, a_k) \notin S(\gamma\varphi, T)$; so $(\gamma(x_1, \ldots, \gamma(x_k)) \notin T$.

Conversely, suppose that T and $-T$ satisfy the above invariance criterion for partial endomorphisms on $\varphi(A)$. Let φ' be any element of $\Phi(\mathcal{A}, \mathcal{R})$. By Theorem 2, there exists a partial endomorphism γ on $\varphi(A)$ such that $\varphi' = \gamma\varphi$. We have

$$\begin{aligned}(a_1, \ldots, a_k) \in S(\Phi, T) \quad &\text{iff} \quad (\varphi(a_1), \ldots, \varphi(a_k)) \in T|_{\varphi(A)}\\ &\text{iff} \quad (\gamma\varphi(a_1), \ldots, \gamma(a_k)) \in T|_{\gamma\varphi(A)}\\ &\text{iff} \quad (a_1, \ldots, a_k) \in S(\gamma\varphi, T).\end{aligned}$$

Thus $S(\varphi, T) = S(\varphi', T)$; hence T is $\mathcal{A}$-reference invariant relative to $\mathcal{R}$. ◇

22.4.2 Theorem 4 (p. 287)

Suppose $\mathcal{A}$, $\mathcal{R}$ are relational structures, $\mathcal{A}$ is irreducible, and $\Phi(\mathcal{A}, \mathcal{R}) \neq \varnothing$. If T is a relation on R that is $\mathcal{A}$-reference invariant relative to $\mathcal{R}$, then $S(T)$ is structure invariant in $\mathcal{A}$. If, further, $T \subseteq \bigcup_{\varphi \in \Phi}\varphi(A)^k$, then T is structure invariant in $\mathcal{R}$.

PROOF. Suppose α is an endomorphism of $\mathscr{A}$. Select any φ in Φ. Note that $\varphi\alpha\varphi^{-1}$ is a partial endomorphism on $\varphi(A)$ in $\mathscr{R}$ and that $\varphi\alpha\varphi^{-1}\varphi = \varphi\alpha$ is in Φ. Thus, if $(a_1, \ldots, a_k) \in S(T)$, then by definition $(\varphi(a_1), \ldots, \varphi(a_k)) \in T$ and by Theorem 3 $(\varphi\alpha(a_1), \ldots, \varphi\alpha(a_k)) = (\varphi\alpha\varphi^{-1}\varphi(a_1), \ldots, \varphi\alpha\varphi^{-1}\varphi(a_k)) \in T$, whence $(\alpha(a_1), \ldots, \alpha(a_k)) \in S(T)$. Thus $S(T)$ is structure invariant.

Suppose $T \subseteq \bigcup_{\varphi \in \Phi}\varphi(A)^k$ and γ is an endomorphism of $\mathscr{R}$. Consider $(t_1, \ldots, t_k) \in T$. By hypothesis, for some $\varphi \in \Phi$ there exists $a_1, \ldots, a_k \in A$ such that $\Phi(a_i) = t_i$, where $(a_1, \ldots, a_k) \in S(T)$. Since, by Theorem 2, $\gamma\varphi \in \Phi$, $(\gamma\varphi(a_1), \ldots, \gamma\varphi(a_k)) = (\gamma(t_1), \ldots, \gamma(t_k)) \in T$, and so T is structure invariant. ◇

22.4.3 Theorem 5 (p. 288)

Suppose $\mathscr{A}$ and $\mathscr{R}$ are relational structures such that $\mathscr{A}$ is irreducible, $\Phi(\mathscr{A}, \mathscr{R}) \neq \varnothing$, and $\mathscr{R}$ is compatible with $\mathscr{A}$.

(i) *T on R is $\mathscr{A}$-reference invariant relative to $\mathscr{R}$ iff T is structure invariant in $\mathscr{R}$.*

(ii) *S on A is structure invariant in $\mathscr{A}$ iff there exists T on R such that T is $\mathscr{A}$-reference invariant relative to $\mathscr{R}$ and $S(T) = S$.*

PROOF.

(i) Suppose T is reference invariant. Let $S(T)$ be the induced relation. Let γ be an endomorphism of $\mathscr{R}$ and $\varphi \in \Phi(\mathscr{A}, \mathscr{R})$. By compatibility, there is an endomorphism α of $\mathscr{A}$ such that $\gamma\varphi = \varphi\alpha$. Consider any $(x_1, \ldots, x_k) = \mathbf{x} \in T$, then there exists $\mathbf{a} \in S(T)$ such that $\varphi(\mathbf{a}) = \mathbf{x}$. By Theorem 4, $S(T)$ is structure invariant; so $\alpha(\mathbf{a}) \in S(T)$, whence $\varphi\alpha(\mathbf{a}) = \gamma\varphi(\mathbf{a}) = \gamma\mathbf{x} \in T$, proving that T is structure invariant.

The converse was proved in Theorem 3.

(ii) Suppose S is structure invariant in $\mathscr{A}$. Define $T = \bigcup_{\varphi \in \Phi}\varphi(S)$. We first show that $T \cap \varphi(A)^k = \varphi(S)$. Obviously, $\varphi(S) \subseteq T \cap \varphi(A)^k$; so for the equality to fail there exists k-ary vectors $\mathbf{a}, \mathbf{b}$ such that $\mathbf{a} \notin S, \mathbf{b} \in S$, and for some $\psi \in \Phi$, $\psi(\mathbf{b}) = \varphi(\mathbf{a})$. By Theorem 3, there is partial endomorphism γ of $\mathscr{R}$ such that $\psi = \gamma\varphi$; and by comparability, there is an endomorphism α of $\mathscr{A}$ such that $\gamma\varphi = \varphi\alpha$. Therefore, $\varphi(\mathbf{a}) = \psi(\mathbf{b}) = \gamma\varphi(\mathbf{b}) = \varphi\alpha(\mathbf{b})$, and so $\mathbf{a} = \alpha(\mathbf{b})$. But since $\mathbf{b} \in S$, by structure invariance $\alpha(\mathbf{b}) \in S$, contrary to assumption.

By definition of $S(\varphi, T)$, $\mathbf{a} \in S(\varphi, T)$ iff $\varphi(\mathbf{a}) \in T$. But by what we have just shown, $\varphi(\mathbf{a}) \in \varphi(S)$, and so $S(\varphi, T) \subseteq S$. Since $\varphi(S) \subseteq T$, $S =$

$S(\varphi, \varphi(S)) \subseteq S(\varphi, T)$, and so $S = S(\varphi, T)$, proving T is reference invariant and $S = S(T)$.

The converse was proved in Theorem 4. 

COROLLARY. *Suppose $\mathscr{A}$ is irreducible, $\Phi(\mathscr{A}, \mathscr{R}) \neq \varnothing$ and, for every $\varphi \in \Phi(\mathscr{A}, \mathscr{R})$, $\varphi(A) = R$.*

(i) *The endomorphisms are automorphisms,*

(ii) *the two groups of automorphisms are isomorphic, and*

(iii) *$\mathscr{R}$ is compatible with $\mathscr{A}$.*

PROOF.

(i) Suppose α is an endomorphism of $\mathscr{A}$, then $\varphi\alpha \in \Phi$. But $\varphi\alpha$ is *onto*; so α must be *onto*. Suppose $\alpha(a) = \alpha(b)$ for some $a, b \in A$ with $a \neq b$, then $\varphi\alpha(a) = \varphi\alpha(b)$, and so $\varphi\alpha$ is not 1:1, contrary to assumption. So α is 1:1 and hence is an automorphism. The arguments in $\mathscr{R}$ are symmetrical since $\mathscr{A}$ and $\mathscr{R}$ are isomorphic.

(ii) Choose $\varphi \in \Phi$ and let α and β be automorphisms of $\mathscr{A}$. It is easy to see that $\varphi\alpha\varphi^{-1}$ is an automorphism of $\mathscr{R}$ and that $\varphi\alpha\beta\varphi^{-1} = (\varphi\alpha\varphi^{-1})(\varphi\beta\varphi^{-1})$. Moreover, the mapping is *onto* since if γ is an automorphism of $\mathscr{R}$, $\varphi^{-1}\gamma\varphi$ is an automorphism of $\mathscr{A}$ and $\varphi(\varphi^{-1}\gamma\varphi)\varphi^{-1} = \gamma$. It is 1:1 since $\varphi\alpha\varphi^{-1} = \varphi\beta\varphi^{-1}$ implies $\alpha\varphi^{-1} = \beta\varphi^{-1}$, which in turn implies $\alpha = \beta$.

(iii) Suppose $\varphi \in \Phi(\mathscr{A}, \mathscr{R})$. Since φ is onto $\mathscr{R}$, any partial endomorphism is an endomorphism and so, by part (i), is an automorphism. Since $\mathscr{A}$ and $\mathscr{R}$ are irreducible, we can define $\alpha = \varphi^{-1}\gamma\varphi$. It is trivial to show that α is an automorphism, and so we are done. ◇

22.5 DEFINABILITY

Our approach has been to say that, in the context of measurement, we are certain it is empirically meaningful to talk about any relation on A that is induced by a numerical relation on R that is $\mathscr{A}$-reference invariant relative to $\mathscr{R}$. In Theorem 3, we established as a necessary condition on the numerical relation that it be invariant under permissible transformations, whereas in Theorem 4 we established a necessary condition on the induced empirical relation that it be invariant under all endomorphisms of $\mathscr{A}$, a property we call structure invariance. Finally, in Theorem 5, we showed that when $\mathscr{R}$ is compatible with $\mathscr{A}$, all three concepts, structure invariance in $\mathscr{A}$, structure invariance in $\mathscr{R}$, and reference invariance, are the same.

The question remains, however, which of the three, if any, actually captures the intrinsic sense of meaningfulness, namely, that the relation on A be defined in terms of the defining relations of $\mathscr{A}$. Since the defining relations are by definition invariant under the endomorphisms of $\mathscr{A}$, any relations defined in terms of these are also invariant under the endomorphisms of $\mathscr{A}$. So again, structure invariance is a necessary condition for a relation to be definable; the only question is whether it is also sufficient. The following counterexample makes clear that it is not sufficient.

Consider the relation R given by the axiom:

$$\text{Either } (\forall a)(\forall b)(aRb \text{ iff } a \succsim b) \text{ or } (\forall a)(\forall b)(aRb \text{ iff } b \succsim a).$$

It is easy to see that R is structure invariant for any weak order $\mathscr{A} = \langle A, \succsim \rangle$, but obviously R cannot be defined in terms of $\succsim$ in the elementary theory of weak orders. By Padoa's principle (Section 21.2.4), we can easily show this. Let $A = \{1,2\}$, $\succsim$ be $\geqslant$, and

$$R_1 = \{(1,1),(2,2),(1,2)\}$$
$$R_2 = \{(1,1),(2,2),(2,1)\}.$$

Note that homogeneity is a property of this elementary theory. Moreover, this kind of counterexample is not ruled out for theories that are not elementary because, as was remarked in Section 21.2.4, Padoa's principle applies to such theories as well.

As an example of a meaningfulness argument let us return once more to the idea of independence of events in probability measurement. In Section 22.2 we discussed relational structures of form $\mathscr{A} = \langle \mathscr{E}, \succsim, \mathscr{B}, \cup_{\mathscr{B}} \rangle$, where $\mathscr{E}$ is interpreted as a family of subsets of X, closed under the union and intersection operations and ordered (in probability) by $\succsim$, $\mathscr{B}$ is the set of disjoint pairs of $\mathscr{E}$, and $\cup_{\mathscr{B}}$ is the restriction of $\cup$ to $\mathscr{B}$.

The usual first-order axioms define structures of *qualitative probability* (Definition 5.4, Volume I, p. 204). Not every such structure has an additive representation (Volume I, p. 205); but even if we restrict attention to those that do there can be distinct representations P, P' such that for some A, B in $\mathscr{E}$, $P(A \cap B) = P(A)(B)$ but $P'(A \cap B) \neq P'(A)P'(B)$. Thus, for such structures, Padoa's principle tells us that $\perp$ cannot be defined in terms of $\succsim$, $\mathscr{B}, \cup_{\mathscr{B}}$ such that $A \perp B$ iff $P(A \cap B) = P(A)P(B)$.

If we add more axioms to the system, including one that is not first-order (the Archimedean axiom), then we can obtain a unique additive representation P and can thus define $\perp$ by the numerical relation $P(A \cap B) = P(A)P(B)$, as discussed in Section 22.2.3. But the unique P, and thus $\perp$, is not defined in the first-order theory. The fact that $\perp$ can be defined in terms of other empirical relations for a special class of structures, using powerful analytic methods, does not mean that it should be so defined; empirically, it is better to regard it as a distinct primitive relation, which under special circumstances has an extension that is determined by that of other relations.

A different syntactic approach is currently under investigation by Narens (2002) that is based on the idea of adding a primitive of meaningfulness to the classical Zermelo-Fraenkel framework for set theory. The work is not yet published and is in too much flux to report here; we merely alert the reader to the fact it will eventually appear.

22.6 MEANINGFULNESS AND STATISTICS

As was remarked in the introduction, the linkage between meaningfulness and statistics has been controversial. Among those who contend that the nature of measures used to record data does (or should) affect the nature of the statistical techniques used to analyze these data are Cliff (1982), Senders (1953, 1958), Siegel (1956), Stevens (1946, 1951, 1959, 1968), and Townsend and Ashby (1984). And those who have argued, more or less sweepingly, that statistical procedures are not directly controlled by the type of measures involved are Anderson (1961), Behan and Behan (1954), Burke (1953/1963), Gaito (1959, 1980), Humphreys (1964), Lord (1953), Luce (1967), and I. R. Savage (1957). Michell (1986) has contended that the entire problem has been misformulated and that various conclusions arise depending upon one's view as to the inherent nature of measurement: representational, operational, or classical. In particular for the representational camp, he alleges "The question never was one of permissible statistics or of meaningfulness. It was always one of legitimate inference" (p. 402). We give our view, which agrees with the latter but not the former of the two quoted sentences.

On the one hand, the link between meaningfulness and statistics is obvious: statistical calculations involve various numerical functions $f(x_1, \ldots, x_k)$, such as the arithmetic mean or the standard deviation, for a set of measurements $x_1, \ldots, x_k$. It seems reasonable to require that the numerical relations defined by such functions be meaningful relative to the underlying empirical structure. But on the other hand, the logic of statistical inference seems to require only distributional assumptions, not assumptions about the way in which the measurements x_i are used to represent empirical relations $S_1, \ldots, S_n$ in the empirical structure $\mathscr{A}$.

An attempt to clarify the linkage between meaningfulness and statistics was made by Adams *et al.* (1965). We agree with their approach and principal conclusions, especially their statement that

> "the practice of ignoring scale type in making statistical tests could lead to the formulation of empirically meaningless hypotheses." (p. 124).

Our emphasis will be somewhat different from theirs; among other things, we shall emphasize the quoted conclusion more than they did, and in particular, focus on the difference between meaningful hypotheses (involving description of *populations*) and meaningful *statistics* (involving description of samples).

The first subsection presents an informal introduction to the topic, using the most commonly encountered descriptive functions as illustrations.

22.6.1 Examples

Case A. The arithmetic mean. This is the well-known function $\bar{x} = (1/k)\Sigma_{i=1}^{k} x_i$. Relations defined in terms of the mean often take the form

(i) $\bar{x} - w = 0$ or

(ii) $\bar{x} - \bar{y} - w = 0.$

Relation (i) specifies a value w for the mean; relation (ii) specifies a value w for the difference between the two means $\bar{x} = (1/k)\Sigma_i x_i$ and $\bar{y} = (1/l)\Sigma_i y_i$.

If we assume an empirical structure $\mathscr{A}$ that maps onto a real interval R, then w as well as x_i, y_i can be regarded as scale values of a homomorphism φ from A into R. Then (i) defines a $(k + 1)$-ary relation on R and (ii) defines a $(k + l + 1)$-ary relation on R. These numerical relations are reference invariant provided that φ is unique up to similarity transformations $\varphi' = r\varphi$, since such transformations map $x \to rx$, $y \to ry$, and $w \to rw$.

Adams *et al.* introduced the term "reference invariance" for these examples since the mean $\bar{x}$ or the difference in means $\bar{x} - \bar{y}$ refers to an invariant empirical object in A whose value is w for homomorphism φ or rw for homomorphism $r\varphi$. In our terminology, the arithmetic mean and the differences among arithmetic means are reference invariant relative to $\mathscr{R}$ for ratio scales.

If φ is unique up to positive linear transformations, then (i) is still reference invariant but (ii) is not; the difference in means is not reference invariant for interval scales. For example, the difference between 50° and 41°F is not the temperature of any *object* at 9°F; changing to °C, we have 50°F → 10°C, 41°F → 5°C, but 9°F → −12.8°C not 5°C.

There is a third relation that is sometimes considered, namely,

(iii) $\bar{x} - \bar{y} = 0.$

This defines a $(k + 1)$-ary relation on R that is reference invariant, even for interval scales. Adams *et al.* describe such invariance in the difference of means as *comparison invariant* for interval scales. This terminology is slightly misleading because the only comparison that can be made is *same* or *different*. Usually one is more interested in hypotheses about the magnitude of the difference in means, and such hypotheses involve relation (ii). For such comparisons it is better to consider the numerical relation:

(iv) $(\bar{x} - \bar{y}) - (v - w) = 0.$

This is $(k + l + 2)$-ary and invariant for positive linear transformations. Thus, differences in means can be considered reference invariant for interval scales, but the "referent" consists of a pair of objects in A with scale values v, w.

For ordinal scales, none of (i)–(iv) is reference invariant in general. In fact, when R is a real interval, the only monotone functions that preserve relations (i), (iii), and (iv) are $x \to rx + s$, and the only ones that preserve (ii) are $x \to rx$.

Case B. The median. If $x_1, \ldots, x_{2k+1}$ are distinct, then the median $m(x_1, \ldots, x_{2k+1})$ is defined to be that x_j such that k of the x_i are $< x_j$ and k are $> x_j$. For any strictly increasing function γ, $\gamma(x_j)$ is such that k of the $\gamma(x_i)$ are $< \gamma(x_j)$ and k are $> \gamma(x_j)$; thus

$$m[\gamma(x_1), \ldots, \gamma(x_{2k+1})] = \gamma[m(x_1, \ldots, x_{2k+1})].$$

Consequently, the $(2k + 2)$-ary relation

$$m(x_1, \ldots, x_{2k+1}) - w = 0,$$

the median, is reference invariant for ordinal (and all stronger) scales.

Case C. The geometric mean. If $x_1, \ldots, x_k$ are positive, then the geometric mean is defined by $\bar{x}_g = \exp[(1/k)\Sigma_{i=1}^{k} \log x_i]$. The analogs to (i)–(iv) are

(i) $\bar{x}_g/w = 1$

(ii) $\bar{x}_g/\bar{y}_g w = 1$

(iii) $\bar{x}_g/\bar{y}_g = 1$

(iv) $\bar{x}_g v/\bar{y}_g w = 1$

Relations (i), (iii), and (iv) are invariant under $x \to tx^r$, so the geometric mean is reference invariant and comparison invariant for log-interval scales. Relation (ii) is invariant only under $x \to x^r$. If these relations are of interest, it may be better to use a different numerical relational structure in which $\log \varphi$ rather than φ is the homomorphism; then Case C reduces to Case A.

We turn next to some familiar measures of dispersion.

Case D. Standard deviation. This is defined by

$$s_x = \left[(1/k) \sum_{i=1}^{k} (x_1 - \bar{x})^2\right]^{1/2}$$

Important relations defined by using s_x are

(i) $s_x/w = 1$

(ii) $s_x/s_y = \tau$.

In relation (i), $x_1, \ldots, x_k$ and w are scale values so we have a $(k + 1)$-ary relation on R. In (ii), $x_1, \ldots, x_k$ and $y_1, \ldots, y_1$ are measured values and τ is a fixed constant; different values of τ define different $(k + 1)$-ary relations on R.

The mapping $x \to rx + t$ sends s_x into rs_x. Hence relation (i) is invariant only for $t = 0$ (ratio scales). The value of the standard deviation is reference invariant for ratio scales. Relations of form (ii), however, are invariant for all r, t. In the terminology of Adams *et al.*, ratios of standard deviations are *absolutely invariant* for interval scales.

Case E. Coefficient of variation. This is the ratio $s_x/\bar{x}$. Its being a constant is invariant for ratio scales so the coefficient of variation is absolutely invariant for ratio scales. The coefficient of variation does not seem to be useful for interval scales.

One thing illustrated by these examples is that it is not the statistic itself that is appropriate or inappropriate, meaningful or meaningless, but rather numerical relations involving the statistic. For example, Case D(i) involving s_x is meaningless for interval scales but Case D(ii) is meaningful. This view is not very different from that taken by Adams *et al.*, or, for that matter, by Stevens. The concepts of absolute, reference, and comparison invariance are based on the meaningfulness of particular sorts of relations; in general, many other sorts of relations may arise, and these particular ones are just useful examples. This point, too, was made by Adams *et al.* (1965, p. 113).

22.6.2 Meaningful Relations Involving Population Means

Let us link statistical measurement concepts in the following way. We have an empirical relational structure $\mathscr{A} = \langle A, S_1, \ldots, S_n \rangle$, a numerical relational structure $\mathscr{R} = \langle R, T_1, \ldots, T_n \rangle$, and a nonempty set of homomorphisms $\Phi(\mathscr{A}, \mathscr{R})$. A population is a probability space $\langle B, \mathscr{E}, P \rangle$, where B is a subset of the empirical domain A, $\mathscr{E}$ is a σ-field of subsets of B, and P is a countably additive probability measure on $\mathscr{E}$. For many purposes it would be sufficient to consider the case in which B is finite with k elements and $P(\{b\}) = 1/k$ for all b in B, but occasionally we shall want to consider some continuous distributions. The most important step in this link is the assumption that $B \subseteq A$, i.e., *the population consists of elements for which measurement homomorphisms* φ *are defined*. We could consider B as an additional unary relation on A, but the homomorphisms in $\Phi(\mathscr{A}, \mathscr{R})$ do not represent this relation in $\mathscr{R}$; in fact, for a multivariate Gaussian population we could have $\varphi(B) = R = \mathrm{Re}^m$, i.e., every m-dimensional real vector could be a value of φ for some b in B.

Assume that $B = \{b_1, \ldots, b_k\}$ is finite and denote $P(\{b_i\}) = p_i$; we have $\Sigma_i p_i = 1$. For interval-scale measurement, the $(k+1)$-ary numerical relation on R defined by

$$\Sigma_i p_i x_i - w = 0$$

is reference invariant and so it corresponds to a meaningful relation in $\mathscr{A}$.

In particular, it defines the following empirically meaningful relation on A: b^* in A is a mean of B iff $\Sigma_i p_i \varphi(b_i) - \varphi(b^*) = 0$. For interval-scale measurement this relation does not depend upon φ. However, for ordinal-scale measurement, the same relation does depend upon φ and so it cannot be defined by the same equation. One could, of course, define a different relation for each φ in $\Phi(\mathscr{A}, \mathscr{R})$ but then none of them would be of much empirical interest.

Similarly, for two populations $\langle B, \mathscr{E}, P \rangle$ and $\langle C, \mathscr{F}, Q \rangle$ we can find a meaningful relation for interval scales using the equation

$$\Sigma_i p_i \varphi(b_i) - \Sigma_i q_i \varphi(c_i) - [\varphi(b^*) - \varphi(c^*)] = 0.$$

The pair (b^*, c^*) is a difference in means for B, C. Again, for ordinal scales, this is either not defined (if φ is unspecified) or uninteresting (if it depends on a particular arbitrary choice of φ).

In the case of infinite populations, we assume that the homomorphisms are $\mathscr{E}$-measurable, and we replace the sum $\Sigma p_i \varphi(b_i)$ by the integral $\int : \varphi(b)\, dP$; otherwise, nothing is changed.

22.6.3 Inferences about Population Means

Next, let us consider what happens when the populations are either very large or infinite so only a sample can be observed. If we know nothing whatsoever about the population distribution for a given homomorphism φ, but we are willing to assume independent and identically distributed measurements yielding sample values $x_1, \ldots, x_k$, then we can still make an inference about the population mean. If the population mean is $x^* = \varphi(b^*)$, then asymptotically, $k^{1/2}(\bar{x} - x^*)/s_x$ has a standard Gaussian distribution. This can be used to reject values of x^* for which the above test statistic is large in absolute value. Now this large-sample argument in no way depends on the meaningfulness or the interest of the hypothesis that b^* is a mean of B; it is perfectly valid whether φ is ordinal or interval.[2] In other words, we do not criticize the calculation of the sample mean $\bar{x}$ with ordinal data for use in a test statistic whose distribution is asymptotically known. Rather, it is the hypothesis that b^* is the mean of B being tested that is subject to criticism.

The argument for comparing two population means by looking at large samples is similar to the one-sample argument just given. Once again, the probability theory argument is valid that the test statistic $\{\bar{x} - \bar{y} - [\varphi(b^*) - \varphi(c^*)]\}/(s_x^2/k + s_y^2/l)^{1/2}$ can be used to reject the hypothesis that (b^*, c^*) is a difference in means for B, C. There is no objection to calculating $\bar{x}, \bar{y}, s_x^2, s_y^2$ for ordinal data. The objection is to the hypothesis itself, namely, that (b^*, c^*) is a difference in means for B, C, which for ordinal scales is either meaningless if φ is unspecified in it, or uninteresting if φ is specified arbitrarily.

22.6.4 Parametric Models for Populations

Large-sample arguments such as those involved in the preceding subsection are not fashionable nowadays. However, one often assumes that a random variable $\mathbf{X} = \varphi(b)$ has a Gaussian distribution with parameters μ, σ^2 and once again assumes independent and identically distributed (Gaussian) measurements $x_1, \ldots, x_k$. In that case the test statistic $(k-1)^{1/2} \times(\bar{x} - \mu)/s_x$ has precisely Student's t distribution with $k - 1$ degrees of freedom for every k and regardless of the value of σ^2.

[2] The approach to a Gaussian distribution as $k \to \infty$ is based on the Central Limit Theorem, and both the existence of the limit distribution and the rate of convergence to it do vary with the distribution of measurement values $\varphi(b)$. Arbitrary monotone transformations can make the convergence very slow or even destroy it altogether if $\varphi(B)$ is unbounded. In practice, these difficulties are not too severe; for large samples one has a good estimate of the shape of the distribution and a good idea of when the sample size k is large enough.

Comparing this with the test in the preceding subsection, we note two changes. The change from $k^{1/2}$ to $(k-1)^{1/2}$ is of no importance; the argument in the previous subsection is an asymptotic one for which $k^{1/2}$ and $(k-1)^{1/2}$ are equivalent. The crucial change is from $x^* = \varphi(b^*)$ to μ. Previously, we had an empirical relation involving an object b^* in B. This relation was meaningful and subject to test with interval scale measurement but not with ordinal scale measurement. Here, however, the hypothesized value of μ receives its meaning not from the measurement process leading to the measured value $\varphi(b^*)$ but from the Gaussian probability model for which it is a parameter. If the assumption that the data are realizations of independent and identically distributed Gaussian random variables is valid, then the hypothesis that their common mean is μ is meaningful, and the t test is valid.

Interestingly, however, the parametric model for the population does not free us of measurement questions; it embroils us in them ever more deeply.

Suppose, first that the numerical values $\varphi(b)$ are generated by ordinal measurement and that one has no prior information about the shape of the population distribution corresponding to the specific measurement procedure that leads to the homomorphism φ. If the Gaussian hypothesis is true for some φ, it is false for all monotone transforms $\gamma\varphi$ except the linear ones. The chance of hitting by accident the one family of linearly related homomorphisms for which φ is Gaussian out of all possible monotonically related homomorphisms is negligible. Therefore, in this case, the underlying distributional assumption can be rejected a priori as having essentially zero probability. The hypothetical statistical reasoning is valid but, since the supposition under which the parameter μ is meaningful is judged false, any conclusions are meaningless.

In point of fact, we doubt that the Gaussian assumption is usually made in this blind fashion; generally, an investigator has some notion that his measurement representation φ is "reasonable," and would resist really violent monotone transformations. But that means that there is additional empirical information captured by φ. The measurement structure is not as simple as it seemed. We shall elaborate this idea below.

Before turning to the measurement structures that underlie parametric population models, we have one more purely ordinal version with which to deal. Suppose that the Gaussian assumption is not applied blindly to an arbitrarily chosen ordinal homomorphism φ but rather φ is purposely selected so as to force the Gaussian shape to fit the sample data. This can be done, for example, by replacing each x value by its percentile rank and then by the standard Gaussian deviate corresponding to that percentile. That done, the t test can be used. However, the actual hypothesis being tested is not about the mean of a particular Gaussian random variable; it concerns the mean of whatever variable results from the transformation. That is, it concerns the median not the mean, and the actual numerical relation being tested is reference invariant for ordinal scales. The t statistic is also nearly invariant under monotone transformations since it depends chiefly on the rank of x^* relative to $x_1, \ldots, x_k$; it approaches perfect invariance as $k \to \infty$. A simpler and therefore preferable approach uses the binomial distribution directly to test hypotheses about the median; its power is nearly as good, even for untransformed samples from a Gaussian distribution.

It is unclear whether other methods can be devised for selecting a transformation of φ that makes the sample approximately Gaussian and that is not reducible to an hypothesis that is reference invariant for ordinal scales.

If the same idea is extended to a comparison of two population means, we immediately run into a snag; it is not possible, in principle, to devise a monotone transformation that forces two different population distributions to be simultaneously Gaussian (Theorem 6.8, I, p. 286). If such a transformation is indeed possible, then it provides a representation for important empirical relations so we are led once more to our next topic: measurement structures underlying parametric families of population distributions.

It is true that if the two populations being compared have identical distributions, then this identity will be preserved for every monotone transformation; and hence, this null hypothesis could be tested by combining the two samples, estimating the transformation that renders the combined distribution Gaussian, and calculating the test statistic based on $\bar{x} - \bar{y}$ using this transformed scale. Nevertheless, if the transformation is developed as described, then the underlying hypothesis is ordinally invariant: For every p the pth percentile of population B is equal to the pth percentile of population C. The test statistic, though nominally based on $\bar{x} - \bar{y}$, is mostly determined by ranks; a better procedure is available that uses only the ranks (Wilcoxon two-sample or Mann-Whitney U).

22.6.5 Measurement Structures and Parametric Models for Populations

Having disposed of the cases in which a Gaussian distribution is forced on ordinal data and therefore does not have any empirical content, we now turn to the use of distributional assumptions with empirical content. We assume a representation function ψ, mapping the empirical objects in A into the reals with the properties that ψ both preserves an empirical order on A and constitutes a random variable with known (say, Gaussian) shape for each population in A. The function ψ is not necessarily a homomorphism for the full relational structure $\mathscr{A}$; it does preserve an aspect of that structure, namely, an order relation on A, and it also provides a representation for the probabilities in populations by means of Gaussian random variables. Such a function is unique up to linear transformations; it is also a monotonic function, possibly nonlinear, of any homomorphism φ on $\mathscr{A}$.

This idea generalizes easily to more than two populations and to various two-parameter families of distribution closed under linear transformations.

Let us make what we have just said quite formal, not with the idea of providing an axiomatization that leads to such a representation but rather to make clear the formal similarity between a distributional axiom system and a measurement one.

DEFINITION 4. *Let* $\mathscr{A} = \langle A, \succsim, S_2, \ldots, S_n \rangle$ *and* $\mathscr{R} = \langle \mathrm{Re}, \geqslant, T_2, \ldots, T_n \rangle$ *be relational structures, in which* $\succsim$ *is a weak order and* $\geqslant$ *is the usual order on* Re. *Let* $\langle B, \mathscr{E}, P \rangle$ *and* $\langle C, \mathscr{F}, Q \rangle$ *be countably additive probability spaces with* B *and* C *both subsets of* A *and* $B \cap C = \varnothing$. *The function* $\psi: A \to \mathrm{Re}$ *is a* Gaussian representation *for* $\langle A, \succsim, \langle B, \mathscr{E}, P \rangle, \langle C, \mathscr{F}, Q \rangle \rangle$ *iff*:

1. *For all* $a, a' \in A$, $a \succsim a'$ *iff* $\psi(a) \geqslant \psi(a')$.

2. ψ *restricted to* B *and* ψ *restricted to* C *each are random variables with Gaussian distributions.*

Just as the axiom that $\langle A, \succsim, \bigcirc \rangle$ has an additive representation into $\langle \mathrm{Re}^+, \geqslant, + \rangle$ has sharp empirical content that we try to capture in another way by means of the axioms of extensive measurement; likewise the axiom that $\langle A, \succsim, \langle B, \mathscr{E}, P \rangle, \langle C, \mathscr{F}, Q \rangle \rangle$ has a Gaussian representation has sharp content characterizing an enormous number of detailed facts about the ordering $\succsim$ on A and the probability functions P and Q on subsets of A.

A Gaussian representation ψ is obviously a homomorphism from $\langle A, \succsim \rangle$ into $\langle \mathrm{Re}, \geqslant \rangle$ (Condition 1) but of course it need not be a homomorphism of $\mathscr{A}$ into $\mathscr{R}$ since the relations $S_2, \ldots, S_n$ and their representations $T_2, \ldots, T_n$ are not even mentioned in the conditions. Really, the S_i are extra baggage; ψ is already determined up to interval-scale transformations by Conditions 1 and 2. If φ is a homomorphism of $\mathscr{A}$ into $\mathscr{R}$, then in particular, it is a homomorphism of $\langle A, \succsim \rangle$ into $\langle \mathrm{Re}, \geqslant \rangle$; so by the theory of ordinal measurement (Theorem 3, Chapter 2, p. 42), $\psi = \gamma\varphi$ where γ is strictly increasing function from $\varphi(A)$ to Re. But γ need not be linear even though ψ is an interval scale and φ may be one also; ψ and φ are tied together only by the ordering $\succsim$, on A. What we have, really, is two different theories: one involving relations $S_2, \ldots, S_n$, the other involving populations $\langle B, \mathscr{E}, P\rangle$ and $\langle C, \mathscr{F}, Q\rangle$, and they are tied together only weakly by $\langle A, \succsim\rangle$.

An example is in order. Consider a renewal process unfolding in time, such as the successive failures of light bulbs in a socket. Let us suppose that the distribution between successive failures is something smooth and unimodal, say gamma, but definitely not Gaussian. Time durations are, of course, extensive in character. Nonetheless, for many statistical purposes one is completely justified in ignoring the additive structures of the usual representation of time and transforming it so that the distribution of transformed times is a Gaussian.

The reader should carefully contrast the status of the Gaussian parameter μ_B in Definition 4 and that of $\varphi(b^*)$ in Section 22.6.2. Both can be thought of legitimately as "the mean of the population B" (or $\langle B, \mathscr{E}, P\rangle$, if we are careful). But they arise in radically different ways. In Section 22.6.2, there is no distribution theory for $\langle B, \mathscr{E}, P\rangle$; the only random variables that arose are those defined on B in the course of a measurement representation for $S_1, \ldots, S_n$, i.e., homomorphisms φ of $\mathscr{A}$ into $\mathscr{R}$. The value $\varphi(b^*)$ is the mean for the homomorphism φ. If we have at least interval-scale measurement for $\Phi(\mathscr{A}, \mathscr{R})$, then the mean is reference invariant: the object b^* in A, corresponding to the mean of population B, is independent of the choice of homomorphism. In Definition 4, by contrast, there is a distributional theory for $\langle B, \mathscr{E}, P\rangle$, but it is unrelated to the relations $S_2, \ldots, S_n$ on A. The number μ_B is a parameter of the theory, which also happens to be the mean of the random variable $\psi|_B$. For any element a in A, the numerical relation

$$\tau = [\psi(a) - \mu_B]/\sigma_B$$

has an empirical interpretation, namely,

$$P(\{b \in B | a \succsim b\}) = \int_{-\infty}^{\tau} (2\pi)^{-1/2} \exp(-z^2/2)\, dz.$$

In particular, $\psi(b^*) = \mu_B$ has the meaning $P(\{b \in B | b^* \succsim b\}) = \frac{1}{2}$. So if we put $a = c^* = \psi^{-1}(\mu_C)$ in the relation we see that the relation

$$\tau = (\mu_C - \mu_B)/\sigma_B$$

among parameters is empirically meaningful. Likewise, relations stating numerical values for $(\mu_C - \mu_B)/\sigma_C$ and for σ_B/σ_C are empirically meaningful; of course, these are just the invariants of the distributional theory under transformation $\psi' = r\psi + s$, which sends μ into $r\mu + s$ and σ into $r\sigma$.

If the representation and interval-scale uniqueness theorems for $\mathscr{A}$ arise from empirically correct axioms [i.e., $\Phi(\mathscr{A}, \mathscr{R})$ is nonempty and all homomorphisms are related by positive linear transformations], then hypotheses about values of population means or differences among population means are meaningful and possibly interesting, and they can be tested by distribution-free or large sample statistical procedures. On the other hand, if the assumptions of Definition 4 are empirically correct, then hypotheses about values of $(\mu_C - \mu_B)/\sigma_B$ or σ_B/σ_C are meaningful and possibly interesting, and they can be tested by well-known statistical methods that center about the F distribution. In either case, the sample statistics $\bar{x}$ and $\bar{y}$ can play an important role.

If the corresponding empirical theories are false, then φ or ψ cannot be constructed, and the hypotheses concerning means cannot be formulated meaningfully, any more than one could hypothesize a value for the eccentricity of the orbit of Mars in a universe in which that orbit was a spiral instead of an ellipse.

We can now see that the controversy over the appropriateness of statistical calculations is generated by a confusion between two types of empirical theories: theories about relations $S_2, \ldots, S_n$, which typically generate sets of homomorphisms that form interval scales or better, and theories about population probabilities, which typically generate different sets of order homomorphisms that also form interval scales. It sometimes happens that both types of theory apply simultaneously, with representations φ and ψ that are monotonically but nonlinearly related; in that case, there are two population means: those for random variable φ and those for ψ, and both must be studied.

Of course, the most satisfactory situation occurs where the probability spaces $\langle B, \mathscr{E}, P\rangle$ and $\langle C, \mathscr{F}, Q\rangle$ are linked to the relations $S_2, \ldots, S_n$ in an empirically interesting way, following laws that pin down the relation between φ and ψ. The simplest situation, which is often assumed to obtain,[3] is one in which φ and ψ are linearly related and the population standard deviations are equal $\sigma_B = \sigma_C$. Another common assumption is the log-Gaussian one: φ is log-interval and ψ is linearly related to $\log \varphi$, with constant standard deviations on the ψ scale. Under these much stronger assumptions there are additional meaningful hypotheses; for example,

[3] If b^* and c^* are means for B, C, and φ provides an additive representation for a concatenation operation $\bigcirc$ on A, then a necessary condition for this simplest situation is the following: If $E \in \mathscr{E}$ and $F \in \mathscr{F}$ satisfy $E \bigcirc c^* = F \bigcirc b^*$, then $P(E) = Q(F)$. We have not tried to analyze the situation completely to find further necessary and/or sufficient conditions.

differences between μ_B and μ_C can be studied in these units of φ or of $\log \varphi$ rather than only relative to σ_B or σ_C.

From our point of view, an important situation is one in which the real focus of interest is on the relations $S_1, \ldots, S_n$, and the probability spaces are introduced merely to account for "error," i.e., for the fact that seemingly constant empirical conditions give rise to varying objects $b \in B$ with varying scale values $\varphi(b)$. Obviously, in this case, hypotheses about means of φ values are the focus of interest and require interval-scale measurement. If one does not like large-sample or distribution-free statistics, one examines the empirical distributions of samples to see whether Definition 4 holds with $\psi = \varphi$.[4] If the equality appears to be true, then it is convenient to replace hypotheses about φ-means with hypotheses about μ-values and to use powerful parametric methods. But if not true, then it is wrong to find a different representation ψ, nonlinearly related to φ and to apply parametric (Gaussian) methods to μ-values for ψ. This gives the correct answer to the wrong question. The ψ representation may itself be interesting, but we started by assuming that the primary interest was in $S_1, \ldots, S_n$. The tails of a Gaussian distribution, though fascinating, must not be allowed to wag the whole relational structure under study.

22.6.6 Meaningful Relations in Uniform Structures

In the preceding section we considered two kinds of models: a measurement homomorphism φ that preserves the structure of relations $\succsim$, $S_2, \ldots, S_n$, and a Gaussian representation ψ that preserves only $\succsim$ but models the various populations by Gaussian random variables. The assumption of a Gaussian form obviously is motivated by mathematical convenience rather than verisimilitude. It is therefore interesting to obtain an analogous representation ψ in which there is no commitment in advance to a definite form for the distribution; rather, the form is determined by the data.

Such results were obtained by Levine (1970, 1972), and some of them were described in a conjoint measurement setting in Chapter 6. We shall give a very brief sketch here, restated in the notation of the preceding subsection. The basic idea is that we have a homomorphism φ of $\mathscr{A}$ into $\mathscr{R}$, plus a family $\mathscr{B}$ of populations (probability spaces in A). The homomorphism φ defines a family of distribution functions on reals. Levine's results show conditions under which a transformation $x \to \gamma(x)$ can be con-

[4] Of course, as we noted earlier, this has probability 0 of being true, a priori, but then so does any specific equality in science.

structed such that the transformed distributions can all be obtained from a fixed distribution by translations (1970) or by linear transformations (1972). Thus we obtain a mapping $\psi = \gamma\varphi$ that represents all the populations in $\mathscr{B}$ either by a one-parameter family of distributions (differing by translation) or by a two-parameter family (differing by linear transformations). The fixed distribution from which the others are obtained is constructed along with the transformation γ; it can have any shape.

Specifically, let $\mathscr{A} = \langle A, \succsim, S_2, \ldots, S_n \rangle$ be an empirical relational structure and let φ be a homomorphism of $\mathscr{A}$ into $\mathscr{R} = \langle \mathrm{Re}, \geqslant, T_2, \ldots, T_n \rangle$. Let $\mathscr{B}$ be a family of countably additive probability spaces $\langle B, \mathscr{E}_B, P_B \rangle$, in which each B is a subset of A such that $\varphi|_B$ is $\mathscr{E}_B$-measurable (i.e., φ defines a random variable on B). We can define the family of distribution functions $\{G_B\}$ by:

$$G_B(x) = P_B(\{b \in B \mid \varphi(b) \leqslant x\}).$$

The family $\{G_B\}$ is said to be *uniform* if there exists a one-to-one strictly increasing function γ of Re onto Re, a distribution H on Re, and constants μ_B, such that for all B

$$G_B[\gamma^{-1}(x)] = H(x - \mu_B).$$

(See Definition 6.9 for a slightly different but essentially equivalent version.)

If the family $\{G_B\}$ is uniform, then we can define a new homomorphism on A by $\psi = \gamma\varphi$ and new distribution functions for the random variables $\psi|_B$:

$$H_B(x) = P_B(\{b \in B | \psi(b) \leqslant x\}).$$

It is now trivial to show that $H_B(x) = H(x - \mu_B)$, i.e., the distributions on the ψ scale are identical except for location parameters μ_B. (If the distribution H is chosen to have mean 0, which is always possible, then μ_B is the mean of $\psi|_B$.)

Levine (1970) and Section 6.7 give qualitative axiom systems sufficient to guarantee that a family $\{G_B\}$ is uniform, and the proofs show how γ and H can be constructed.

Similarly, we say the family $\{G_B\}$ is affine if there exist γ, H as above, and constants μ_B, σ_B, such that for all B

$$G_B[\gamma^{-1}(x)] = H[(x - \mu_B)/\sigma_B].$$

That is, on the scale $\psi = \gamma\varphi$, the transformed distributions H_B differ only by location and size parameters. This is axiomatized in Levine (1972).

When the family $\mathscr{B}$ contains at least three populations and is uniform, then interval-scale uniqueness for $\psi = \gamma\varphi$ is generally obtained (there is a special case with weaker uniqueness). Thus numerical relations such as $\mu_B - \psi(b^*) = 0$ or $\mu_B - \mu_C = \psi(b^*) - \psi(c^*)$ are reference invariant relative to uniform structures of populations. Once more, as in the preceding section, questions about φ means and questions about location parameters are quite independent; either can be the focus of interest, and either can be approached statistically. The principal difference from the last section is that the ψ-theory is based on qualitative assumptions and is far more likely to be empirically valid since it allows any shape of distribution.

There is a similar interval-scale uniqueness theorem for affine families, albeit obtained under somewhat stronger conditions; it leads to exactly the same sorts of meaningful numerical relations as in the two-parameter Gaussian case.

Although the uniform and affine theories provide for parametric models of populations that are more flexible and thus more likely to be of greater intrinsic interest than the Gaussian model, they do not permit the familiar statistical methods involving the F distribution. Thus, they are less likely to be used for the sake of convenient small-sample inference when the real focus of interest is the means of the φ distributions. Actually, in the uniform case, there is a very simple and powerful procedure for testing comparison hypotheses of the form $\mu_B - \mu_C = \psi(b^*) - \psi(c^*)$; it uses the Mann-Whitney U statistic. This procedure is distribution-free but highly parametric; it assumes that the ψ representation is given. Insofar as this procedure becomes popular, we are apt to see it misused. Someone interested in differences between φ means but not willing to assume the Gaussian form, will calculate confidence intervals for location-parameter differences, using Mann-Whitney U, without realizing that they are really assuming an underlying uniform family of populations with representation $\psi = \varphi$. It would probably be possible to extend the Mann-Whitney methods to the affine case, again based on the assumption that the ψ representation is known.

22.7 DIMENSIONAL INVARIANCE

The most interesting examples of numerical relations are the numerical laws that relate the values from several different measurement homomorphisms. A numerical law $f(x_1, \ldots, x_k) = 0$ cannot be invariant if all of the

x_i undergo independent ratio-scale transformations (except for the trivial cases in which f is always 0 or never 0). As discussed in Section 22.2.4, a numerical law involving independent representations $x_1, \ldots, x_k$ of either ratio or interval type may also involve scale-dependent[5] parameters $\theta_1, \ldots, \theta_l$. We say that the law $f[\mathbf{x}, \boldsymbol{\theta}] = 0$, $\mathbf{x} = (x_1, \ldots, x_l)$, $\boldsymbol{\theta} = (\theta_1, \ldots, \theta_l)$ is reference invariant if the empirical relation $S(\boldsymbol{\varphi}, f)$ defined by

$$\mathbf{a} \in S(\boldsymbol{\varphi}, f) \qquad \text{iff} \qquad f[\boldsymbol{\varphi}(\mathbf{a}), \boldsymbol{\theta}(\boldsymbol{\varphi})] = 0$$

is the same for every $\boldsymbol{\varphi}$. Here we use the abbreviated notation in which $\mathbf{a}$ is a k-tuple of empirical objects from different empirical structures, $\mathbf{x} = \boldsymbol{\varphi}(\mathbf{a})$ is a k-tuple of numerical values, $\boldsymbol{\varphi}$ is a k-tuple of homomorphisms $(\varphi_1, \ldots, \varphi_k)$, and $\boldsymbol{\theta}(\boldsymbol{\varphi})$ is a vector of parameters that varies with the choice of $\boldsymbol{\varphi}$.

Most numerical laws of physics exhibit two special features, which require further discussion. First, though there are many measurement scales, they do not transform independently because the various physical attributes are interconnected by a number of multiplicative relations. A measurement representation that incorporates all of these relations has few degrees of freedom. In mechanics we conventionally adopt mass, length, and time as basic attributes; the units of all other representations are determined by the choice of units for these three. These constraints reflect the fact there are some numerical laws that do not require any scale-dependent parameters to be meaningful. An example is the law relating the period of a simple pendulum t to the length of the pendulum l and the acceleration of gravity, g: $t^2 - Clg^{-1} = 0$, where C is a constant, characteristic for all pendulums with the same angular excursion. If we change the units of length and time, $l' = \beta l$ and $t' = \gamma t$, then the unit of acceleration is determined: $g' = (\beta\gamma^{-2})g$. Hence $t^2 - Clg^{-1} = 0$ iff $(t')^2 - Cl'(g')^{-1} = 0$.

A second important feature of numerical physical laws is the fact that the scale-dependent parameters that do arise are usually multipliers, which means that they can be incorporated into the existing multiplicative structure as additional physical quantities. A good example is Hooke's law, which relates the force F on a spring to its length l. In Section 22.2.5 we wrote this in the form $F - C\theta l = 0$, where C is a constant, characteristic of

[5] The word *scale* has a double meaning in practice; it is often used to refer to a specific homomorphism, though technically it refers to the entire collection of them. Thus, the correct term here should be homomorphism-dependent but because "scale-dependent parameters" is such a familiar term we let it stand.

the spring, and θ is a scale-dependent parameter. (Recall that if a^* is a standard spring, φ_1 is the length homomorphism, and φ_2 is the force homomorphism, then $\theta = \varphi_2(a^*)/\varphi_1(a^*)$.) An alternative approach is to incorporate $C\theta$ into the measurement structure itself, calling it the rectilinear compliance of the spring, sometimes denoted C_M (see Volume I, p. 540), with units of force/length, more conventionally, mass/sec^2 or MT^{-2} (since force is MLT^{-2} and length is L in the usual dimensional terms with the mass-length-time basis). Hooke's law becomes $F - C_M l = 0$, and this is invariant under changes of units of mass, length, and time in the same manner as the pendulum law discussed above.

To systematize these ideas about the reference invariance of numerical laws in physics, we employ a special numerical relational structure $\mathscr{R}$, called a *structure of physical quantities*. This structure permits both the presentation of operations and order relations having a real unit representation (Section 20.4.2) in each physical attribute and the representation of multiplicative relations between attributes; and the permissible transformations for such a representation are precisely the changes of units in the basic attributes (usually mass, length, and time). Such permissible transformations are called *similarities*. Meaningful numerical relations in $\mathscr{R}$ are precisely those that are invariant under similarities; they are called *dimensionally invariant relations*.

This sort of analysis of meaningfulness of numerical laws of physical quantities was perhaps implicit in Chapter 10, in which the structure of physical quantities was developed axiomatically, the representation of physical attributes by such a structure was sketched, and the theory of dimensionally invariant numerical relations was developed. It was made more explicit by Luce (1978). In the following subsections, we present an abbreviated and more concrete treatment of the representing structure, then develop the empirical relational structure for physical attributes and the representation and uniqueness theorems in some detail. Finally, we show how dimensionally invariant relations on $\mathscr{R}$ are used to represent structure-invariant empirical relations; we relate this to the concept of physically similar systems (Section 10.10.2).

22.7.1 Structures of Physical Quantities

We shall not repeat the axiomatic development in Section 10.2, but for clarity and completeness, we present informally the fundamental ideas for a three-dimensional structure of physical quantities with a specific basis, that the reader should think of as mass, length, and time. The generalization to structures of other dimensionality will be obvious.

The set R is $\mathrm{Re} \times \mathrm{Re}^3$, i.e., all 4-tuples (τ, p, q, r), where τ is real and p, q, r are rational. In the representation of the next section, τ will be the "numerical value" given rise to by the homomorphism φ, and p, q, r will be the exponents in the multiplicative relation between φ and the three basic homomorphisms. For example, a length measurement l will be a 4-tuple of form $(\tau, 0, 1, 0)$ whereas a force measurement F will be of form $(\tau, 1, 1, -2)$; these reflect the usual dimensional designations L and MLT^{-2}.

"Multiplication" in R, denoted $*$, is defined by multiplying the τ component values and adding component-wise the p, q, and r values (adding exponents); that is,

$$(\tau, p, q, r) * (\gamma', p', q', r') = (\tau\tau', p + p', q + q', r + r').$$

For example, in Hooke's law, rectilinear compliance, C_M has the form $(\tau, 1, 0, -2)$, so the product of a rectilinear compliance with a length $C_M * l$, has the exponents of a force. The multiplicative identity $(1, 0, 0, 0)$ is denoted $\mathbf{1}$; any element $x = (\tau, p, q, r)$ in R has a multiplicative inverse $x^{-1} = (\tau^{-1}, -p, -q, -r)$ such that $x * x^{-1} = \mathbf{1}$. Hooke's law can thus be written as $C_M * l * F^{-1} = \mathbf{1}$.

Each subset of R in which p, q, r are fixed and τ varies throughout Re is called a *dimension*. The ordering $\geqslant$ and the operation $+$ are carried over from $\langle \mathrm{Re}, \geqslant, + \rangle$ to each dimension; thus $x \geqslant y$ or $x + y = z$ are defined only when x, y, z in R all lie in the same dimension. The dimension in which $p = q = r = 0$ is a special one called the *dimensionless quantities*. It is convenient, and not especially misleading, to treat $(\tau, 0, 0, 0)$ and the real number τ as interchangeable. Thus, $\mathbf{1}$ is identified with 1 and $\mathbf{0} = (0, 0, 0, 0)$ with 0. Elements of this dimension are often denoted by Greek pi, with subscripts or other affixes. The physical quantity $\pi_1 = C_M * l * F^{-1}$ is dimensionless and so subtraction, $\pi_1 - 1$, is defined, and Hooke's law can be written as $\pi_1 - 1 = 0$. It turns out that this is typical: Buckingham's "Pi" Theorem, proved in Section 10.3, asserts that every dimensionally invariant relation on physical quantities $x_1, \ldots, x_k$ can be written as $h(\pi_1, \ldots, \pi_s) = 0$, where each π_i is a dimensionless product of x_j's and h is defined by an ordinary real-valued function of s real variables.

The dimension that contains the physical quantity x is denoted by $[x]$. We shall sometimes use Q, Q_i, etc., as a notation for dimensions; thus, if $x \in Q$, $[x] = Q$. Obviously, for any x, y in R either $[x] = [y]$ (x, y lie in the same dimension) or $[x] \cap [y] = \varnothing$; thus, with respect to the equivalence relation of lying in the same dimension, $[x]$ is just the equivalence class containing x.

The *structure of physical quantities with basis* can be thought of as the 7-tuple $\mathscr{R} = \langle R, *, \geqslant, +, Q_M, Q_L, Q_T \rangle$, where R, $*$, $\geqslant$, and $+$ were explained and Q_M, Q_L, Q_T are the three dimensions with respective exponents triples (1,0,0), (0,1,0), and (0,0,1).

For any m, l, t in Re^+ we can define an automorphism $\eta = \eta(m, l, t)$ of $\mathscr{R}$ onto itself by the formula

$$\eta(\tau, p, q, r) = (m^p l^q t^r \tau, p, q, r).$$

For x in Q_M, x has form $(\tau, 1, 0, 0)$, and $\eta(x) = (m\tau, 1, 0, 0)$ is again in Q_M; η is multiplication by m on Q_M. Similarly, η is multiplication by l or t on Q_L or Q_T. Any automorphism must preserve $\geqslant$ and $+$ on each basic dimension and therefore must consist of multiplication on each. Any automorphism must map 1 into itself and can thus be shown to be the identity mapping for dimensionless quantities. Finally, since

$$(\tau, p, q, r) = (\tau, 0, 0, 0) * (1, 1, 0, 0)^p * (1, 0, 1, 0)^q * (1, 0, 0, 1)^r$$

(where pth powers, etc., are defined in terms of $*$) and any automorphism preserves $*$, it follows that every automorphism of $\mathscr{R}$ has the form just specified. Such an automorphism $\eta(m, l, t)$ is called *similarity*.

Clearly, the restriction to similarities is a result of specifying the basis Q_M, Q_L, Q_T; if this were excluded from the relational structure, automorphisms would include changes of basis as well as similarities. In Definition 10.2, we handled this a bit differently. Rather than explicitly select a basis, we simply restricted similarities to those automorphisms that map each dimension into itself. The two formulations are equivalent.

A k-ary relation on R is called *dimensionally invariant* if it is mapped into itself by every similarity of $\mathscr{R}$.

The only relations that we shall consider on R are ones in which each place is filled by an element from some one dimension, i.e., a subset of a Cartesian product $\times_i Q_i$, where $Q_1, \ldots, Q_k$ are k dimensions of $\mathscr{R}$ (not necessarily distinct).

Suppose we have a set of k dimensions, $Q_1, \ldots, Q_k$, such that we can form s dimensionless quantities from them by means of an $s \times k$ matrix of rational numbers (ρ_{ij}), i.e., if $x_i \in Q_i$, $i = 1, \ldots, k$, then

$$\pi_i = \prod_{j=1}^{k} x_j \rho_{ij}$$

is dimensionless. Then, for any real function h with s arguments, define the following dimensionally invariant relation T:

$$(x_1, \ldots, x_k) \in T \quad \text{iff} \quad h(\pi_1, \ldots, \pi_s) = 0.$$

Here we identify π_i with a real number, its τ component. Note that the dimensionless character of τ depends only on the dimensionless Q_j and the matrix (ρ_{ij}), not on the choice of $x_j \in Q_j$. T is vacuous if the π_i are not dimensionless. The dimensional invariance of T follows because any similarity is the identity map on dimensionless quantities, so $\eta(\pi_i) = \pi_i$, $i = 1, \ldots, s$.

Conversely, Buckingham's Pi Theorem (Section 10.3.2) assures us that every dimensionally invariant relation on R is obtained in this manner.

22.7.2 Triples of Scales

In developing a theory for the relational structure underlying the dimensional representation used in physics, we begin with binary conjoint structures whose components can be thought of as structures with ratio-scale representations. These ratio scales can arise either because the components are extensive structures, as was assumed in Chapter 10, or more generally because they are homogeneous PCSs with unit representations, as discussed in Section 20.4.2. One way to go about their study is via the theory of distributive triples, which was worked out in Section 20.2.7. For the reader who is satisfied with that development, the remainder of this section can be skipped and reading should be resumed at 22.7.3.

The earlier development was somewhat more restrictive than is really necessary, and we carry out the more general theory here. It follows closely the ideas and some of the results in Falmagne and Narens (1983) but with minor modifications. Basically what we show is that the empirical condition of distributivity can be replaced by a conceptual condition of compatibility among the admissible transformations of three scales of homomorphisms: on the two components of the conjoint structure and on that structure itself.

Let $C = \langle A \times X, \succsim \rangle$ be a conjoint structure (Section 19.6.1), and suppose for numerical maps φ_A: $A \to (\text{onto}) I_A$, φ_X: $X \to (\text{onto}) I_X$, and M: $A \times X \to (\text{onto}) I$, where I_A, I_X, and I are real intervals; we have the numerical representation: for all a, b in A and x, y in X,

$$ax \succsim by \quad \text{iff} \quad M[\varphi_A(a), \varphi_X(x)] \geqslant M[\varphi_A(b), \varphi_X(y)].$$

In an abuse of notation that will, however, simplify the format, we write a for $\varphi_A(a)$, etc. in the arguments of M expressions.

Any strictly increasing map of either I, I_A, I_X onto itself is an order automorphism of that component, and the corresponding sets of order automorphisms of the numerical representation M, called scales, are denoted respectively, $\mathscr{S}$, $\mathscr{S}_A$, $\mathscr{S}_X$. Sometimes constraints are imposed on the possible pairs of component scale transformations that can be made, i.e., only the set $R \subseteq \mathcal{S}_A \times \mathcal{S}_X$ is available. As usual, we write fRg for $(f, g) \in R$.

For $f \in \mathcal{S}_A$, $g \in \mathcal{S}_X$, with fRg, a mapping $M_{f,g}: I_A \times I_x \to (\text{onto})I$ may exist and be order preserving in the sense that

$$ax \succsim by \qquad \text{iff} \qquad M_{f,g}[f(a), g(x)] \geqslant M_{f,g}[f(b), g(y)].$$

Since two order preserving mappings are necessarily related by a strictly increasing function, we see there exists $H_{f,g}: I \to (\text{onto})I$ such that for all $a \in I_A$ and $x \in I_x$

$$M_{f,g}[f(a), g(x)] = H_{f,g}[M(x, y)]. \tag{2}$$

Falmagne and Narens (1983) began with a more general family of numerical codes than this and treated the property embodied in Equation (2) as a condition to be studied. Note that in such a general family this restriction, which they called "order meaningfulness," is tantamount to demanding that reference invariance hold for the representations. Because we have no need for the more general framework, we confine attention to representations satisfying Equation (2).

The questions we explore are the consequences of imposing two further conditions. One is to suppose that the underlying scales with which we are dealing are ratio scales in the sense that $\mathcal{S}_A$ and $\mathcal{S}_X$ are each isomorphic to a ratio scale. For $\mathcal{S}_A$ this means that there is a strictly increasing mapping $h: I_A \to (\text{onto})\text{Re}^+$ such that for each $f, f^* \in \mathcal{S}_A$ there exists $r \in \text{Re}^+$ such that

$$f^* = h^{-1} r h(f), \tag{3}$$

and, conversely, given any r and f, the f^* defined by Equation (3) is a member of $\mathcal{S}_A$.

The second constraint has to do with the basic idea of dimensional invariance in this context. There are a number of ways in which it can be formulated, but perhaps the simplest to understand is as a kind of compatibility among order automorphisms f and g. Consider the requirement that (f, g) be an order automorphism of $\mathcal{S}$. One way to formulate what that means is: for all $a, b \in I_A$, $x, y \in I_X$, $f \in \mathcal{S}_A$, $g \in \mathcal{S}_x$, if fRg, then

$$M(a, x) \geqslant M(b, y) \qquad \text{iff} \qquad M[f(a), g(x)] \geqslant M[f(b), g(y)]. \tag{4a}$$

It is easy to verify that this is equivalent to the existence of an order automorphism $P_{f,g}$ on I such that for all $a, b \in I_A$ and $x, y \in I_X$:

$$M[f(a), g(x)] = P_{f,g}[M(a, x)]. \tag{4b}$$

We shall refer to this condition, embodied either as Equation (4a) or (4b), as *compatibility of scales*. The condition formulated in Equation (4b) was suggested by Luce (1959) in a single-factor context as a "principle of theory construction." Basically, what is being said is that the three classes of scales are interconnected so that admissible (i.e., $f \in \mathcal{S}_A$, $g \in \mathcal{S}_X$, and fRg) transformations on the components are equivalent to an admissible transformation (i.e., $P_{f,g}$ in $\mathcal{S}$) on the conjoint structure. Another way of describing it is that the order automorphisms of the components form factorizable order automorphisms of the conjoint structure. This condition has been criticized; see Rozeboom (1962) and Luce (1962).

Several other invariance concepts have been suggested at one time or another. For example, Falmagne and Narens (1983) introduced the idea of the representation being *isotone* iff for order automorphisms f and g with fRg, there exists an order automorphism $m_{f,g} \in \mathcal{S}$ such that under the usual restriction on the symbols

$$M_{f,g}(a, x) = m_{f,g}[M(a, x)]. \tag{5}$$

They also point out that, in this context, dimensional invariance (see Section 10.3.2) can be formulated as: for all $a, b \in I_A$, $x, y \in I_X$, $f, f^* \in \mathcal{S}_A$, $g, g^* \in \mathcal{S}_X$, with fRg and f^*Rg^*:

$$\begin{aligned} M_{f,g}(a, x) \geqslant M_{f,g}(b, y) \quad &\text{iff} \\ M_{f,g}[f^*(a), g^*(x)] \geqslant M_{f,g}[f^*(b), g^*(y)]& \end{aligned} \tag{6a}$$

which is equivalent to the existence of a family of scales F_{f,g,f^*,g^*} on I such that

$$M_{f,g}[f^*(a), g^*(x)] = F_{f,g,f^*,g^*}[M_{f,g}(a, x)].$$

Equations (4), (5), and (6) all seem vaguely similar, each attempting to capture some notion of invariance. Careful examination shows them to be different: Equation (4) is entirely a constraint on M for various values of the argument; Equation (5) relates $M_{f,g}$ to M at the same value of the arguments; and Equation (6) is entirely a constraint on $M_{f,g}$ at various values of the arguments. Since, however, Equation (2) is assumed to hold throughout and it relates $M_{f,g}$, evaluated at $f(a)$, $g(x)$, to M, evaluated at

a, x, it is not too surprising that the three concepts, compatible, isotonic, and dimensionally invariant scales, are related. Indeed, they are the same concept.

THEOREM 6. *Suppose* $C = \langle A \times X, \succsim \rangle$ *is a conjoint structure;* $\mathscr{S}_A$, $\mathscr{S}_X$, *and* $\mathscr{S}$ *are scales on intervals* I_A, I_X, *and* I, *respectively;* $R \subseteq \mathscr{S}_A \times \mathscr{S}_X$; *and for each* $f \in \mathscr{S}_A$ *and* $g \in \mathscr{S}_X$ *with* fRg, *there is* $M_{f,g} \in \mathscr{S}$ *that preserves* C. *Suppose, further,* $\mathscr{S}_A$, $\mathscr{S}_X$ *and* R *are such that identities and inverses are included, and if* fRg *and* f^*Rg^*, *then* $f(f^*)Rg(g^*)$. *Then Equations* (4), (5), *and* (6) *are equivalent.*

So, in what follows, it does not matter which of these concepts we use, but it is certainly convenient to have a single term for it; we use *compatibility*.

The major result of the section is that if compatibility holds and if the component scales are isomorphic to ratio scales, then the conjoint structure has a representation in terms of the products of powers of these isomorphisms. That is the basic building block of the structure of physical quantities (see Sections 10.2 and 22.7.3).

THEOREM 7. *Suppose the hypotheses of Theorem* 6 *obtain. If, in addition, scales are compatible and if* $\mathscr{S}_A$ *has an isomorphism* h *to a ratio scale* [*see Equation* (2)] *and* $\mathscr{S}_X$ *has an isomorphism* u *to a ratio scale, then for some constants* μ *and* ν, $h^{\mu}u^{\nu}$ *is a representation of the conjoint structure.*

Note that the hypothesis that the scales are isomorphic to ratio scales sneaks in the assumption of homogeneity.

22.7.3 Representation and Uniqueness Theorem for Physical Attributes

Our general purpose is to show how dimensionally invariant relations in $\mathscr{R}$ induce meaningful empirical relations and vice-versa. To accomplish this, we need to formulate the empirical relational structure $\mathscr{A}$ in which the meaningful relations are induced and to establish the representation and uniqueness theorem for $\Phi(\mathscr{A}, \mathscr{R})$. The uniqueness theorem must state that the only permissible transformations are the similarities of $\mathscr{R}$. Such an empirical structure and its representation in $\mathscr{R}$ were described in Section 10.9 (Embedding physical attributes in a structure of physical quantities). We repeat it here but base it on distributive triples (Section 20.2.7) or compatible triples (Section 22.7.2) rather than on laws of similitude and exchange.

The empirical relational structure $\mathscr{A}$ will be called a *structure of physical attributes*. It is an "umbrella" structure, involving a finite number of physical attributes (ordered sets), some exhibiting ratio-scale measurement and all interrelated as distributive or compatible triples.

We assume that there are u weak orders $\langle A_i, \succsim_i \rangle$, $i = 1, 2, \ldots, u$ that are called *physical attributes*. Associated with the first $t < u$ are additional relations, perhaps concatenation operations, S_{ij}, j in an index set $J(i)$, on A_i such that

$$\mathscr{A}_i = \langle A_i, \succsim_i, S_{ij} \rangle_{j \in J(i)}$$

are totally ordered relational structures with ratio scale representations on Re^+. We know a good deal about how this can come about (see Section 20.2). One very special case includes the several extensive structures that arose in classical physics.

Some of the attributes are conjoint structures. These are specified by a relation C on U^3, where $U = \{1, 2, \ldots, u\}$. If $(i, j, k) \in C$, then $\mathscr{A}_k$ is a conjoint structure with components $\mathscr{A}_i$ and $\mathscr{A}_j$ in the sense that $A_k = A_i \times A_j$ and $\mathscr{A}_k$, $\mathscr{A}_j$, and $\mathscr{A}_k$ form a compatible triple.[6] We make the following assumptions about C:

(1) If (i, j, k) and $(i', j', k) \in C$, then $i = i'$ and $j = j'$.

(2) If k is such that for all $i, j \in U$, $(i, j, k) \notin C$, then either $k \leqslant t$ or there exist $i, j \leqslant t$ such that $(i, k, j) \in C$.

We can describe these assumptions in the following way. We say $\mathscr{A}_k$ is *decomposable* if there are i and j such that $(i, j, k) \in C$, otherwise $\mathscr{A}_k$ is *indecomposable*. The first assumption says that a decomposable attribute is not decomposable in any other way. The second one says that an indecomposable attribute is either a ratio-scale structure or it is a member of a compatible triple whose other two attributes are ratio scalable.

The first assumption seems, on the face of it, too restrictive. For example, suppose we have the attributes mass, velocity, energy and momentum with the usual measures m, v, E, and p. Then, since $p = mv$ and $E = \frac{1}{2}mv^2$, we can either think of E as decomposable into m and v or into p and v. If both decompositions are included in the qualitative structure, then it is necessary to impose some sort of an assumption to force internal consistency of the exponents that arise from Theorem 7. The alternative is to include just one of the decompositions and to allow the others to arise naturally in the representation. The latter is simpler to present.

[6] Technically, we should say there is an isomorphism between A_k and $A_i \times A_j$, but we will simply substitute $A_i \times A_j$ for A_k.

We define the C-basis of such a structure to consist of the following ratio scalable attributes:

(i) those that are indecomposable, and

(ii) those that are part of a triple, one member of which is decomposable and not ratio scalable.

If the C-basis has m-members, which by relabeling can be taken to be $1, 2, \ldots, m \leqslant t$, then we say the structure of physical attributes is *m-dimensional*.

Having described the structure, we can now introduce the representation theorem. Up until now, we have considered φ_i as a real-valued function on A_i that preserves $\succsim_i$, is additive over $\bigcirc_i$ when the operation exists, and enters into conjoint representations as a multiplier after being raised to some rational power. Now we point out the order-preserving, additive, and multiplicative properties can all be defined in a structure of physical quantities, $\mathscr{R}$, provided φ_i maps A_i into a dimension Q_i of that structure. The ordering $\geqslant$ and operation $\oplus_i$ are defined within Q_i, and the rational powers and multiplications can be taken in the sense of the operation $*$.

THEOREM 8. *Let*

$$\mathscr{A} = \langle A_1, \ldots, A_u, \succsim_1, \ldots, \succsim_u, \{S_{1j}\}_{j \in J(1)}, \ldots, \{S_{tj}\}_{j \in J(t)}, \{C\} \rangle$$

be an m-dimensional structure of physical attributes with C-basis $\mathscr{A}_i$, $i = 1, \ldots, m$. *Let*

$$\mathscr{R} = \langle (\mathrm{Re}^+)^{m+1}, *, +, \geqslant, Q_1, \ldots, Q_m \rangle$$

be the m-dimensional structure of physical quantities with basis $Q_1, \ldots, Q_m$; *then, there exist functions*

$$\varphi_i \colon A_i \to (\mathrm{Re}^+)^{m+1}, \qquad i = 1, \ldots, u,$$

such that:

(i) *For each* $i = 1, \ldots, t$ *there exist numerical relational structures* $\mathscr{R}_i = \langle R_i, \geqslant, R_{ij} \rangle_{j \in J(i)}$, *where* R_i *is a dimension of* $\mathscr{R}$ *and* R_{ij} *are relations on* R_i *of the same order as* S_{ij}, *such that* φ_i *is a homomorphism of* $\mathscr{A}_i$ *onto* $\mathscr{R}_i$ *and for* $i = 1, \ldots, m$, $R_i = Q_i$.

(ii) *For* $i = t + 1, \ldots, u$, φ_i *is a homomorphism of* $\mathscr{A}_i$ *onto a dimension of* $\mathscr{R}$.

(iii) *For each compatible triple* $(i, j, k) \in C$, *there exist unique nonzero numbers* ρ, ρ' *such that for all* $a_i \in A_i$, $a_j \in A_j$

$$\varphi_k[(a_i, a_j)] = \varphi_i(a_i)^{\rho} * \varphi_j(a_j)^{\rho'}.$$

Moreover, φ_i', $i = 1, \ldots, u$ *are other functions satisfying* (i)–(iii) *iff there is a similarity* η *of* $\mathscr{R}$ *such that for* $i = 1, \ldots, u$, $\varphi_i' = \eta\varphi_i$.

The proof of Theorem 8, which is fairly straightforward once one has Theorem 7 and the choice of the basis, is left as Exercise 10.

This theorem reformulates and corrects Theorem 10.11. That theorem does not completely clarify the nature of the consistency required of the conjoint triples. Here we have by-passed the consistency problem by our choice of C, which uniquely defines a basis. It should be noted that by making other choices for C, different bases are selected. It reformulates Theorem 10.11 in several ways. First, it invokes compatible automorphism groups rather than laws of exchange and similitude, thereby making the results more general and easier to prove. It makes more explicit the nature of the representation, and in particular it makes explicit the identification of the two bases. Only by establishing this correspondence do we obtain uniqueness up to similarities.

Theorem 8, together with Theorem 5, yields the fact that dimensionally invariant relations in the image of the structure of physical attributes correspond to meaningful relations in that structure. We say *meaningful* because Theorem 5 holds and so all three invariance concepts agree. This result permits us to view the basic Pi Theorem of Buckingham (Theorem 10.4) in a new light. It simply shows how meaningful relations of such a structure can be defined explicitly in terms of the defining relations of the structure of physical attributes. Specifically, the defining relations tell us which multiplicative combinations of ratio and conjoint measurements are dimensionless. And the theorem says that any meaningful relations can be expressed as a numerical relation holding among some of these dimensionless (numerical) quantities.

One might hope that in some similar way the general concept of a meaningful relation in a measurement structure could be explicitly expressed in terms of the defining relations of that structure as reflected in its numerical representation. But that has yet to be done.

22.7.4 Physically Similar Systems

The correspondence of dimensional invariance and meaningfulness just mentioned does not cover the most important ground. To clarify the result and its limitations, we consider again how Hooke's law is formulated.

Suppose our physics is such that the dimensions available are:

Name	Symbol for dimension	Exponent in basis		
		M	L	T
Mass	Q_M	1	0	0
Length	Q_L	0	1	0
Time	Q_T	0	0	1
Velocity	Q_v	0	1	−1
Acceleration	Q_a	0	1	−2
Energy	Q_E	1	2	−2
Momentum	Q_p	1	1	−1
Force	Q_F	1	1	−2
Density	Q_d	1	−3	0

Hooke's law involves the physical dimensions of force Q_F with exponents $(1, 1, -2)$, length Q_L with exponents $(0, 1, 0)$, and compliance Q_C with exponents $(1, 0, -2)$. The relation T is defined by

$$(x, y, z) \in T \quad \text{iff} \quad \text{for } x \in Q_F, \quad y \in Q_C, \quad z \in Q_L,$$
$$x^{-1} * y * z - 1 = 0.$$

This is certainly dimensionally invariant (see the discussion in 22.7.1). The induced relation $S(T)$ is defined by

$$(a_i, a_j, a_k) \in S(T) \quad \text{iff} \quad (\varphi_i(a_i), \varphi_j(a_j), \varphi_k(a_k)) \in T.$$

Unfortunately, $S(T)$ is vacuous because there is no attribute in the list of nine that maps onto the dimension Q_C i.e., $\varphi_j(a_j) \in Q_C$ is always false. Thus $S(T)$ is meaningful but trivially so. It certainly does not correspond in meaning to Hooke's law.

The particular problem raised by Hooke's law could be remedied by adding an attribute in which the objects are springs and the ordering is just that of rectilinear compliance; there would be a conjoint mapping involving that attribute and the extensive attributes force and length. But then we could give yet another example, demanding yet another attribute. The underlying problem is that many physical laws involve numerous scale-dependent parameters, as discussed in the introduction to this section; these parameters are part of the structure $\mathscr{R}$, but they are not in the range of homomorphisms mapping $\mathscr{A}$ into $\mathscr{R}$.

A solution to this problem was suggested in Section 10.10.2. We consider a set of dimensions $Q_1, \ldots, Q_k$ in R that represent the physical attributes that covary for different configurations of a single "physical system." For example, Q_{F} and Q_{L} represent force and length, which covary for one particular spring, the configurations of which satisfy Hooke's law. We have, further, the set of dimensions $Q_{k+1}, \ldots, Q_{k+l}$ that represents the scale-dependent parameters of the "physical systems." In Hooke's law, this is just Q_{C}, which is a constant physical quantity for any particular spring. The relation T defined over $Q_1 \times \cdots \times Q_k \times Q_{k+1} \times \cdots \times Q_{k+1}$ is still dimensionally invariant, but we no longer try to use the homomorphism φ alone to induce $S(T)$; instead, we define S by means of φ, which maps a product of k attributes into $Q_1 \times \cdots \times Q_k$, together with a *system measure* θ, which maps subsets of the same k-fold product ("physical systems") into $Q_{k+1} \times \cdots \times Q_{k+l}$. For Hooke's law and the example of a 9-attribute structure given above, we have (φ_i, φ_j) mapping length $\times$ force into $Q_{\mathrm{L}} \times Q_{\mathrm{F}}$ and a system measure θ that assigns a value in Q_{C} to each spring. Each spring is not regarded as an element in yet another (tenth) attribute but rather a subset of (length, force) combinations.

In the following, we shall use $\mathscr{A}_1, \ldots, \mathscr{A}_k$ to refer to any k (not necessarily distinct) attributes of a structure of physical attributes $\mathscr{A}$. (We change notation because we no longer need to distinguish between basic and nonbasic attributes, and we want to avoid introducing double subscripts.) The Cartesian product of the domains these attributes is $\times_{i=1}^{k} A_i$, and a typical element is $\mathbf{a} = (a_1, \ldots, a_k)$. If φ is a homomorphism of $\mathscr{A}$ into the structure of physical quantities $\mathscr{R}$, we shall let $Q_1, \ldots, Q_k$ be the dimensions corresponding to $\mathscr{A}_1, \ldots, \mathscr{A}_k$, and we shall also denote by φ the component-wise mapping:

$$\varphi\colon \mathop{\times}_{i=1}^{k} A_i \to \mathop{\times}_{i=1}^{k} Q_i.$$

We write $\mathbf{x} = \varphi(\mathbf{a})$ or, component-wise, $x_i = \varphi_i(a_i)$.

Let $Q_{k+1}, \ldots, Q_{k+l}$ be l additional dimensions of $\mathscr{R}$ and let θ be a function:

$$\theta\colon \mathop{\times}_{i=1}^{k} A_i \to \mathop{\times}_{i=k+1}^{k+l} Q_i.$$

Finally, let T be a subset of $\times_{i=k+1}^{k+l} Q_i$, i.e., T is a $(k+1)$-ary relation on R with dimension Q_i in place i, $i = 1, \ldots, k+1$. The induced relation,

$S = S(\varphi, \boldsymbol{\theta}, T)$ is defined by

$$\mathbf{a} \in S \qquad \text{iff} \qquad (\varphi(\mathbf{a}), \boldsymbol{\theta}(\mathbf{a})) \in T.$$

This mode of definition is not helpful unless a dimensionally invariant T on R yields an induced relation $S(T)$, independent of the homomorphism φ. How can this occur? First, if φ is changed, $\boldsymbol{\theta}$ must be changed with it: this can be accomplished by writing $\boldsymbol{\theta}(\mathbf{a})$ as a function $\theta^*(\varphi(\mathbf{a}))$, where θ^* is a fixed function from $\times_{i=1}^{k} Q_i \to \times_{i=k+1}^{k+1} Q_i$. We now regard S as induced by φ, $\boldsymbol{\theta}^*$, and T, and we ask when $S = S(\theta^*, T)$ depends only on the relation T and function θ^* defined in $\mathscr{R}$ but not on φ. Next, suppose that η is a similarity of $\mathscr{R}$, that T is dimensionally invariant, and that φ and $\eta\varphi$ induce the same relation $S(\varphi, \theta^*, T) = S(\eta\varphi, \theta^*, T)$. Then

$$\begin{aligned} (\eta\varphi(\mathbf{a}), \eta\theta^*\varphi(\mathbf{a})) \in T \qquad &\text{iff} \qquad (\varphi(\mathbf{a}), \theta^*(\mathbf{a})) \in T \\ &\text{iff} \qquad (\eta\varphi(\mathbf{a}), \theta^*\eta\varphi(\mathbf{a})) \in T \end{aligned}$$

where the first equivalence is dimensional invariance of T, and the second one is invariance of S. Thus it is plausible to impose the further requirement that $\eta\theta^* = \theta^*\eta$, i.e., the function θ^* commutes with every similarity. Such a function is called a system measure (Vol. I, p. 509), and we shall also refer to $\boldsymbol{\theta} = \theta^*\varphi$ as a *system measure* if θ^* commutes with all similarities.

These conditions suffice: if $T \subseteq \times_{i=1}^{k+1} Q_i$ is dimensionally invariant, and

$$\theta^*\colon \mathop{\times}_{i=1}^{k} Q_i \to \mathop{\times}_{i=k+1}^{k+l} Q_i$$

is a system measure, then the relation $S(\varphi, \theta^*, T)$ on $\times_{i=1}^{k} A_i$ is independent of φ, and, hence, θ^*, T induce a meaningful empirical relation on $\mathscr{A}$.

Finally, we turn to the intrinsic characterization of empirical relations on $\mathscr{A}$ that are defined in the above manner. Here we discuss the automorphisms of $\mathscr{A}$ and physically similar systems in $\mathscr{A}$.

A *physical system* in $S(\theta^*, T)$ is identified by a constant value of the function θ^* or system measure; thus, for any fixed homomorphism φ, we partition S into equivalence classes or physical systems by writing

$$\mathbf{a} \sim \mathbf{b} \qquad \text{iff} \qquad \theta^*\varphi(\mathbf{a}) = \theta^*\varphi(\mathbf{b}).$$

This partition is independent of φ since if $\varphi' = \eta\varphi$, where η is a similarity, the commutativity of θ^* and η leads to

$$\theta^*\varphi(\mathbf{a}) = \theta^*\varphi(\mathbf{b}) \qquad \text{iff} \qquad \theta^*\varphi'(\mathbf{a}) = \theta^*\varphi'(\mathbf{b}).$$

Suppose that $\mathbf{a} \sim \mathbf{b}$ and $\mathbf{a}', \mathbf{b}'$ are two other elements of S such that their φ values are related to those of $\mathbf{a}, \mathbf{b}$ by a similarity of η: $\varphi(\mathbf{a}') = \eta\varphi(\mathbf{a})$ and $\varphi(\mathbf{b}') = \eta\varphi(\mathbf{b})$. Then $\mathbf{a}' \sim \mathbf{b}'$, as is easily shown. Thus a similarity induces a partial mapping of one physical system into another. When the mapping φ is onto $\times_{i=1}^{k} Q_i$, then any similarity induces a full map of each physical system in S onto another. More precisely, a similarity η induces a map $\lambda = \varphi^{-1}\eta\varphi$ of $\times_{i=1}^{k} A_i$ onto itself, and λ maps any physical system in S on another. *The relation S is therefore the union of families of physically similar systems.*

A converse to this result can be established using Theorem 10.13. If φ maps $\times_{i=1}^{k} A_i$ onto $\times_{i=1}^{k} Q_i$, and S is the union of families of physically similar systems, and the stability group of S is rational (see Vol. I, p. 511) then $S = S(\theta^*, T)$ for some system measure θ^*: $\times_{i=1}^{k} Q_i \to \times_{i=k+1}^{k+1} Q_i$ and some dimensionally invariant $(k + 1)$-ary relation $T \subseteq \times_{i=l+1}^{k+l} Q_i$.

If the similarity η induces a map $\lambda = \varphi^{-1}\eta\varphi$ of the full structure of physical attributes $\mathscr{A}$ onto itself, then λ is an automorphism in the sense that λ is an automorphism of each separate attribute $\langle A_i, \succsim_i \rangle$ or $\langle A_i, \succsim_i, \bigcirc_i \rangle$, and for each conjoint mapping $(i, j, k) \in C$,

$$(\lambda(a_i), \lambda(a_j)) = \lambda(a_i, a_j).$$

Moreover, every such compatible automorphism is related to a similarity: since the mapping $\varphi' = \varphi\lambda$ is a homomorphism of $\mathscr{A}$ into $\mathscr{R}$, and consequently, by the uniqueness part of Theorem 8, $\varphi\lambda = \eta\varphi$. This last formula is meaningful even when φ is not *onto*. Hence we can generalize the notion of a family of physically similar systems to include any case in which any system is mapped onto any other system by an automorphism of $\mathscr{A}$. Any S that is the union of families of similar systems in this broader sense (and that satisfies the technical requirement of rational stability group) is obtained as $S(\theta, T)$ for a suitable choice of $Q_{k+1}, \ldots, Q_{k+l}, \theta^*$, and dimensionally invariant T.

In short, similar physical systems are those that are mapped onto one another by automorphisms of the structure of physical attributes. A family of such systems (or a union of several families) constitutes a meaningful empirical relation on $\mathscr{A}$, defined by means of a system measure that assigns scale-dependent parameter vectors of physical quantities $\boldsymbol{\theta}(\mathbf{a}) = \theta^*(\varphi(\mathbf{a}))$ to each system a and a dimensionally invariant relation T that is defined over pairs of form $(\varphi(\mathbf{a}), \theta^*(\mathbf{a}))$.

22.7.5 Fundamental versus Index Measurement

In the natural sciences, and especially in the physical ones, the use of fundamental measurement structures is pervasive. New uses are discovered often, for example, measures of charge, temperature, mass, time, and distance are applied in a variety of ways in biophysics and physiology. When measurement is more problematic, as it has been in the social and behavioral sciences, we are strongly tempted to imitate the methods of the sciences allied to physics by finding new applications of these same physical measures. Indeed, it is sometimes extremely useful to do so. Much progress has been made in cognitive psychology by the systematic analysis of response times: the duration of intervals marked by occurrences of stimulus presentation and discrete responses (see Luce, 1986a; Posner, 1978; Sternberg, 1969/1970; Welford, 1980). And the study of color perception depends so heavily on physical measurement that it is often regarded as a branch of optics (see Chapter 15), though more than physics is involved.

In a number of situations, however, physical measurements are used in a way that does not engage the full measurement structures that underlie them. Especially important are uses in which the physical measure is thought to be an order-preserving index of some hypothetical underlying quantity, which itself has not received a full measurement analysis. To examine the logic of such "index" measurement, we begin with a concrete example familiar to psychologists.

Hunger is believed to be an important variable influencing behavior; it appears to be a variable that varies in intensity. One might hope, therefore, that it can be measured quantitatively. Various physical measures have been suggested as indices of hunger: amount of food ingested, initial rate of ingestion, force exerted to overcome a restraint to reach food, percentage reduction from normal body weight, time since last food ingestion, etc. One problem with these indices is that in some circumstances they are not monotonically related. Ignoring that problem or bypassing it by confining attention to a limited set of circumstances, one could adopt some particular index, say, hours of deprivation, as a "measure" or, perhaps, "operational definition" of hunger.

Physical measurement of time intervals is a ratio scale, but this does *not* mean that the time since last food, as a measure of hunger, is also a ratio scale. The ratio character of time measurement is based on the numerical representation of time intervals so that the value associated with the concatenation of adjacent intervals is the sum of values associated with those intervals. This measurement is carried out in practice by counting the oscillatory behavior of natural oscillators such as some types of atoms. But the subdivision of a time interval since last food ingestion into small intervals defined by the motion of a physical oscillator has, so far as we know, nothing empirically to do with hunger. If we have no reason to incorporate such physical oscillations into an empirical structure for hunger, then there is nothing empirical about the representation of hunger that limits which monotonic transformations of time we can use as indices of hunger. We lose nothing empirically by indexing hunger by, for example, the logarithm or the square of the measured time. The same is not true of time itself since we are using + to represent concatenation of time intervals. For the measurement of hunger by time deprivation alone we have, at best, ordinal information. (These remarks apply just as well if we were to use as a measure of hunger a (log interval) representation derived from conjoint measurement such as sugar concentration in the blood.)

These observations in no way deny the probable usefulness of such indices of hunger. One way to advance knowledge is to obtain detailed observations, often presented as a numerical plot of two variables, one of which is the index. For example, a plot of some measure of the intensity with which a task is performed as a joint function of the time since last food and the amount (or quality) of the food offered as a reward for performance may be informative in learning how to incorporate the hypothetical variable *hunger* into a motivational theory (see Example 1, p. 316 of Vol. I). One possible direction is to determine whether additive conjoint scales are compatible with such plotted functions. If they are, then that representation might constitute a true interval scale of "hunger." There is absolutely no reason to anticipate that the additive contribution of the variable "time since last food" would be proportional to that time. Indeed, it may not even prove to be monotonic, in which case time of deprivation is an imperfect index of hunger. Alternatively, such plots might suggest other generalizations about deprivation and reward, leading to a quite different theory, one in which detailed numerical representation of the plots might not be of primary interest. Such a theory might be physiologically based, in which case the measure "time of deprivation" would be just another instance of the use of physical measures in novel ways. But then the measurement would in fact be of time not "hunger." An interval- or ratio-scale measure of a useful intervening variable that deserves the appellation *hunger* might or might not emerge from such a theory.

Put another way, the emergence of a fundamentally measured quantity is a by-product of an explicit theory into which that variable is incorporated. In this work we have been mostly concerned with theories simple enough to be stated using informal set theory and axiomatics. In practice, measurement sometimes emerges from much more complex theories. The intuition that hunger can be represented numerically is pretheoretical. Actual measurement of such a pretheoretical variable may come from a rather simple theory such as additive conjoint structure or from something more complex; or the hypothetical variable may dissolve into many related concepts as the theory develops. The use of physical indices for such pretheoretical variables may be a useful initial step toward such a theory, but it must not be confused with the fundamental measurement of the variables.

One reason for confusion is the prevalence of "pointer" measurement in the physical sciences in which length or angle is used as a measure of temperature, current, mass, and many other attributes. Consider, for example, the use of springs to provide measurement of weight by means of displacements of length. Are we not saying that because length increments are measured on a ratio scale, the spring provides a ratio-scale measurement of weight? Are we not committing the same misapplication of uniqueness of one scale to another that we warned against in the hunger example? The differences are considerable, and the answer is *no*.

In the weight case, we have three distinct ingredients, most of which are lacking in the hunger example. First, there is an extensive theory of length measurement leading to ratio-scale representations. Second, there is an extensive theory of mass measurement leading to ratio-scale representations. And third, there are two physical laws that relate the measures of length and weight via the mass involved. Newton's law connects the force called *weight* via the acceleration, due to gravity, to mass, so that in a fixed gravitational field weight is proportional to mass. And Hooke's law connects the force applied to a spring to its displacement, again as a direct proportion within certain limits of elastic deformation of the spring. Thus the theory asserts (within certain limits) that weight and spring displacement are proportional. So the measurement procedure that uses a spring to measure weight directly is valid according to this theory. This example is typical of the instances of pointer measurement in physical science.

No comparable structure exists in the hunger example. There is a ratio scale of time, but there is no independent theory for the measurement of hunger, except for the pretheoretical conjecture that hunger is a monotonic function of hours of deprivation. Furthermore, there is no analogue of Newton's and Hooke's laws, that is, no empirical relation between the notion of hunger and the concatenations that pertain to time measurement.

The distinction between fundamental measurement and numerical indices is particularly relevant to the discussions concerning the measurement of intelligence and of personality traits. The dictum that "intelligence is just what intelligence tests measure" identifies a hypothetical, pretheoretic variable (e.g., intelligence) with a particular index, the number of correct answers on an intelligence test, corrected for age. The theory of measurement developed in these volumes can only in a limited way provide methods suited to measuring hunger or intelligence. To be applied, there must first be a detailed empirical and theoretical analysis of the relevant phenomena. During that process, the theory of measurement can help clarify the status of different numerical assignments whether index or fundamental in character.

22.8 PROOFS

22.8.1 Theorem 6 (p. 315)

Suppose $\mathscr{C} = \langle A \times X, \succsim \rangle$ is a conjoint structure with scales $\mathscr{S}_A$, $\mathscr{S}_X$, and $\mathscr{S}$ on intervals associated with A, X, and $A \times X$; R is included in $\mathscr{S}_A \times \mathscr{S}_X$; and for each $f \in \mathscr{S}_A$ and $g \in \mathscr{S}_X$ with fRg, there exists $M_{f,g} \in \mathscr{S}$ that preserves $\mathscr{C}$. Moreover, the scales are closed under inverses and composition, and R is such that if fRg and $f * Rg^*$, then $f(f^*)Rg(g^*)$. Then the properties of being compatible [*Equation* (4)], *isotone* [*Equation* (5)], *and dimensionally invariant* [*Equation* (6)] *are equivalent to each other.*

PROOF. Equation (4) implies Equation (6):
Letting $f' = f^{-1}f^*$ and $g' = g^{-1}g^*$, then

$$\begin{aligned} M_{f,g}[f^*(a), g^*(x)] &= H_{f,g}\{M[f'(a), g'(x)]\} && \text{[by (2)]} \\ &= H_{f,g}\{P_{f',g'}[M(a,x)]\} && \text{[by (4)].} \end{aligned}$$

Similarly,

$$M_{f,g}(a,x) = H_{f,g}\{M[f^{-1}(a), g^{-1}(x)]\} = H_{f,g}\{P_{f^{-1},g^{-1}}[M(a,x)]\}.$$

Taking inverses and denoting function composition by $\cdot$ we see Equation (6) obtains with

$$Q_{f,g;f^*,g^*} = H_{f,g} \cdot P_{f',g'} \cdot [P_{f^{-1},g^{-1}}]^{-1} \cdot [H_{f,g}]^{-1}.$$

Equation (6) implies Equation (5):

$$\begin{aligned} M_{f,g}(a,x) &= H_{f,g}\{M[f^{-1}(a), g^{-1}(x)]\} && \text{[by (2)]} \\ &= H_{f,g}\{F_{\iota,\iota;f^{-1},g^{-1}}[M_{\iota,\iota}(a,x)]\} && \text{[by (6)]} \\ &= m_{f,g}M(a,x), \end{aligned}$$

where $m_{f,g} = H_{f,g}(Q_{\iota,\iota;f^{-1},g^{-1}})$, which is Equation (5).

Equation (5) implies Equation (4):

$$\begin{aligned} H_{f,g}[M(a,x)] &= M_{f,g}[f(a), g(x)] && \text{[by (2)]} \\ &= m_{f,g}\{M[f(a), g(x)]\} && \text{[by (5)]}, \end{aligned}$$

whence

$$M[f(a), g(x)] = [m_{f,g}]^{-1} H_{f,g}[M(a,x)],$$

which is Equation (4) with $P_{f,g} = [m_{f,g}]^{-1}(H_{f,g})$. ◇

22.8.2 Theorem 7 (p. 315)

Under the hypotheses of Theorem 6, if the scales are compatible and if those on A and X are isomorphic to ratio scales under isomorphisms h and u, respectively, then for some constants μ and ν, the conjoint structure has the representation $h^{\mu}u^{\nu}$.

PROOF. We first show that the Thomsen condition holds in $\mathscr{C}$. Suppose $M(a, x) = M(b, y)$ and $M(b, z) = M(c, x)$. Because $\mathscr{S}_X$ is isomorphic to a ratio scale, it is homogeneous. Thus, there exist g, g^* in $\mathscr{S}_X$ such that $g(x) = z$ and $g^*(z) = g(y)$. Observe that using the commutativity of ratio scales

$$x = g^{-1}(z) = g^{-1}g^{*-1}g(y) = g^{*-1}(y). \tag{7}$$

By compatibility,

$$\begin{aligned} M(a,z) &= M[\iota(a), g(x)] \\ &= P_{\iota,g}[M(a,x)] && \text{(4b)} \\ &= P_{\iota,g}[M(b,y)] && \text{(hypothesis)} \\ &= P_{\iota,g}\{M[\iota(b), g^{-1}g^*(z)]\} && \text{(choice of } g, g^*) \\ &= P_{\iota,g}P_{\iota,g^{-1}g^*}[M(b,z)] && \text{(4b)} \\ &= P_{\iota,g}P_{\iota,g^{-1}g^*}[M(c,x)] && \text{(hypothesis)} \\ &= P_{\iota,g}P_{\iota,g^{-1}g^*}[M[\iota(c), g^{*-1}(y)]] && \text{(7)} \\ &= P_{\iota,g}P_{\iota,g^{-1}g^*}P_{\iota,g^{*-1}}[M(c,y)]. \end{aligned}$$

Observe, by three applications of Equation (4b),

$$\begin{aligned} M(c,y) &= M[c, gg^{-1}g^*g^{*-1}(y)] \\ &= P_{\iota,g}P_{\iota,g^{-1}g^*}P_{\iota,g^{*-1}}[M(c,y)], \end{aligned}$$

so $M(a, z) = M(c, y)$.

By Theorem 6.2, we know that M has a multiplicative representation:

$$M(a, x) = F[v(a)p(x)].$$

Using Equation (4b) again,

$$v[f(a)p[g(x)]] = P_{f,g}f[v(a)p(x)]. \tag{8}$$

Let $L_{f,g} = f^{-1}P_{f,g}F$, $v_f = vfv^{-1}$, $p_g = pgp^{-1}$, $v(a) = s$, $p(x) = t$. Then Equation (8) is equivalent to

$$v_f(s)p_g(t) = L_{f,g}(st).$$

If we set $s = 1$ and then $t = 1$ to express v_f and p_g in terms of $L_{f,g}$, then the resulting functional equation is known to be solved by

$$\begin{aligned} L_{f,g}(t) &= k(f, g)t^{m(f,g)} \\ v_f(t) &= k'(f)t^{m(f,g)} \\ p_g(t) &= k''(g)t^{m(f,g)}. \end{aligned}$$

Because of the dependencies in the last two equations, necessarily $m(f, g) = m$, a constant. Thus, on the first component, we have

$$v[f(a)] = k'(f)v(a)^m,$$

and by the assumption about u,

$$u[f(a)] = c(f)u(a).$$

Eliminating $f(a)$

$$v^{-1}k'(f)v(a) = u^{-1}c(f)u(a).$$

If we set $v(a) = s$, $k'(f) = K[c(f)]$, where K is strictly increasing, then we can write

$$\begin{aligned} v^{-1}[K(c)s^m] &= u^{-1}\{c[u(a)]\} \\ &= u^{-1}\{c[uv^{-1}(s)]\}. \end{aligned}$$

Letting $w = uv^{-1}$, this becomes

$$w\{K(c)s^m\} = cw(s).$$

Selecting c_0 so that $K(c_0) = 1$, we see

$$w(s^m) = c_0 w(s);$$

and setting $s = 1$,

$$w[K(c)] = cw(1).$$

Thus,

$$w[K(c)s^m] = w[H(c)]w(s^m)/w(1)c_0.$$

The well known solution is of the form

$$w(t) = Kt^{1/\mu},$$

and substituting back, we find,

$$v(s) = Cu(s)^{\mu}.$$

The other component is similar, and the conclusion follows. ◇

22.9 REPRISE: UNIQUENESS, AUTOMORPHISMS, AND CONSTRUCTABILITY

22.9.1 Alternative Representations

As we have repeatedly emphasized throughout these volumes, any structure having a numerical representation has many. There are two distinct senses in which this is true. One is illustrated by the observation that an exponential transformation converts an additive representation in $\langle \text{Re}, \succsim, + \rangle$ to a multiplicative one in $\langle \text{Re}^+, \succsim, \cdot \rangle$. Indeed, $\langle \text{Re}, \succsim, + \rangle$ is isomorphic under any strictly increasing mapping f to a distinct numerical structure with the associative operation $x \oplus y = f^{-1}[f(x) + f(y)]$. Each of this infinity of possibilities represents the empirical structure as well as any other, and the choice among them is entirely conventional or pragmatic, none being logically superior nor empirically preferable. The second type of nonuniqueness, which we discuss in the next subsection, arises from automorphisms or endomorphisms of a particular representing structure.

Although the choice of numerical structure may be partially conventional, it is far from unimportant; as expositors or appliers of measurement, we cannot be altogether cavalier in selecting representations for a number of highly pragmatic reasons. For example, in cases in which an additive representation exists, it is undoubtedly wise to use addition, assuming that choice is not in contradiction to some more important consideration, because it is consistent with counting, which is one of the first mathematical skills learned. Moreover, modern work in complexity theory (Rabin, 1977) shows that addition of natural numbers is from a computability point of view, far simpler, in a well specified technical sense, than multiplication. Put another way, there can be issues of computation that are simpler in one representation than another. So we can think of the choice of the representing structure as a convention about how we want to carry out certain calculations. Further, since people do not shift from one convention to another easily without getting confused, it is advisable to stick to the most familiar. But neither of these reasons is decisive: witness the use of addition for extensive measures and multiplication for conjoint ones in physics. Physicists could also have elected to use an additive representation for the conjoint structures, requiring them to work with logarithms of extensive measures instead of their powers. Probably the main disadvantage would not be that but rather the potential confusion in knowing whether a particular use of + refers to combining within or across dimensions. There is no ambiguity in physics; it is used only within a dimension.

When there is a standard, conventional choice, one typically makes no further reference to the nonuniqueness that is due to the existence of isomorphic numerical alternatives of this sort. The problem remains when no convention exists, however, as we saw when we dealt with nonassociative concatenation structures (Chapter 19) or more general relational structures (Section 20.2). In those cases it was resolved by accepting a different, more abstract conventional choice, namely, select the representation(s) in which Archimedean-ordered (sub)groups of automorphisms of the structure are represented as either addition or multiplication. Again the choice is conventional but not arbitrary because it leads to familiar representations of admissible transformations. Let us consider automorphisms more fully.

22.9.2 Nonuniqueness and Automorphisms

Of these two types of nonuniqueness, both of which entail conventional choices for the representation, the second, which concerns the several possible representations of a structure into a fixed numerical structure or, equally, of the structure onto itself, is by far the more interesting because it reflects properties of the empirical structure itself. In extensive structures, such nonuniqueness is made manifest in the arbitrary choice of unit in ratio scales; and in conjoint structures, in the arbitrary choice of both unit and zero in interval scales. It simply does not matter to which mass we assign the value unity, but once we agree upon that convention, the entire mass scale is completely determined. Note that it truly is a matter of convention in a homogeneous structure such as the classical theory of mass because the only way to single out any element as special is by convention; none is distinguished by its properties. (In a discrete theory such as quantum theory there is, of course, a natural unit, namely the minimal element.) In some ways, the choice of unit in the homogeneous case appears to exhibit much the same conventional character as does that of alternative isomorphic representations, but it differs in one important respect.

Consider, for example, an extensive structure $\mathscr{A} = \langle A, \succsim, \bigcirc \rangle$ that is isomorphic to $\mathscr{R} = \langle \mathrm{Re}^+, \geqslant, + \rangle$, and let φ be one such isomorphism. Because of the uniqueness theorem for such measurement, we know that for each $r > 0$, $r\varphi$ is another isomorphism and that any two isomorphisms onto $\mathscr{R}$ are so related. Now, what is interesting and important is that the mapping $\alpha(r) = \varphi^{-1} r \varphi$, which takes $\mathscr{A}$ onto itself, is actually an automorphism of $\mathscr{A}$, that is, an exact structure-preserving map of $\mathscr{A}$ onto itself. Moreover, if α is an automorphism of $\mathscr{A}$, then the mapping $\varphi\alpha$ of $\mathscr{A}$ onto $\mathscr{R}$ is itself an isomorphism, and so of the form $r(\alpha)\varphi$, in which $r(\alpha) = \varphi\alpha\varphi^{-1}$. So we have the relation $\varphi\alpha = r\varphi$, in which $\alpha = \alpha(r)$ and $r = r(\alpha)$. In other words, the nonuniqueness captured by saying extensive measures form a ratio scale can, at least in this case, be given an intrinsic characterization in terms of automorphisms of the empirical structure with itself. This fact, as we have seen in Chapters 13, 19, 20, and the present one is of considerable importance.

As we know, the set of all automorphisms of a structure forms a mathematical group under function composition, with the identity map the identity of the group. Of course, automorphisms other than the identity do not always exist (e.g., the probability structures of Chapter 5 have none). In this situation, it is sometimes useful to consider something a bit more general than the automorphisms. A structure-preserving map of $\mathscr{A}$ *into* itself (i.e., a homomorphism of $\mathscr{A}$ into $\mathscr{A}$) is called an *endomorphism* of $\mathscr{A}$. Of course, each automorphism is an endomorphism, but the converse is not generally true. There are a number of situations, in particular when the representation is into but not onto the real numbers, when endomorphisms are more useful than automorphisms. For one thing, they often exist in abundance when there are no nontrivial automorphisms.

Automorphisms have had two distinct but related roles in measurement theory. The first is in providing an intrinsic characterization of families of numerical structures that are candidates for representing qualitative measurement structures. The key to this work is the (nonobvious) fact, explored in Chapter 20, that for structures exhibiting both finite uniqueness (i.e., specifying the values at some N points uniquely determines the isomorphism) and a lot of symmetry (homogeneity of degree at least 1), the possibilities for the automorphism group are rather limited. The three most familiar cases are, of course: ordinal, in which the transformation group is all strictly increasing functions; interval, in which it is all positive affine transformations (on Re^+, $x \to rx^s$, $r > 0$, $s > 0$); and ratio, in which it is all similarity transformations (on Re^+, $x \to rx$, $r > 0$). For finitely unique, homogeneous structures that can be mapped onto the real numbers or are concatenation structures, the ratio, the interval, and some cases lying between these two are the only possibilities. It was results such as this that led to studying all homogeneous structures in which the automorphisms with no fixed point (and which are called translations) form an Archimedean-ordered group. This class of structures was seen to provide a natural extension of the dimensional structure of physics. Thus, for those cases in which the automorphism groups are very rich, a great deal is known. But as the automorphisms become sparse, so too does our knowledge.

The work just mentioned illustrates one of the advantages of an abstract, axiomatic approach to measurement. It has clarified concepts such as uniqueness and scale type about which there was some confusion until recently.

22.9.3 Invariance under Automorphisms

The other role of automorphisms, which is common throughout mathematics, has to do with relations that are invariant under automorphisms of a structure. Observe that, by definition, an automorphism α must have the property that for each j in J and each $a_1, \ldots, a_{k(j)}$ in A,

$$(a_1, \ldots, a_{k(j)}) \in S_j \quad \text{iff} \quad (\alpha(a_1), \ldots, \alpha(a_{k(j)})) \in S_j.$$

By definition the primitives or defining relations of the structure are invariant under the automorphisms of the structure. It therefore follows that anything that can be explicitly formulated just in terms of the primitives must also have the property of being invariant under the automorphisms of the structure. This observation has led to considerable interest in the classes of "things" that exhibit such invariance. For example, in Section 10.3 we were interested in relations that are dimensionally invariant in the sense of being invariant under so-called similarity transformations of the structure of physical quantities. As we just saw in Section 22.7, this invariance can in fact be viewed as a special case of invariance under automorphisms. And in geometry during the 19th century, it was proposed that a geometrical object be defined as a relation of the structure that is invariant under the automorphisms of the geometry in question (Section 12.2.5).

This invariance approach to meaningfulness, to understanding something about what it means to accept a concept as being within a structure or a theory, appears to give a nontrivial necessary condition for at least some homogeneous structures. But it is certainly unsatisfactory in the nonhomogeneous case. The difficulty can be summarized as follows: In a structure such as probability that has no nontrivial automorphisms, invariance under automorphisms lacks any bite. On the other hand, we know by the example given in Section 22.5 that even for homogeneous structures invariance is broader than definability, and we do not know if other easily applied restrictions on invariance can be found that characterize definability in ordered relational structures.

A different stance is possible. One could accept the invariance definition of meaningful as adequate for measurement structures and conclude that, though almost everything in a highly irregular structure is meaningful, little about it is of interest. For example, we can make IQ measurement conventionally into an absolute scale (only the trivial automorphism) by adopting the usual scoring conventions. A statement that "x is twice as intelligent as y" becomes meaningful in light of these conventions: it simply means that the pair of performances of x and y on the test is a member of some large but well-defined set of pairs such that the left-hand member of any pair gets a score twice that of the right-hand member. The difficulty is that such an assertion is of very little scientific interest.

This case should be contrasted with probability, which is also an absolute scale but which also exhibits a great deal of regularity. In that structure a statement that $P(A) = \frac{1}{3}$ is not only meaningful but of interest in some contexts.

There probably are important connections between the problem of defining meaningfulness and that of formulating satisfactory theorems about the representation and uniqueness of structures when there are few or no automorphisms.

22.9.4 Constructability of Representations

As usually stated, representation theorems simply assert the existence of isomorphisms or homomorphisms. Just as in the case of existence theorems for systems of equations, algebraic, differential, and functional, representation theorems provide reassurance but little help to anyone who needs to construct the representation. So the question arises, does the proof of a representation provide aid in finding actual homomorphisms? We make several observations on the matter.

First, it is often possible and sometimes easy to prove the existence of a homomorphism by using the Cantor-Birkhoff result (Theorem 2, Chapter 2). The basic technique is to show how to construct a countable order-dense subset. The construction draws on the axioms of the structure. Once established, this technique yields an order-preserving representation that, by taking limits, can be extended to the entire structure. In terms of that function, one can construct numerical relations that correspond to the empirical ones. For example, suppose $\langle A, \succsim, \bigcirc \rangle$ is a structure that satisfies all of the axioms of an extensive structure except for associativity. As we saw in Exercises 7–9 in Chapter 19 (or Narens and Luce, 1976), it is possible to construct a countable, order-dense subset, yielding the existence of a homomorphism φ of $\langle A \succsim \rangle$ into $\langle \mathrm{Re}^+, \geqslant \rangle$. Define the numerical operation $\odot$ as follows: for each x, y in Re^+ for which there exist a, b in A such that $\varphi(a) = x$, $\varphi(b) = y$ and $a \bigcirc b$ is defined, let

$$x \odot y = \varphi[\varphi^{-1}(x) \bigcirc \varphi^{-1}(y)] = \varphi(a \bigcirc b).$$

It is not difficult to prove that $\odot$ has just the correct properties and so φ represents $\langle A, \succsim, \bigcirc \rangle$ in $\langle \mathrm{Re}^+, \geqslant, \odot \rangle$.

Although reassuring, this result is not very helpful because the Cantor-Birkhoff construction does not provide a practical way to approximate the value of the homomorphism at any point to a prescribed level of accuracy. The reason is that although an arbitrarily chosen element from A can be placed in intervals bounded by elements from the countable dense subset, one does not know in advance how far out in the count one must go to achieve the prescribed level of accuracy. Moreover, the process is inherently unstable in the sense that deleting or adding an element to the countable order-dense subset can have a profound effect on the numerical values assigned.

These difficulties do not arise in extensive measurement when we base it on standard sequences. Then the representation of any element can be approximated with a prescribed degree of accuracy in a number of steps proportional to the reciprocal of the error, and the ratio of two values so estimated is independent of the element chosen to be the unit. The advantages of having such a constructive technique are considerable.

Alternatively, if one knew the operation $\odot$ in advance, it might be possible to devise an approximation procedure based on solving systems of inequalities, as was done in Chapter 9 for finite, additive conjoint structures. But there we knew that the operation was + whereas in the case of generalized structures the operation is unknown, and it too has to be constructed.

The challenge to the theorist is increased by demanding a useful, constructive method for finding the representation. Often the problem of constructability of a representation arises simultaneously with establishing the degree of uniqueness of the representation, though not the class of permissible transformations. That is, for some N, N-point uniqueness is established, but the degree of homogeneity is not. In particular, if the values of the isomorphism are selected for N arbitrary but distinct points of the structure, and we show how to determine it for any $N + 1$st point, then we have simultaneously proved N-point uniqueness and provided a constructive method. From a purely mathematical point of view, it is often an extra burden to provide such constructions because they are considerably more complex to follow than those based on the Cantor-Birkhoff result.

Successful examples of such constructions were provided in Sections 2.2.4, 16.7, and 19.5.4. We have failed, though not for lack of trying, to find such a construction for the axiomatization of a general metric with additive segments (Sections 14.2 and 14.3) first given by Beals and Krantz (1967). The reason for this difference is subtle. In the standard-sequence constructions that have worked, a specifiable set of inequalities is sufficient to determine the value of an element at a prescribed level of accuracy. In the metric axiomatization, one needs to isolate elements lying on an additive segment between two other elements. The approaches to this so far attempted require a potentially infinite set of inequalities. What is needed is a formal method that makes use of only a small part of the iso-similarity contour that is being approximated, but that has not been achieved.

Is this effort worth the pain? Can use be made of a method of constructing approximations to the homomorphism and numerical relations? We do not know the answer. It is clear that the construction of extensive measurement was, in fact, the basis of all early measurement of length, mass, and time. Often, now, combinations of physical laws are used to reduce measurement in one domain to that in another thereby bypassing direct use of standard sequences. Such indirect measurement is certainly not a possibility now for structures with inherently nonadditive representations, because the numerical operation is not fully specified. Additional, specific laws that will be reflected as functional equations must be uncovered before that route can be followed. Moreover, nothing has yet been done with the constructive methods provided for nonassociative structures. It may well be that these methods of approximation will require somewhat more complex empirical constructions than those used for extensive structures, and so whether they will actually be carried out depends upon how badly one wants either to test a theory or to measure something precisely.

When we turn to structures with neither operations nor a means of defining them, as for conjoint structures, the kinds of constructive methods involving standard sequences no longer exist. Mathematically, the tack taken in the case of homogeneous, Archimedean-ordered translations is to use the homogeneity of the translations to map the structure onto its translation group and then use Hölder's theorem to map that into the positive real numbers. This method is potentially constructive only if it is possible to characterize the translation group in a constructive fashion. For positive concatenation structures, that is possible via the *n*-copy operators, but in general we do not have an explicit way to construct that group.

For one class of situations, constructing a representation is especially easy. In these cases, the underlying structure includes some rich numerical information, and the representation theorem asserts that a homomorphism is some explicit function, with only a few free parameters, of the given numerical properties. In that case, only the parameters need be estimated from the data. We encountered one example of such a system in Section 3.14.1 where the underlying objects were probability distributions, the attribute was riskiness, and the representation was a weighted average of mean and variance. A more complex example in the same domain appears in Luce and Weber (1986). Their representation involves seven parameters of quantities calculated separately for the positive, zero, and negative outcomes. Another example can be found in Chapter 15 on color measurement, where, again, the stimuli are described in terms of distributions over wavelength. These more specific representations are both more testable and more easily rejected, and they probably should be our goal in those cases in which (partial) numerical descriptions of the objects being measured are naturally available. But caution needs to be exercised to avoid confusion regarding the status of this numerical information. We have already discussed a related issue in Section 22.7.5 on index measurement.

It should be noted in closing that, as outlined in Section 21.8.3, there exist in the foundations literature some attempts to make more explicit what is meant by a construction, but these have not yet had much impact on research into the foundations of measurement.

EXERCISES

1. Let $U = \{(x, y) | x - (3/2)y \geqslant 0\}$, $A = (1, 2, 3, 3\frac{1}{2}\}$, $S_1 = U_{|A}$, $\mathscr{A} = \langle A, S_1 \rangle$, $R = \{1, 2, 3\}$, $T_1 = U|_R$, $\mathscr{R} = \langle R, T_1 \rangle$. Show that $\Phi(\mathscr{A}, \mathscr{R})$ contains exactly one element φ and that φ is not 1:1. (22.2.2)

2. Prove Theorem 1. (22.2.2)

3. Suppose $\mathscr{A}, \mathscr{R}$ are relational structures of the same type and let $\Phi(\mathscr{A}, \mathscr{R})$ be the set of homomorphisms from $\mathscr{A}$ into $\mathscr{R}$. Let $\approx$ denote the congruence relation of $\mathscr{A}$ and let $\mathbf{a}$ denote the equivalence class under $\approx$

containing a. Prove:

(i) The mapping η: $a \to \mathbf{a}$ is a homomorphism of $\mathscr{A}$ onto $\mathscr{A}/\approx$.

(ii) If $\boldsymbol{\varphi}$ is a homomorphism of $\mathscr{A}/\approx$ into $\mathscr{R}$, then

(a) $\boldsymbol{\varphi}$ is 1:1.

(b) $\boldsymbol{\varphi}\eta$ is in $\Phi(\mathscr{A}, \mathscr{R})$.

(c) $\boldsymbol{\varphi}\eta(a) = \boldsymbol{\varphi}\eta(b)$ iff $a \approx b$.

(iii) Conversely, if $\varphi \in \Phi(\mathscr{A}, \mathscr{R})$ is such that $a \approx a'$ implies $\varphi(a) = \varphi(a')$, then $\boldsymbol{\varphi}(\mathbf{a}) = \varphi(a)$ defines a homomorphism $\boldsymbol{\varphi}$ of $\mathscr{A}/\approx$ into $\mathscr{R}$ such that $\boldsymbol{\varphi}\eta = \varphi$.

(iv) The relation $=$ on R is $\mathscr{A}/\approx$-reference invariant relative to $\mathscr{R}$, the induced relation $S(=)$ is just equality on $A/\approx$, and $S(\boldsymbol{\varphi}\eta, =)$ is $\approx$ on A. (22.2.6)

4. Suppose $\mathscr{A}, \mathscr{R}$ are relational structures with nonempty $\Phi(\mathscr{A}, \mathscr{R})$, that S_1 is a weak order $\succsim$, and T_1 is $\geqslant$. Prove that $\approx$ includes $\sim$. (22.2.6)

5. In the measurement structure of Exercise 1, show that $=$ on R is $\mathscr{A}$-reference invariant relative to $\mathscr{R}$ and find $S(=)$. (22.2.6)

6. Prove that if $\Phi(\mathscr{A}, \mathscr{R})$ contains exactly one element, then that element is either an irreducible homomorphism or an isomorphism. (Hint: if $a \approx a'$ and $\varphi(a) \neq \varphi(a')$, construct φ' with the values at a and a' interchanged.) (22.2.6)

7. Prove that if $=$ on R is $\mathscr{A}$-reference invariant relative to $\mathscr{R}$, then either every homomorphism is irreducible or every homomorphism is an isomorphism. (22.2.6)

8. Prove Theorem 2. (22.3.1)

9. Construct an example of a measurement structure for which there is a relation on A that is structure invariant in $\mathscr{A}$ and that is not induced by any relation T on R that is $\mathscr{A}$-reference invariant relative to $\mathscr{R}$. (Hint: pay attention to Theorem 5.) (22.3.3)

10. Prove Theorem 8. (22.7.3)

References

Aczél, J. (1966). *Lectures on functional equations and their applications*. New York: Academic Press.

Aczél, J. (Ed.) (1984). *Functional equations: History, applications, and theory*. Dordrecht, Netherlands: Reidel.

Adams, E. W., Fagot, R. F., & Robinson, R. E. (1965). A theory of appropriate statistics. *Psychometrika*, **30**, 99–127.

Adams, E. W., Fagot, R. F., & Robinson, R. E. (1970). On the empirical status of axioms in theories of fundamental measurement. *Journal of Mathematical Psychology*, **7**, 379–409.

Albeverio, S., Fenstad, J. E., Hoegh-Krohn, R., & Lindstrom, T. (1986). *Nonstandard methods in stochastic analysis and mathematical physics*. New York: Academic Press.

Allais, M. (1953). Le comportement de l'homme rationnel devant le risque: Critique des postulats et axiomes de l'école Américaine. *Econometrica*, **21**, 503–546.

Allais, M., & Hagen, O. (Eds.) (1979). *Expected utility hypothesis and the Allais paradox*. Dordrecht, Netherlands: Reidel.

Allen, R. D. G. (1956). *Mathematical economics*. London: Macmillan.

Alper, T. M. (1984). *Groups of homeomorphisms of the real line*. Unpublished AB thesis, Harvard University, Cambridge, MA.

Alper, T. M. (1985). A note on real measurement structures of scale type (m, m + 1). *Journal of Mathematical Physchology*, **29**, 73–81.

Alper, T. M. (1987). A classification of all order-preserving homeomorphism groups of the reals that satisfy finite uniqueness. *Journal of Mathematical Psychology*, **31**, 135–154.

Anderson, N. H. (1961). Scales and statistics: Parametric and nonparametric. *Psychological Bulletin*, **58**, 305–316.

Arrow, K. J. (1982). Risk perception in psychology and economics. *Economic Inquiry*, **20**, 1–9.

Barwise, J. (Ed.) (1977). *Handbook of mathematical logic*. Amsterdam: North-Holland.

Beals, R., & Krantz, D. H. (1967). Metrics and geodesics induced by order relations. *Mathematische Zeitschrift*, **101**, 285–298.

Behan, F. L., & Behan, R. A. (1954). Football numbers (continued). *American Psychologist*, **9**, 262–263.
Beth, E. W. (1953). On Padoa's method in the theory of definition. *Indagationes Mathematicae*, **15**, 330–339.
Borcherding, K., Brehmer, B., Vlek, C., & Wagenaar, W. A. (Eds.) (1984). *Research perspectives on decision making under uncertainty*. Amsterdam: North-Holland.
Burke, C. J. (1953). Additive scales and statistics. *Psychological Review*, **60**, 73–75. (Reprinted in M. H. Marx (Ed.), *Theories in contemporary psychology* (pp. 147–149). New York: MacMillan, 1963).
Busemann, H. (1955). *The geometry of geodesics*. New York: Academic Press.
Cameron, P. J. (1989). Groups of order-automorphisms of the rationals with prescribed scale type. *Journal of Mathematical Psychology*, **33**, 163–171.
Campbell, N. R. (1920). *Physics: The elements*. London & New York: Cambridge University Press. (Reprinted as *Foundations of science: The philosophy of theory and experiment*. New York: Dover, 1957).
Campbell, N. R. (1928). *An account of the principles of measurement and calculation*. London: Longmans, Green.
Chang, C. C., & Keisler, H. J. (1977). *Model theory* (2nd ed.). Amsterdam: Elsevier/North-Holland.
Chew, S. H. (1983). A generalization of the quasilinear mean with applications to measurement of income inequality and decision theory resolving the Allais paradox. *Econometrica*, **51**, 1065–1092.
Cliff, N. (1982). What is and isn't measurement. In G. Keran (Ed.), *Statistical and methodological issues in psychology and social science research* (pp. 3–38). Hillsdale, NJ: Erlbaum.
Cohen, M., & Narens, L. (1979). Fundamental unit structures: a theory of ratio scalability. *Journal of Mathematical Psychology*, **20**, 193–232.
Cohen, M. A. (1988). Dedekind completion of positive concatenation structures: Necessary and sufficient conditions. *Journal of Mathematical Psychology*, **32**, 64–71.
Davies, P. M., & Coxon, A. P. M. (Eds.) (1982). *Key texts in multidimensional scaling*. Exeter, NH: Heinemann Educational Books.
Dedekind, R. (1872/1901). *Stetigkeit und irrationale Zahlen*. Brunswick, 1872. [*Essays on the theory of numbers*. New York: Dover, 1963 (reprint of the English translation).]
Dhombres, J. (1979). *Some aspects of functional equations* (Lecture Notes). Department of Mathematics, Chulalongkorn University, Bangkok, Thailand.
Droste, M. (1987). Ordinal scales in the theory of measurement. *Journal of Mathematical Psychology*, **31**, 60–82.
Edwards, W., & Tversky, A. (Eds.) (1967). *Decision making*. Harmondsworth, England: Penguin Books.
Ehrenfeucht, A. (1961). An application of games to the completeness problem for formalized theories. *Fundamenta Mathematicae*, **49**, 129–141.
Ehrenfeucht, A., & Mostowski, A. (1957). Models of axiomatic theories admitting automorphisms. *Fundamenta Mathematicae*, **43**, 50–68.
Ellsberg, D. (1961). Risk, ambiguity, and the Savage axioms. *Quarterly Journal of Economics*, **75**, 643–669.
Falmagne, J.-C., & Narens, L. (1983). Scales and meaningfulness of quantitative laws. *Synthese*, **55**, 287–325.
Ferguson, A., Meyers, C. S. (Vice Chairman), Bartlett, R. J. (Secretary), Banister, H., Bartlett, F. C., Brown, W., Campbell, N. R., Craik, K. J. W., Drever, J., Guild, J., Houstoun, R. A., Irwin, J. O., Kaye, G. W. C., Philpott, S. J. F., Richardson, L. F., Shaxby, J. H., Smith, T., Thouless, R. H., & Tucker, W. S. (1940). Quantitative estimates of sensory events. The

advancement of science. *Report of the British Association for the Advancement of Science*, **2**, 331–349.

Fishburn, P. C. (1981a). Subjective expected utility: A review of normative theories. *Theory and Decision*, **13**, 139–199.

Fishburn, P. C. (1981b). Restricted thresholds for interval orders: A case of nonaxiomatizability by a universal sentence. *Journal of Mathematical Psychology*, **24**, 276–283.

Fraïssé, R. J. (1955). Sur quelques classifications des relations, basées sur des isomorphismes restreints. I. Étude générale. II. Application aux relations d'ordre, et construction d'exemples montrant que ces classifiations sont distinctes. *Publications Scientifiques de l'Université d'Alger, Série A*, **2**, 15–60, 273–295.

Fuchs, L. (1963). *Partially ordered algebraic systems*. Reading, MA: Addison-Wesley.

Gaito, J. (1959). Nonparametric methods in psychological research. *Psychological Reports*, **5**, 115–125.

Gaito, J. (1980). Measurement scales and statistics: Resurgence of an old misconception. *Psychological Bulletin*, **87**, 564–567.

Gilboa, I. (1987). Expected utility with purely subjective nonadditive probabilities. *Journal of Mathematical Economics*, **16**, 65–88.

Glass, A. M. W. (1981). *Ordered permutation groups*. London & New York: Cambridge University Press.

Gleason, A. M. (c1959). Unpublished research.

Gödel, K. (1930). Die Vollständigkeit der Axiome des logischen Funktionenkalküls. *Monatshefte für Mathematik und Physik*, **37**, 349–360.

Hilbert, D. (1899). *Grundlagen der Geometrie* (8th ed., with revisions and supplements by P. Bernays, 1956). Stuttgart: Teubner.

Hölder, O. (1901). Die Axiome der Quantität und die Lehre vom Mass. *Berichte über die Verhandlungen der Königlich Sächsischen Gesellschaft der Wissenschaften zu Leipzig, Mathematische-Physische Klasse*, **53**, 1–64.

Holman, E. (1971). A note on additive conjoint measurement. *Journal of Mathematical Psychology*, **8**, 489–494.

Humphreys, L. (1964). Review of *Introduction to psychological research*, W. A. Scott & M. Wertheimer, 1962. *Contemporary Psychology*, **9**, 76.

Jameson, D., & Hurvich, L. M. (1955). Some quantitative aspects of an opponent-colors theory. I. Chromatic responses and spectral saturation. *Journal of the Optical Society of America*, **45**, 546–552.

Kahneman, D., & Tversky, A. (1979). Prospect theory: An analysis of decision under risk. *Econometrica*, **47**, 263–291.

Kaplan, M. F., & Schwartz, S. (Eds.) (1975). *Human judgment and decision processes*. New York: Academic Press.

Keisler, H. J. (1977). Fundamentals of model theory. In J. Barwise (Ed.), *Handbook of mathematical logic* (pp. 47–103). Amsterdam: North-Holland.

Kelley, J. L. (1955). *General topology*. Princeton, NJ: Van Nostrand.

Kleene, S. C. (1952). *Introduction to metamathematics*. Princeton, NJ: Van Nostrand.

Kleene, S. C. (1967). *Mathematical logic*. New York: Wiley.

Krantz, D. H. (1973). Measurement-free tests of linearity in biological systems. *IEEE Transactions on Systems, Man, and Cybernetics*, **SMC-3**, 266–271.

Kreisel, G. (1954). Applications of mathematical logic to various branches of mathematics. *Applications scientifique de la logique mathématique* (pp. 37–50). Paris: Gauthier–Villars.

Levine, M. V. (1970). Transformations that render curves parallel. *Journal of Mathematical Psychology*, **7**, 410–443.

Levine, M. V. (1972). Transforming curves into curves with the same shape. *Journal of Mathematical Psychology*, **9**, 1–16.

Lord, F. M. (1953). On the statistical treatment of football numbers. *American Psychologist*, **8,** 750–751.
Löwenheim, L. (1915). Über Möglichkeiten im Relativkalkül. *Mathematische Annalen*, **76,** 447–470.
Luce, R. D. (1959). On the possible psychophysical laws. *Psychological Review*, **66,** 81–95.
Luce, R. D. (1962). Comments on Rozeboom's criticisms of "On the possible psychophysical laws." *Psychological Review*, **69,** 548–551.
Luce, R. D. (1967). Remarks on the theory of measurement and its relation to psychology. In *Les modèles et la formalization du comportement* (pp. 27–42). Paris: Éditions du Centre National de la Recherche Scientifique.
Luce, R. D. (1971). Similar systems and dimensionally invariant laws. *Philosophy of Science*, **38,** 157–169.
Luce, R. D. (1978). Dimensionally invariant numerical laws correspond to meaningful qualitative relations. *Philosophy of Science*, **45,** 1–16.
Luce, R. D. (1986a). *Response times*. London & New York: Oxford University Press.
Luce, R. D. (1986b). Uniqueness and homogeneity of ordered relational structures. *Journal of Mathematical Psychology*, **30,** 391–415.
Luce, R. D. (1987). Measurement structures with Archimedean ordered translation groups. *Order*, **4,** 165–189.
Luce, R. D. (1988). Rank-dependent, subjective-expected utility representations. *Journal of Risk and Uncertainty*, **1**, 305–332.
Luce, R. D., & Cohen, M. (1983). Factorizable automorphisms in solvable conjoint structures. I. *Journal of Pure and Applied Algebra*, **27,** 225–261.
Luce, R. D., & Narens, L. (1983). Symmetry, scale types, and generalizations of classical physical measurement. *Journal of Mathematical Psychology*, **27,** 44–85.
Luce, R. D., & Narens, L. (1985). Classification of concatenation structures by scale type. *Journal of Mathematical Psychology*, **29,** 1–72.
Luce, R. D., & Narens, L. (1992). Intrinsic Archimedeanness and the continuum. In C.W. Savage & P. Ehrlich (Eds.) *Philosophical and Foundational Issues in Measurement Theory*. Hillsdale, NJ: Lawrence Erlbaum Associates. pp. 15–38.
Luce, R. D., & Weber, E. U. (1986). An axiomatic theory of conjoint, expected risk. *Journal of Mathematical Psychology*, **30,** 188–205.
Machina, M. J. (1987). Choice under uncertainty: Problems solved and unsolved. *Economic Perspectives*, **1,** 121–154.
Makkai, M. (1977). Admissible sets and infinitary logic. In J. Barwise (Ed.), *Handbook of mathematical logic* (pp. 233–281). Amsterdam: North-Holland.
Manders, K. (1979). The theory of all substructures of a structure. *Journal of Symbolic Logic*, **44,** 583–598.
McKinsey, J. C. C., & Suppes, P. (1955). On the notion of invariance in classical mechanics. *British Journal for Philosophy of Science*, **5,** 290–302.
Mendelson, E. (1964). *Introduction to mathematical logic*. New York: Van Nostrand.
Michell, J. (1986). Measurement scales and statistics: A clash of paradigms. *Psychological Bulletin*, **100,** 398–407.
Moler, N., & Suppes, P. (1968). Quantifier-free axioms for constructive plane geometry. *Compositio Mathematica*, **20,** 143–152.
Monk, J. D. (1975). *Mathematical logic*. London & New York: Springer-Verlag.
Narens, L. (1974). Measurement without Archimedean axioms. *Philosophy of Science*, **41,** 374–393.
Narens, L. (1976). Utility-uncertainty trade-off structures. *Journal of Mathematical Psychology*, **13,** 296–322.
Narens, L. (1980). A note on Weber's law for conjoint structures. *Journal of Mathematical Psychology*, **21,** 88–92.

Narens, L. (1981a). A general theory of ratio scalability with remarks about the measurement-theoretic concept of meaningfulness. *Theory and Decision*, **13,** 1–70.

Narens, L. (1981b). On the scales of measurement. *Journal of Mathematical Psychology*, **24,** 249–275.

Narens, L. (1985). *Abstract measurement theory*. Cambridge, MA: MIT Press.

Narens, L. (2002). *Theories of Meaningfulness*. Mahwah, NJ: Lawrence Erlbaum Associates.

Narens, L., & Luce, R. D. (1976). The algebra of measurement. *Journal of Pure and Applied Algebra*, **8,** 197–233.

Narens, L., & Luce, R. D. (1986). Measurement: The theory of numerical assignments. *Psychological Bulletin*, **99,** 166–180.

Niederée, R. (1987). On the reference to real numbers in fundamental measurement: A model-theoretic approach. In E. E. Roskam & R. Suck (Eds.), *Progress in mathematical psychology* (Vol. 1, pp. 3–23). New York, Amsterdam: Elsevier.

Padoa, A. (1902). Un nouveau système irreductible de postulats pour l'algèbre. *Compte Rendu du deuxième Congres International de Mathématiciens,* pp. 249–256.

Pfanzagl, J. (1968). *Theory of measurement*. New York: Wiley.

Pfanzagl, J. (1971). *Theory of measurement*. New York: Wiley.

Pollatsek, A., & Tversky, A. (1970). A theory of risk. *Journal of Mathematical Psychology*, **7,** 540–553.

Posner, M. I. (1978). *Chronometric explorations of mind*. Hillsdale, NJ: Erlbaum.

Presburger, M. (1930). Über die Völlstandigkeit eines gewissen Systems der Arithmetik ganzer Zahlen in welchem die Addition als einzige Operation hervortritt. *Comptes Rendus, 1er Congres des Mathématiciens des Pays Slaves 1929 (Warsaw),* 92–101, 395.

Quiggin, J. (1982). A theory of anticipated risk. *Journal of Economic Behavior and Organizations*, **3,** 323–343.

Rabin, M. O. (1977). Decidable theories. In J. Barwise (Ed.), *Handbook of mathematical logic* (pp. 595–629). Amsterdam: North-Holland.

Ramsay, J. O. (1975). Review of *Foundations of measurement*, Volume I. *Psychometrika*, **40,** 257–262.

Ramsay, J. O. (1976). Algebraic representation in physical and behavioral sciences. *Synthese*, **32,** 419–453.

Ramsey, F. P. (1931). *The foundations of mathematics and other logical essays*. New York: Harcourt, Brace.

Roberts, F. S. (1968). *Representations of indifference relations*. Unpublished doctoral dissertation, Stanford University, Stanford, CA.

Roberts, F. S. (1969). Indifference graphs. In F. Harary (Ed.), *Proof techniques in graph theory* (pp. 139–146). New York: Academic Press.

Roberts, F. S. (1980). On Luce's theory of meaningfulness. *Philosophy of Science*, **47,** 424–433.

Roberts, F. S. (1984a). On the theory of meaningfulness of ordinal comparisons. *Measurement*, **2,** 35–38.

Roberts, F. S. (1984b). Applications of the theory of meaningfulness to order and matching experiments. In E. Degree & J. Van Buggenhaut (Eds.), *Trends in mathematical psychology* (pp. 283–292). Amsterdam: Elsevier/North-Holland.

Roberts, F. S., & Franke, C. H. (1976). On the theory of uniqueness in measurement. *Journal of Mathematical Psychology*, **14,** 211–218.

Roberts, F. S., & Rosenbaum, Z. (1985). Some results on automorphisms of valued digraphs and the theory of scale type in measurement. In Y. Alavi et al. (Eds.), *Proceedings of the Fifth International Conference on Graph Theory and Its Applications* (pp. 659–669). New York: Wiley.

Roberts, F. S., & Rosenbaum, Z. (1988). Tight and loose value automorphisms. *Discrete Applied Mathematics*, **22,** 169–179.

Robinson, A. (1951). *On the metamathematics of algebra.* Amsterdam: North-Holland.

Robinson, A. (1961). Non-standard analysis. *Proceedings of the Koninklijke Nederlandse Akademie van Wetenschappen*, **64**, 432–440.

Robinson, A. (1963). *Introduction to model theory and to the metamathematics of algebra.* Amsterdam: North-Holland.

Robinson, R. E. (1963). *A set theoretical approach to empirical meaningfulness of empirical statements* (Tech. Rep. 55, Psych. & Educ. Ser.). Stanford CA: Stanford University, Institute for Mathematical Studies in the Social Sciences.

Rosen, J. (1975). *Symmetry discovered.* London & New York: Cambridge University Press.

Rosen, J. (1983). *A symmetry primer for scientists.* New York: Wiley.

Rosenstein, J. G. (1982). *Linear orderings.* New York: Academic Press.

Rozeboom, W. W. (1962). The untenability of Luce's principle. *Psychological Review*, **69**, 542–547, 552.

Ryll-Nardzewski, C. (1952). The role of the axiom of induction in elementary arithmetic. *Fundamenta Mathematicae*, **39**, 239–263.

Savage, I. R. (1957). Nonparametric statistics. *Journal of the American Statistical Association*, **52**, 331–344.

Scheffé, H. (1959). *The analysis of variance.* New York: Wiley.

Schoemaker, P. J. H. (1980). *Experiments on decisions under risk: The expected utility hypothesis.* Beston, MA: Kluwer-Nijhoff.

Scott, D., & Suppes, P. (1958). Foundational aspects of theories of measurement. *Journal of Symbolic Logic*, **23**, 113–128.

Segal, U. (1987). The Ellsberg paradox and risk aversion: An anticipated utility approach. *International Economic Review*, **28**, 175–202.

Senders, V. L. (1953). A comment on Burke's additive scales and statistics. *Psychological Review*, **60**, 423–424.

Senders, V. L. (1958). *Measurement and statistics.* London & New York: Oxford University Press.

Shepard, R. N. (1962a). The analysis of proximities: Multidimensional scaling with an unknown distance function. I. *Psychometrika*, **27**, 125–140.

Shepard, R. N. (1962b). The analysis of proximities: Multidimensional scaling with an unknown distance function. II. *Psychometrika*, **27**, 219–246.

Shepard, R. N. (1974). Representation of structure in similarity data: Problems and prospects. *Psychometrika*, **39**, 373–421.

Sheperdson, J. C. (1965). Non-standard models for fragments of number theory. In J. W. Addison, L. Henkin, & A. Tarski (Eds.), *The theory of models* (pp. 342–358). Amsterdam: North-Holland.

Shoenfield, J. R. (1967). *Mathematical Logic.* Reading, MA: Addison-Wesley.

Siegel, S. (1956). *Nonparametric statistics.* New York: McGraw-Hill.

Skala, H. J. (1975). *Non-Archimedean utility theory.* Dordrecht, Netherlands: Reidel.

Śkołem, T. (1920). Logisch-kombinatorische Untersuchungen über die Erfüllbarkeit oder Beweisbarkeit mathematischer Sätze nebst einem Theorem über dichte Mengen. *Skrifter utgitt av Videnskabsselskabet, I. Mathematisk Naturvidenskapelig Klasse,* **4**, 36 pp.

Śkołem, T. (1955). Peano's axioms and models of arithmetic. In T. Śkołem, G. Hasenjaegar, G. Kreisel, A. Robinson, H. Wang, & J. Łoś (Eds.), *Mathematical interpretations of formal systems* (pp. 1–14). Amsterdam: North-Holland.

Sternberg, S. (1969). Memory scanning: Mental processes revealed by reaction-time experiments. *American Scientist*, **57**, 421–457. Reprinted in J. S. Antrobus (Ed.), *Cognition and affect* (pp. 13–58). Boston, MA: Little, Brown, 1970.

Stevens, S. S. (1946). On the theory of scales of measurement. *Science*, **103**, 677–680.

Stevens, S. S. (1951). Mathematics, measurement and psychophysics. In S. S. Stevens (Ed.), *Handbook of experimental psychology* (pp. 1–49). New York: Wiley.

Stevens, S. S. (1959). Measurement, psychophysics and utility. In C. W. Churchman & P. Ratoosh (Eds.), *Measurement: Definitions and theories* (pp. 18–63). New York: Wiley.

Stevens, S. S. (1968). Measurements, statistics, and the schemapiric view. *Science*, **161**, 849–856.

Stevens, S. S. (1974). S. S. Stevens. In G. Lindzey (Ed.), *A history of psychology in autobiography* (Vol. 6, pp. 395–420). Englewood Cliffs, NJ: Prentice-Hall.

Stevens, S. S., & Davis, H. (1938). *Hearing: Its psychology and physiology*. New York: Wiley.

Suppes, P. (1957). *Introduction to logic*. New York: Van Nostrand.

Suppes, P. (1959). Measurement, empirical meaningfulness, and three-valued logic. In C. W. Churchman & P. Ratoosh (Eds.), *Measurement: Definitions and theories* (pp. 129–143). New York: Wiley.

Suppes, P. (1960). *Axiomatic set theory*. New York: Van Nostrand. (Revised edition, Dover, 1972)

Suppes, P. (1962). Models of data. In E. Nagel, P. Suppes, & A. Tarski (Eds.), *Logic, methodology and philosophy of science: Proceedings of the 1960 international congress* (pp. 252–261). Stanford, CA: Stanford University Press. [Czechoslovakian translation: Modely dat. In K. Berka & L. Tondl (Eds.), *Teorie modelu a modelovani* (pp. 223–235). Dordrecht, Netherlands: Reidel, 1967; German translation: *Modelle von Daten*. In M. Galzer & W. Heidelberger (Eds.), *Zur Logik empirischer Theorien* (pp. 191–204). Berlin and New York: de Gruyter, 1983].

Suppes, P., & Zinnes, J. L. (1963). Basic measurement theory. In R. D. Luce, R. R. Bush, & E. Galanter (Eds.), *Handbook of mathematical psychology* (Vol. 1, pp. 1–76). New York: Wiley.

Szmielew, W. (1983). *From affine to Euclidean geometry: An axiomatic approach*. Boston: Reidel. (Translated by Mr. Moszynska from the Polish edition. Warszawa: PWN-Polish Scientific Publishers, 1981.)

Tait, W. W. (1959). A counterexample to a conjecture of Scott and Suppes. *Journal of Symbolic Logic*, **24**, 15–16.

Tarski, A. (1930a). Fundamentale Begriffe der Methodologie der deduktiven Wissenschaften. I. *Monatshefte für Mathematik und Physik*, **37**, 361–404. [Reprinted as Fundamental concepts of the methodology of the deductive sciences. In (J. H. Woodger, Trans.), *Logic, semantics, metamathematics: Papers from 1923 to 1938 by Alfred Tarski* (pp. 60–109). London & New York: Oxford University Press (Clarendon), 1956.]

Tarski, A. (1930b). O pojeciu prawdy w odniesieniu do sformalizowanych nauk dedukcyjnych. *Ruch Filozoficzny*, **12**, 210–311. [German translation: Der Wahrheitsbegriff in den formalisierten Sprachen. *Studia Philosophica*, 261–405 (1936) (reprint dated 1935). Reprinted as The concept of truth in formalized languages. In (J. H. Woodger, Trans.), *Logic, semantics, metamathematics: Papers from 1923 to 1938 by Alfred Tarski* (pp. 152–278). London & New York: Oxford University Press (Clarendon), 1956.]

Tarski, A. (1951). *A decision method for elementary algebra and geometry* (2nd ed., revised). Berkeley, CA: University of California Press.

Tarski, A. (1954). Contributions to the theory of models. I. II. *Indagationes Mathematicae*, **16**, 572–581, 582–588.

Tarski, A. (1955). Contributions to the theory of models. III. *Indagationes Mathematicae*, **17**, 56–64.

Titiev, R. J. (1972). Measurement structures in classes that are not universally axiomatizable. *Journal of Mathematical Psychology*, **9**, 200–205.

Titiev, R. J. (1980). Computer-assisted results about universal axiomatizability and the three-dimensional city-block metric. *Journal of Mathematical Psychology*, **22**, 209–217.

Townsend, J. T., & Ashby, G. T. (1984). Measurement scales and statistics: The misconception misconceived. *Psychological Bulletin*, **96**, 394–401.

Troelstra, A. S. (1977). Aspects of constructive mathematics. In J. Barwise (Ed.), *Handbook of mathematical logic* (pp. 973–1052). Amsterdam: North-Holland.

Tversky, A. (1969). Intransitivity of preferences. *Psychological Review*, **76**, 31–48.

Tversky, A., & Kahneman, D. (1986). Rational choice and the framing of decisions. *Journal of Business*, **59**, S251–S278.

Vaught, R. (1954). Remarks on universal classes of relational systems. *Indagationes Mathematicae*, **16**, 589–591.

Weber, M., & Camerer, C. (1987). Recent developments in modelling preferences under risk. *Operations Research Spektrum*, **9**, 129–151.

Weitzenhoffer, A. M. (1951). Mathematical structures and psychological measurements. *Psychometrika*, **16**, 387–406.

Welford, A. T. (1980). *Reaction times*. London: Academic Press.

Weyl, H. (1952). *Symmetry*. Princeton, NJ: Princeton University Press.

Yaari, M. E. (1987). The dual theory of choice under risk. *Econometrica*, **55**, 95–115.

Author Index

Numbers in italics refer to the pages on which the complete references are listed. The letter n following a page number indicates that the entry is cited in a footnote to that page.

N

P

Q

R

S

T

V

W

Y

Z

Subject Index

The letter n following a page number indicates that the entry is cited in a footnote to that page.

T

U

V

W

A CATALOG OF SELECTED

DOVER BOOKS

IN SCIENCE AND MATHEMATICS

Astronomy

BURNHAM'S CELESTIAL HANDBOOK, Robert Burnham, Jr. Thorough guide to the stars beyond our solar system. Exhaustive treatment. Alphabetical by constellation: Andromeda to Cetus in Vol. 1; Chamaeleon to Orion in Vol. 2; and Pavo to Vulpecula in Vol. 3. Hundreds of illustrations. Index in Vol. 3. 2,000pp. 6⅛ x 9¼.
Vol. I: 0-486-23567-X
Vol. II: 0-486-23568-8
Vol. III: 0-486-23673-0

EXPLORING THE MOON THROUGH BINOCULARS AND SMALL TELESCOPES, Ernest H. Cherrington, Jr. Informative, profusely illustrated guide to locating and identifying craters, rills, seas, mountains, other lunar features. Newly revised and updated with special section of new photos. Over 100 photos and diagrams. 240pp. 8¼ x 11. 0-486-24491-1

THE EXTRATERRESTRIAL LIFE DEBATE, 1750–1900, Michael J. Crowe. First detailed, scholarly study in English of the many ideas that developed from 1750 to 1900 regarding the existence of intelligent extraterrestrial life. Examines ideas of Kant, Herschel, Voltaire, Percival Lowell, many other scientists and thinkers. 16 illustrations. 704pp. 5⅜ x 8½. 0-486-40675-X

THEORIES OF THE WORLD FROM ANTIQUITY TO THE COPERNICAN REVOLUTION, Michael J. Crowe. Newly revised edition of an accessible, enlightening book recreates the change from an earth-centered to a sun-centered conception of the solar system. 242pp. 5⅜ x 8½. 0-486-41444-2

A HISTORY OF ASTRONOMY, A. Pannekoek. Well-balanced, carefully reasoned study covers such topics as Ptolemaic theory, work of Copernicus, Kepler, Newton, Eddington's work on stars, much more. Illustrated. References. 521pp. 5⅜ x 8½.
0-486-65994-1

A COMPLETE MANUAL OF AMATEUR ASTRONOMY: TOOLS AND TECHNIQUES FOR ASTRONOMICAL OBSERVATIONS, P. Clay Sherrod with Thomas L. Koed. Concise, highly readable book discusses: selecting, setting up and maintaining a telescope; amateur studies of the sun; lunar topography and occultations; observations of Mars, Jupiter, Saturn, the minor planets and the stars; an introduction to photoelectric photometry; more. 1981 ed. 124 figures. 25 halftones. 37 tables. 335pp. 6½ x 9¼. 0-486-40675-X

AMATEUR ASTRONOMER'S HANDBOOK, J. B. Sidgwick. Timeless, comprehensive coverage of telescopes, mirrors, lenses, mountings, telescope drives, micrometers, spectroscopes, more. 189 illustrations. 576pp. 5⅝ x 8¼. (Available in U.S. only.)
0-486-24034-7

STARS AND RELATIVITY, Ya. B. Zel'dovich and I. D. Novikov. Vol. 1 of *Relativistic Astrophysics* by famed Russian scientists. General relativity, properties of matter under astrophysical conditions, stars, and stellar systems. Deep physical insights, clear presentation. 1971 edition. References. 544pp. 5⅝ x 8¼. 0-486-69424-0

Chemistry

THE SCEPTICAL CHYMIST: THE CLASSIC 1661 TEXT, Robert Boyle. Boyle defines the term "element," asserting that all natural phenomena can be explained by the motion and organization of primary particles. 1911 ed. viii+232pp. 5⅜ x 8½.
0-486-42825-7

RADIOACTIVE SUBSTANCES, Marie Curie. Here is the celebrated scientist's doctoral thesis, the prelude to her receipt of the 1903 Nobel Prize. Curie discusses establishing atomic character of radioactivity found in compounds of uranium and thorium; extraction from pitchblende of polonium and radium; isolation of pure radium chloride; determination of atomic weight of radium; plus electric, photographic, luminous, heat, color effects of radioactivity. ii+94pp. 5⅜ x 8½. 0-486-42550-9

CHEMICAL MAGIC, Leonard A. Ford. Second Edition, Revised by E. Winston Grundmeier. Over 100 unusual stunts demonstrating cold fire, dust explosions, much more. Text explains scientific principles and stresses safety precautions. 128pp. 5⅜ x 8½. 0-486-67628-5

THE DEVELOPMENT OF MODERN CHEMISTRY, Aaron J. Ihde. Authoritative history of chemistry from ancient Greek theory to 20th-century innovation. Covers major chemists and their discoveries. 209 illustrations. 14 tables. Bibliographies. Indices. Appendices. 851pp. 5⅜ x 8½. 0-486-64235-6

CATALYSIS IN CHEMISTRY AND ENZYMOLOGY, William P. Jencks. Exceptionally clear coverage of mechanisms for catalysis, forces in aqueous solution, carbonyl- and acyl-group reactions, practical kinetics, more. 864pp. 5⅜ x 8½.
0-486-65460-5

ELEMENTS OF CHEMISTRY, Antoine Lavoisier. Monumental classic by founder of modern chemistry in remarkable reprint of rare 1790 Kerr translation. A must for every student of chemistry or the history of science. 539pp. 5⅜ x 8½. 0-486-64624-6

THE HISTORICAL BACKGROUND OF CHEMISTRY, Henry M. Leicester. Evolution of ideas, not individual biography. Concentrates on formulation of a coherent set of chemical laws. 260pp. 5⅜ x 8½. 0-486-61053-5

A SHORT HISTORY OF CHEMISTRY, J. R. Partington. Classic exposition explores origins of chemistry, alchemy, early medical chemistry, nature of atmosphere, theory of valency, laws and structure of atomic theory, much more. 428pp. 5⅜ x 8½. (Available in U.S. only.) 0-486-65977-1

GENERAL CHEMISTRY, Linus Pauling. Revised 3rd edition of classic first-year text by Nobel laureate. Atomic and molecular structure, quantum mechanics, statistical mechanics, thermodynamics correlated with descriptive chemistry. Problems. 992pp. 5⅜ x 8½. 0-486-65622-5

FROM ALCHEMY TO CHEMISTRY, John Read. Broad, humanistic treatment focuses on great figures of chemistry and ideas that revolutionized the science. 50 illustrations. 240pp. 5⅜ x 8½. 0-486-28690-8

Engineering

DE RE METALLICA, Georgius Agricola. The famous Hoover translation of greatest treatise on technological chemistry, engineering, geology, mining of early modern times (1556). All 289 original woodcuts. 638pp. 6¾ x 11. 0-486-60006-8

FUNDAMENTALS OF ASTRODYNAMICS, Roger Bate et al. Modern approach developed by U.S. Air Force Academy. Designed as a first course. Problems, exercises. Numerous illustrations. 455pp. 5⅜ x 8½. 0-486-60061-0

DYNAMICS OF FLUIDS IN POROUS MEDIA, Jacob Bear. For advanced students of ground water hydrology, soil mechanics and physics, drainage and irrigation engineering and more. 335 illustrations. Exercises, with answers. 784pp. 6⅛ x 9¼. 0-486-65675-6

THEORY OF VISCOELASTICITY (Second Edition), Richard M. Christensen. Complete consistent description of the linear theory of the viscoelastic behavior of materials. Problem-solving techniques discussed. 1982 edition. 29 figures. xiv+364pp. 6⅛ x 9¼. 0-486-42880-X

MECHANICS, J. P. Den Hartog. A classic introductory text or refresher. Hundreds of applications and design problems illuminate fundamentals of trusses, loaded beams and cables, etc. 334 answered problems. 462pp. 5⅜ x 8½. 0-486-60754-2

MECHANICAL VIBRATIONS, J. P. Den Hartog. Classic textbook offers lucid explanations and illustrative models, applying theories of vibrations to a variety of practical industrial engineering problems. Numerous figures. 233 problems, solutions. Appendix. Index. Preface. 436pp. 5⅜ x 8½. 0-486-64785-4

STRENGTH OF MATERIALS, J. P. Den Hartog. Full, clear treatment of basic material (tension, torsion, bending, etc.) plus advanced material on engineering methods, applications. 350 answered problems. 323pp. 5⅜ x 8½. 0-486-60755-0

A HISTORY OF MECHANICS, René Dugas. Monumental study of mechanical principles from antiquity to quantum mechanics. Contributions of ancient Greeks, Galileo, Leonardo, Kepler, Lagrange, many others. 671pp. 5⅜ x 8½. 0-486-65632-2

STABILITY THEORY AND ITS APPLICATIONS TO STRUCTURAL MECHANICS, Clive L. Dym. Self-contained text focuses on Koiter postbuckling analyses, with mathematical notions of stability of motion. Basing minimum energy principles for static stability upon dynamic concepts of stability of motion, it develops asymptotic buckling and postbuckling analyses from potential energy considerations, with applications to columns, plates, and arches. 1974 ed. 208pp. 5⅜ x 8½. 0-486-42541-X

METAL FATIGUE, N. E. Frost, K. J. Marsh, and L. P. Pook. Definitive, clearly written, and well-illustrated volume addresses all aspects of the subject, from the historical development of understanding metal fatigue to vital concepts of the cyclic stress that causes a crack to grow. Includes 7 appendixes. 544pp. 5⅜ x 8½. 0-486-40927-9

ROCKETS, Robert Goddard. Two of the most significant publications in the history of rocketry and jet propulsion: "A Method of Reaching Extreme Altitudes" (1919) and "Liquid Propellant Rocket Development" (1936). 128pp. 5⅜ x 8½. 0-486-42537-1

STATISTICAL MECHANICS: PRINCIPLES AND APPLICATIONS, Terrell L. Hill. Standard text covers fundamentals of statistical mechanics, applications to fluctuation theory, imperfect gases, distribution functions, more. 448pp. 5⅜ x 8½. 0-486-65390-0

ENGINEERING AND TECHNOLOGY 1650–1750: ILLUSTRATIONS AND TEXTS FROM ORIGINAL SOURCES, Martin Jensen. Highly readable text with more than 200 contemporary drawings and detailed engravings of engineering projects dealing with surveying, leveling, materials, hand tools, lifting equipment, transport and erection, piling, bailing, water supply, hydraulic engineering, and more. Among the specific projects outlined-transporting a 50-ton stone to the Louvre, erecting an obelisk, building timber locks, and dredging canals. 207pp. 8⅜ x 11¼. 0-486-42232-1

THE VARIATIONAL PRINCIPLES OF MECHANICS, Cornelius Lanczos. Graduate level coverage of calculus of variations, equations of motion, relativistic mechanics, more. First inexpensive paperbound edition of classic treatise. Index. Bibliography. 418pp. 5⅜ x 8½. 0-486-65067-7

PROTECTION OF ELECTRONIC CIRCUITS FROM OVERVOLTAGES, Ronald B. Standler. Five-part treatment presents practical rules and strategies for circuits designed to protect electronic systems from damage by transient overvoltages. 1989 ed. xxiv+434pp. 6⅛ x 9¼. 0-486-42552-5

ROTARY WING AERODYNAMICS, W. Z. Stepniewski. Clear, concise text covers aerodynamic phenomena of the rotor and offers guidelines for helicopter performance evaluation. Originally prepared for NASA. 537 figures. 640pp. 6⅛ x 9¼. 0-486-64647-5

INTRODUCTION TO SPACE DYNAMICS, William Tyrrell Thomson. Comprehensive, classic introduction to space-flight engineering for advanced undergraduate and graduate students. Includes vector algebra, kinematics, transformation of coordinates. Bibliography. Index. 352pp. 5⅜ x 8½. 0-486-65113-4

HISTORY OF STRENGTH OF MATERIALS, Stephen P. Timoshenko. Excellent historical survey of the strength of materials with many references to the theories of elasticity and structure. 245 figures. 452pp. 5⅜ x 8½. 0-486-61187-6

ANALYTICAL FRACTURE MECHANICS, David J. Unger. Self-contained text supplements standard fracture mechanics texts by focusing on analytical methods for determining crack-tip stress and strain fields. 336pp. 6⅛ x 9¼. 0-486-41737-9

STATISTICAL MECHANICS OF ELASTICITY, J. H. Weiner. Advanced, self-contained treatment illustrates general principles and elastic behavior of solids. Part 1, based on classical mechanics, studies thermoelastic behavior of crystalline and polymeric solids. Part 2, based on quantum mechanics, focuses on interatomic force laws, behavior of solids, and thermally activated processes. For students of physics and chemistry and for polymer physicists. 1983 ed. 96 figures. 496pp. 5⅜ x 8½. 0-486-42260-7

Mathematics

FUNCTIONAL ANALYSIS (Second Corrected Edition), George Bachman and Lawrence Narici. Excellent treatment of subject geared toward students with background in linear algebra, advanced calculus, physics and engineering. Text covers introduction to inner-product spaces, normed, metric spaces, and topological spaces; complete orthonormal sets, the Hahn-Banach Theorem and its consequences, and many other related subjects. 1966 ed. 544pp. 6⅛ x 9¼. 0-486-40251-7

ASYMPTOTIC EXPANSIONS OF INTEGRALS, Norman Bleistein & Richard A. Handelsman. Best introduction to important field with applications in a variety of scientific disciplines. New preface. Problems. Diagrams. Tables. Bibliography. Index. 448pp. 5⅜ x 8½. 0-486-65082-0

VECTOR AND TENSOR ANALYSIS WITH APPLICATIONS, A. I. Borisenko and I. E. Tarapov. Concise introduction. Worked-out problems, solutions, exercises. 257pp. 5⅜ x 8¼. 0-486-63833-2

AN INTRODUCTION TO ORDINARY DIFFERENTIAL EQUATIONS, Earl A. Coddington. A thorough and systematic first course in elementary differential equations for undergraduates in mathematics and science, with many exercises and problems (with answers). Index. 304pp. 5⅜ x 8½. 0-486-65942-9

FOURIER SERIES AND ORTHOGONAL FUNCTIONS, Harry F. Davis. An incisive text combining theory and practical example to introduce Fourier series, orthogonal functions and applications of the Fourier method to boundary-value problems. 570 exercises. Answers and notes. 416pp. 5⅜ x 8½. 0-486-65973-9

COMPUTABILITY AND UNSOLVABILITY, Martin Davis. Classic graduate-level introduction to theory of computability, usually referred to as theory of recurrent functions. New preface and appendix. 288pp. 5⅜ x 8½. 0-486-61471-9

ASYMPTOTIC METHODS IN ANALYSIS, N. G. de Bruijn. An inexpensive, comprehensive guide to asymptotic methods–the pioneering work that teaches by explaining worked examples in detail. Index. 224pp. 5⅜ x 8½ 0-486-64221-6

APPLIED COMPLEX VARIABLES, John W. Dettman. Step-by-step coverage of fundamentals of analytic function theory–plus lucid exposition of five important applications: Potential Theory; Ordinary Differential Equations; Fourier Transforms; Laplace Transforms; Asymptotic Expansions. 66 figures. Exercises at chapter ends. 512pp. 5⅜ x 8½. 0-486-64670-X

INTRODUCTION TO LINEAR ALGEBRA AND DIFFERENTIAL EQUATIONS, John W. Dettman. Excellent text covers complex numbers, determinants, orthonormal bases, Laplace transforms, much more. Exercises with solutions. Undergraduate level. 416pp. 5⅜ x 8½. 0-486-65191-6

RIEMANN'S ZETA FUNCTION, H. M. Edwards. Superb, high-level study of landmark 1859 publication entitled "On the Number of Primes Less Than a Given Magnitude" traces developments in mathematical theory that it inspired. xiv+315pp. 5⅜ x 8½. 0-486-41740-9

CALCULUS OF VARIATIONS WITH APPLICATIONS, George M. Ewing. Applications-oriented introduction to variational theory develops insight and promotes understanding of specialized books, research papers. Suitable for advanced undergraduate/graduate students as primary, supplementary text. 352pp. 5⅜ x 8½. 0-486-64856-7

COMPLEX VARIABLES, Francis J. Flanigan. Unusual approach, delaying complex algebra till harmonic functions have been analyzed from real variable viewpoint. Includes problems with answers. 364pp. 5⅜ x 8½. 0-486-61388-7

AN INTRODUCTION TO THE CALCULUS OF VARIATIONS, Charles Fox. Graduate-level text covers variations of an integral, isoperimetrical problems, least action, special relativity, approximations, more. References. 279pp. 5⅜ x 8½. 0-486-65499-0

COUNTEREXAMPLES IN ANALYSIS, Bernard R. Gelbaum and John M. H. Olmsted. These counterexamples deal mostly with the part of analysis known as "real variables." The first half covers the real number system, and the second half encompasses higher dimensions. 1962 edition. xxiv+198pp. 5⅜ x 8½. 0-486-42875-3

CATASTROPHE THEORY FOR SCIENTISTS AND ENGINEERS, Robert Gilmore. Advanced-level treatment describes mathematics of theory grounded in the work of Poincaré, R. Thom, other mathematicians. Also important applications to problems in mathematics, physics, chemistry and engineering. 1981 edition. References. 28 tables. 397 black-and-white illustrations. xvii + 666pp. 6⅛ x 9¼. 0-486-67539-4

INTRODUCTION TO DIFFERENCE EQUATIONS, Samuel Goldberg. Exceptionally clear exposition of important discipline with applications to sociology, psychology, economics. Many illustrative examples; over 250 problems. 260pp. 5⅜ x 8½. 0-486-65084-7

NUMERICAL METHODS FOR SCIENTISTS AND ENGINEERS, Richard Hamming. Classic text stresses frequency approach in coverage of algorithms, polynomial approximation, Fourier approximation, exponential approximation, other topics. Revised and enlarged 2nd edition. 721pp. 5⅜ x 8½. 0-486-65241-6

INTRODUCTION TO NUMERICAL ANALYSIS (2nd Edition), F. B. Hildebrand. Classic, fundamental treatment covers computation, approximation, interpolation, numerical differentiation and integration, other topics. 150 new problems. 669pp. 5⅜ x 8½. 0-486-65363-3

THREE PEARLS OF NUMBER THEORY, A. Y. Khinchin. Three compelling puzzles require proof of a basic law governing the world of numbers. Challenges concern van der Waerden's theorem, the Landau-Schnirelmann hypothesis and Mann's theorem, and a solution to Waring's problem. Solutions included. 64pp. 5⅜ x 8½. 0-486-40026-3

THE PHILOSOPHY OF MATHEMATICS: AN INTRODUCTORY ESSAY, Stephan Körner. Surveys the views of Plato, Aristotle, Leibniz & Kant concerning propositions and theories of applied and pure mathematics. Introduction. Two appendices. Index. 198pp. 5⅜ x 8½. 0-486-25048-2

TENSOR CALCULUS, J.L. Synge and A. Schild. Widely used introductory text covers spaces and tensors, basic operations in Riemannian space, non-Riemannian spaces, etc. 324pp. 5⅜ x 8¼. 0-486-63612-7

ORDINARY DIFFERENTIAL EQUATIONS, Morris Tenenbaum and Harry Pollard. Exhaustive survey of ordinary differential equations for undergraduates in mathematics, engineering, science. Thorough analysis of theorems. Diagrams. Bibliography. Index. 818pp. 5⅜ x 8½. 0-486-64940-7

INTEGRAL EQUATIONS, F. G. Tricomi. Authoritative, well-written treatment of extremely useful mathematical tool with wide applications. Volterra Equations, Fredholm Equations, much more. Advanced undergraduate to graduate level. Exercises. Bibliography. 238pp. 5⅜ x 8½. 0-486-64828-1

FOURIER SERIES, Georgi P. Tolstov. Translated by Richard A. Silverman. A valuable addition to the literature on the subject, moving clearly from subject to subject and theorem to theorem. 107 problems, answers. 336pp. 5⅜ x 8½. 0-486-63317-9

INTRODUCTION TO MATHEMATICAL THINKING, Friedrich Waismann. Examinations of arithmetic, geometry, and theory of integers; rational and natural numbers; complete induction; limit and point of accumulation; remarkable curves; complex and hypercomplex numbers, more. 1959 ed. 27 figures. xii+260pp. 5⅜ x 8½. 0-486-63317-9

POPULAR LECTURES ON MATHEMATICAL LOGIC, Hao Wang. Noted logician's lucid treatment of historical developments, set theory, model theory, recursion theory and constructivism, proof theory, more. 3 appendixes. Bibliography. 1981 edition. ix + 283pp. 5⅜ x 8½. 0-486-67632-3

CALCULUS OF VARIATIONS, Robert Weinstock. Basic introduction covering isoperimetric problems, theory of elasticity, quantum mechanics, electrostatics, etc. Exercises throughout. 326pp. 5⅜ x 8½. 0-486-63069-2

THE CONTINUUM: A CRITICAL EXAMINATION OF THE FOUNDATION OF ANALYSIS, Hermann Weyl. Classic of 20th-century foundational research deals with the conceptual problem posed by the continuum. 156pp. 5⅜ x 8½. 0-486-67982-9

CHALLENGING MATHEMATICAL PROBLEMS WITH ELEMENTARY SOLUTIONS, A. M. Yaglom and I. M. Yaglom. Over 170 challenging problems on probability theory, combinatorial analysis, points and lines, topology, convex polygons, many other topics. Solutions. Total of 445pp. 5⅜ x 8½. Two-vol. set. Vol. I: 0-486-65536-9 Vol. II: 0-486-65537-7

INTRODUCTION TO PARTIAL DIFFERENTIAL EQUATIONS WITH APPLICATIONS, E. C. Zachmanoglou and Dale W. Thoe. Essentials of partial differential equations applied to common problems in engineering and the physical sciences. Problems and answers. 416pp. 5⅜ x 8½. 0-486-65251-3

THE THEORY OF GROUPS, Hans J. Zassenhaus. Well-written graduate-level text acquaints reader with group-theoretic methods and demonstrates their usefulness in mathematics. Axioms, the calculus of complexes, homomorphic mapping, *p*-group theory, more. 276pp. 5⅜ x 8½. 0-486-40922-8

Physics

OPTICAL RESONANCE AND TWO-LEVEL ATOMS, L. Allen and J. H. Eberly. Clear, comprehensive introduction to basic principles behind all quantum optical resonance phenomena. 53 illustrations. Preface. Index. 256pp. 5⅜ x 8½. 0-486-65533-4

QUANTUM THEORY, David Bohm. This advanced undergraduate-level text presents the quantum theory in terms of qualitative and imaginative concepts, followed by specific applications worked out in mathematical detail. Preface. Index. 655pp. 5⅜ x 8½. 0-486-65969-0

ATOMIC PHYSICS (8th EDITION), Max Born. Nobel laureate's lucid treatment of kinetic theory of gases, elementary particles, nuclear atom, wave-corpuscles, atomic structure and spectral lines, much more. Over 40 appendices, bibliography. 495pp. 5⅜ x 8½. 0-486-65984-4

A SOPHISTICATE'S PRIMER OF RELATIVITY, P. W. Bridgman. Geared toward readers already acquainted with special relativity, this book transcends the view of theory as a working tool to answer natural questions: What is a frame of reference? What is a "law of nature"? What is the role of the "observer"? Extensive treatment, written in terms accessible to those without a scientific background. 1983 ed. xlviii+172pp. 5⅜ x 8½. 0-486-42549-5

AN INTRODUCTION TO HAMILTONIAN OPTICS, H. A. Buchdahl. Detailed account of the Hamiltonian treatment of aberration theory in geometrical optics. Many classes of optical systems defined in terms of the symmetries they possess. Problems with detailed solutions. 1970 edition. xv + 360pp. 5⅜ x 8½. 0-486-67597-1

PRIMER OF QUANTUM MECHANICS, Marvin Chester. Introductory text examines the classical quantum bead on a track: its state and representations; operator eigenvalues; harmonic oscillator and bound bead in a symmetric force field; and bead in a spherical shell. Other topics include spin, matrices, and the structure of quantum mechanics; the simplest atom; indistinguishable particles; and stationary-state perturbation theory. 1992 ed. xiv+314pp. 6⅛ x 9¼. 0-486-42878-8

LECTURES ON QUANTUM MECHANICS, Paul A. M. Dirac. Four concise, brilliant lectures on mathematical methods in quantum mechanics from Nobel Prize-winning quantum pioneer build on idea of visualizing quantum theory through the use of classical mechanics. 96pp. 5⅜ x 8½. 0-486-41713-1

THIRTY YEARS THAT SHOOK PHYSICS: THE STORY OF QUANTUM THEORY, George Gamow. Lucid, accessible introduction to influential theory of energy and matter. Careful explanations of Dirac's anti-particles, Bohr's model of the atom, much more. 12 plates. Numerous drawings. 240pp. 5⅜ x 8½. 0-486-24895-X

ELECTRONIC STRUCTURE AND THE PROPERTIES OF SOLIDS: THE PHYSICS OF THE CHEMICAL BOND, Walter A. Harrison. Innovative text offers basic understanding of the electronic structure of covalent and ionic solids, simple metals, transition metals and their compounds. Problems. 1980 edition. 582pp. 6⅛ x 9¼. 0-486-66021-4